Tropical Glaciers

Tropical glaciers are both highly sensitive indicators of global climate and fresh water reservoirs in some fast developing regions while the bursting of proglacial lakes can be a serious hazard. This book gives a theoretical and practical analysis of tropical glaciology including a useful definition of tropical glacier climate regimes and the analysis of the main glaciological key variables. The Rwenzori and the Cordillera Blanca are investigated as examples of tropical glacierized mountains. The fluctuations of their glaciers since the end of the Little Ice Age are reconstructed and the probable climatic reasons are discussed. The evidence of great expansions of mountain glaciers throughout the tropics on several occasions during the Quaternary is summarized, examined and then applied and contrasted.

GEORG KASER gained his Ph.D. in Meteorology, Geophysics and Physical Geography from the University of Innsbruck. He is an Associate Professor at the Institute for Geography at the University of Innsbruck. His research fields include glaciology, boundary layer climatology above snow and ice surfaces, the Alps and tropical high mountains. He has published many papers and travelled extensively for his research.

HENRY OSMASTON received his D. Phil. from Oxford University for work in the Rwenzori Mountains of Uganda. Having held positions as Senior Lecturer and Deputy Head of Geography and Geology at the University of Bristol, he is now an Honorary Research Fellow in the School of Geographical Sciences there. His research in glaciology, climatology, ecology and land-use has focused on mountainous regions of the tropics and sub-tropics.

INTERNATIONAL HYDROLOGY SERIES

The **International Hydrological Programme** (IHP) was established by the United Nations Educational, Scientific and Cultural Organisation (UNESCO) in 1975 as the successor to the International Hydrological Decade. The long-term goal of the IHP is to advance our understanding of processes occurring in the water cycle and to integrate this knowledge into water resources management. The IHP is the only UN science and educational programme in the field of water resources, and one of its outputs has been a steady stream of technical and information documents aimed at water specialists and decision-makers.

The **International Hydrology Series** has been developed by the IHP in collaboration with Cambridge University Press as a major collection of research monographs, synthesis volumes and graduate texts on the subject of water. Authoritative and international in scope, the various books within the Series all contribute to the aims of the IHP in improving scientific and technical knowledge of fresh water processes, in providing research know-how and in stimulating the responsible management of water resources.

Oblique air photo of Mt. Stanley, Rwenzori Mountains, Uganda (right) and Democratic Republic of the Congo (left) taken from the south. This is a mid-twentieth century view of the glaciers during their retreat from an advance in the nineteenth century. The Savoia Glacier in the centre foreground appears relatively exaggerated due to its closeness, but has since (in 2001) diminished by about a half, as has the Elena Glacier just to the right. In the centre is the Stanley Plateau, partly hidden by Savoia and Elena Peaks. The distant group of the three highest peaks are Margherita (5009 m, right), Alexandra (nearest) and Albert (farthest). The Congo/Uganda boundary passes through Margherita, then runs southwards just left of the main ridge-line, with Congo on the west (left) and Uganda on the east (right). Photo by Harward MacLachlan 1952.

INTERNATIONAL HYDROLOGY SERIES

Tropical Glaciers

Georg Kaser *and* Henry Osmaston

CAMBRIDGE UNIVERSITY PRESS
Cambridge, New York, Melbourne, Madrid, Cape Town, Singapore, São Paulo

Cambridge University Press
The Edinburgh Building, Cambridge CB2 2RU, UK

Published in the United States of America by Cambridge University Press, New York

www.cambridge.org
Information on this title: www.cambridge.org/9780521633338

First published 2002
This digitally printed first paperback version (with corrections) 2006

A catalogue record for this publication is available from the British Library

Library of Congress Cataloguing in Publication data

Kaser, Georg, 1953–
    Tropical glaciers / Georg Kaser, Henry Osmaston.
        p.   cm. (International Hydrology Series)
    Includes bibliographical references and index (p.   ).
    ISBN 0 521 63333 8
    1. Glaciers–Tropics. 2. Mountains–Tropics. I. Title.

GB2594.985 .K37 2001
551.31′2′0913–dc21   2001025505

ISBN-13 978-0-521-63333-8 hardback
ISBN-10 0-521-63333-8 hardback

ISBN-13 978-0-521-02096-1 paperback
ISBN-10 0-521-02096-4 paperback

Contents

Figures and tables

TABLES

Preface

In the 1970s I loved to climb the steep walls and cornice-laden crests of the Cordillera Blanca and Huayhuash in the Andes of Peru. My colleagues in Innsbruck knew this when a letter from Peruvian glaciologists came, via the Austrian Embassy, to our Institute in 1988. So, they decided I had to go and follow the call. Beside some other glaciological problems, a fast-growing glacial lake in the Cordillera Blanca threatened the town of Carhuaz and our Peruvian colleagues called for help. During the four weeks in August 1988 I had the deep feeling that glaciology was something far beyond mere science for the first time in my life. It was not just the glacial lake itself but also the manifold immediate impact of glaciers on the life and economy of the people of the Callejon de Huaylas. Also being a consultant to institutions and staff of a government which was already engaged in a civil war with different groups of *guerilleros* was an exceptional experience.

Things became worse in Peru and the planned continuous cooperation did not start because the *guerilleros* increased their activities, particularly in the proposed study area. Yet, many interesting questions about tropical glaciers had arisen during the visit to the Cordillera Blanca and I thought the glaciers on the Rwenzori mountains in East Africa might give me some answers. After two visits, civil war made further investigation impossible there too but my interest in tropical glaciers had increased. So, I continued to collect all information which was available in the literature, and I started to develop ideas about how different tropical glaciers are from the well-known mid-latitude glaciers. Finally, in 1995, we went back for fieldwork in the Cordillera Blanca. Now, tropical glaciers have become objects of intense interest in the context of both discussions on global change and fast-growing societies, especially in low geographical latitudes. Knowledge on the topic is developing fast, thus the present book is just an intermediate report, a pause for one who is under the spell of the complexity of glacial systems, particularly in the tropical Andes.

During my years of research on tropical high mountains I have met many people who have accompanied me along parts of the way.

Alcides Ames was waiting for me at Lima airport in August 1988. Since then we have gone many ways together – in his Cordillera Blanca, through offices in Huaraz and Lima, through archives and data sets in Innsbruck, on conferences and meetings all over the world. His and his wife Francisca's house in Huaraz has ever been open to me, and his enthusiasm and love for his mountains and glaciers were contagious to me and, several times, have encouraged me to go on. I dedicate the book to Francisca and Alcides.

The team around César Portocarrero of the Unidad de Glaciologia y Recursos Hidricos in Huaraz has given every possible support and help during my work in Peru.

Bernd Noggler has sustained endless mud and hours of interrogation by Ugandan soldiers on Rwenzori and when digging snow pits in the high altitudes of the Cordillera Blanca glaciers. His diploma thesis gave an essential basis to my investigations on the glacier fluctuations of Rwenzori.

The reliable help of Christian Georges was essential for the realization of this book. He answered my countless wishes for graphics, maps and data analysis with much creativity and skill. His simultaneously developing diploma thesis on glacier fluctuations in the Cordillera Blanca aroused many stimulating and constructive discussions.

Many years ago, Mike Kuhn introduced me to glaciology and has continued to be an amicable adviser.

Hanns Kerschner has delivered me from many everyday burdens at the Institute. He has spent much time listening to my worries and concerns and has contributed essential ideas and criticisms to my work.

In the final stage of the manuscript I had the luck to meet with Henry Osmaston. Initially taking up his offer for proofreading the text, I then entered into intensive discussions and a fruitful cooperation with him. This resulted in his contribution to this book on the former glacier fluctuations in the world tropics, especially East Africa and the Rwenzori.

Not least I am very thankful to Liz Morris, the president of the International Commission on Snow and Ice (ICSI), and to

Mike Bonell, chief of the Section on Hydrological Processes and Climate of UNESCO who had the idea and made it possible to publish this book in the Hydrological Series. Matt Lloyd and Simon Mitton of Cambridge University Press made many efforts and gave help with the editorial work.

To all these persons I am particularly thankful.

There are still many other people to whom I am grateful for their help when organizing the expeditions, their help during fieldwork, for carrying heavy loads, for solving computer problems, for many discussions on the topic, besides those who reminded me of the joys of normal life from time to time.

Georg Kaser

SUB-PREFACE

I started work in Uganda as a forester in 1949 and from the first I was attracted by the mountains, especially the Rwenzori, the 'Mountains of the Moon', with their wonderful scenery and strange plants. Fortunately they were for several years part of my charge so I explored them frequently both on duty and on leave. I first studied their botany, then their recent botanical history through the pollen analysis of bogs; finally I wanted to see still further back so studied their glacial history. I was amazed at the widespread evidence of this, and was fortunate that excellent air photos became available, as it is very difficult terrain to travel over. Later I extended my interests to Kilimanjaro and, when I left Uganda in 1963 soon after independence, I used these topics for my doctoral thesis, thus converting myself from a forester into a geographer at Bristol University, and also writing the first mountaineering guide to the Rwenzori.

Later I studied glaciers and glaciation elsewhere, particularly in the Himalaya and Karakoram, partly because work in Uganda was very difficult during the terrible civil wars, so I am happy to be able to return to my interests there now.

Very many people have helped me during this half century, too many to list, and I am grateful to them all. Particularly though I remember Forest Guard Bakurisoni who with me built the Bigo and Elena Glacier Huts in 1951, and for three years faced snow, floods and leopards to traverse the centre of the range every month to measure the rainfall in a network of gauges; also my Bakonzo porters and headmen, who accompanied me on difficult journeys in the Rwenzori, cheerful even when it rained as it so often did, patient with my madness in loading them with plants and samples of mud or rock. My most recent thanks must go to Georg for having the courage to ask such an academic dinosaur to contribute to his book, and to my colleague Ian Evans for his efforts to encourage my modernization. I am grateful to WWF for funding its preparation as part of their support for the Rwenzori National Park. I am very grateful to Sally Thomas and Carol Miller (Cambridge) and especially Janice Robertson (copy-editor) for their help in preparing the book for publication. Finally of course what should I have done without Anna, my wife, who accompanied me on some of these travels, holding me by the heels when I delved head-down in a pit, waited patiently for my return innmerable other times, and endured the thousands of hours that I have spent writing about them.

Henry Osmaston

Note: Prologue, Parts I and II are by G. Kaser; Part III is by H.A. Osmaston; Prospect is by G. Kaser and H.A. Osmaston.

Prologue

The story of a glacial lake

In August 1988, I was in Callejon de Huaylas and in the Cordillera Blanca in Peru, together with Dr John M. Reynolds from the Department of Geological Sciences, Plymouth Polytechnic (UK), as a guest of the Unidad de Glaciologia e Hidrologia–Hidrandina S.A.–Huaraz. The Unidad de Glaciologia e Hidrologia emerged from the Comision del Control de las Lagunas de la Cordillera Blanca, which started its work at the beginning of the 1950s, following disastrous ruptures of lakes. Today's Unidad de Glaciologia y Recursos Hidricos (UGRH) is part of the state corporation of energy Electroperú S.A.

The aim of the visit in 1988 was to prepare a glaciological–climatological–hydrological project in order to examine the extent, the causes and the possible consequences of the observed glacier retreat. Furthermore, it created a link to the tradition of the Institut für Geographie der Universität Innsbruck, which has decisively influenced the exploration and the research of the glaciers in the Cordillera Blanca and the Cordillera Huayhuash under its former department head Prof. Hans Kinzl (among others Kinzl, 1935[1], 1937[1], 1940[1,2,3], 1941, 1943, 1944, 1949, 1950, 1954, 1955[1,2], 1965, 1970[1]; Kinzl and Wagner, 1938; Kinzl et al., 1942; Kinzl et al., 1964).

During our visit we climbed from the hamlet Hualcán (about 3000 m a.s.l., situated above the town Carhuaz) up to the tongue of one glacier, named after its inventory number 513 A (according to the first lake inventory in 1953 (Fernández, 1957)) (Fig. 0.1).

The development of a lake at the foot of the western flank of the Nevado Hualcán (6150 m a.s.l.) had been observed over the past few years. The lake had developed from the retreat and the disintegration of the glacier tongue, which was covered with debris (Fig. 0.2).

Lake 513 A was 'discovered' in 1980 when a lateral proglacial lake situated higher was being visited as a matter of routine. In 1985 the lake was surveyed by colleagues of the UGRH. The glacier tongue filled a basin that was formed by the base rock (granodiorite) measuring about 250 m in width and about 750 m in length. At the front, the glacier had formed

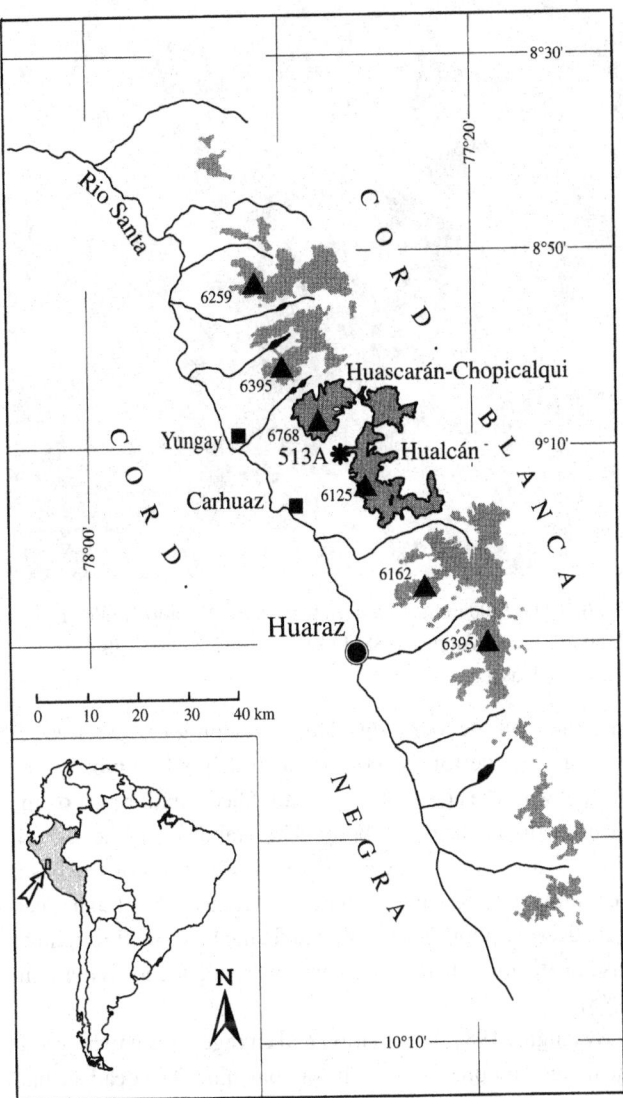

Fig. 0.1 The Nevado Hualcán, Lake 513 A and the town of Carhuaz in the Cordillera Blanca. To the north of this area are the Nevado Huascarán and the town of Yungay which was destroyed in 1970.

Fig. 0.2 The 1700 m high western face of the Nevado Hualcán (6150 m a.s.l.) in August 1988. Lake 513 A is forming at its foot. Photo: G. Kaser.

a moraine on the rock ridge. Due to the progressive retreat of the glaciers, the tongue covered with debris had mostly lost contact with the active glacier. It still filled a major part of the basin with its dead ice body. The stability of the frontal moraine arose from a dead ice core. Large parts of the basin were already filled with a glacier lake. It measured 120 m deep at its deepest point. Its runoff had formed its way through the base of the loose frontal moraine, overlying the ice (Huamán, 1985).

By August 1988 the situation had changed dramatically and the image that presented itself was alarming. The dead ice had melted even further and the surface area of the lake had increased. The various lake basins had deepened by up to 10 m (Fig. 0.3).

The dam of the moraine had subsided by 1.3 m per year and had slumped 4 m altogether. The lake surface level was, therefore, only 1 m underneath the crown of the frontal moraine,

which dammed an approximate 1.5 million m^3 of water (Figs. 0.4, 0.5, 0.6). A small arm of the glacier pushed its tongue into the lake and covered part of it with drifting icebergs (Fig. 0.7), and a steep, 1700 m high wall of ice, disintegrating into seracs from which ice avalanches fell daily, threatened above the lake (Fig. 0.2). The 12 000 inhabitants of the town Carhuaz, 2000 m below, were in immediate danger (Reynolds *et al.*, 1988).

In a crisis meeting in the offices of the Unidad de Glaciologia y Hidrologia in Huaraz, we discussed what might trigger off a catastrophe and the likely extent of such a catastrophe.

There seemed to be three possible causes for the sudden outburst of about $1.5 \times 10^6 \, m^3$ of water:

1. The dead ice core which supports the frontal moraine might not survive the next rainy season, a period of strong melting processes. It seemed highly probable that the frontal moraine would then not withstand the extreme pressure.

2. The whole dead ice body might melt at an increasing rate. It could reach a volume that would no longer be large enough to remain on the bottom of the lake. Just like an iceberg, it would suddenly float upwards, produce a big wave and break through the frontal moraine. It could also sweep away the frontal moraine as part of the ice body swimming on the surface. This type of 'hydraulic fracture' of a dead ice barrier was described theoretically by Nye (1976). It was, however, not clear how large the proportion of debris and rock would be. Therefore, the average density of the dead ice and consequently its buoyancy force in water could not be determined.

3. A big ice avalanche from the western face of the Nevado Hualcán could spill into the lake and, because of the waves produced, the moraine rampart could break. Earthquakes, which are frequent and very often large in the Cordillera Blanca, could quite likely trigger off such an ice avalanche.

In all three cases it had to be expected that, due to its instability, the frontal moraine could be partly destroyed and that would consequently trigger off a sudden rupture of the lake. The erosion of the steep outer slope of the damming moraine is extreme during the bursting of such dams. Experience shows that the breachings of proglacial moraine lakes are among the biggest mudflow catastrophes in high mountain areas (Haeberli, 1992). The water masses streaming out of Lake 513 A would have very likely entrained, on their way down, much loose morainic material. Fig. 0.8 (right) shows a cautious estimate of the extent of a possible breach of the lake and the following mudflow (Kaser, 1988, 1989, 1994).

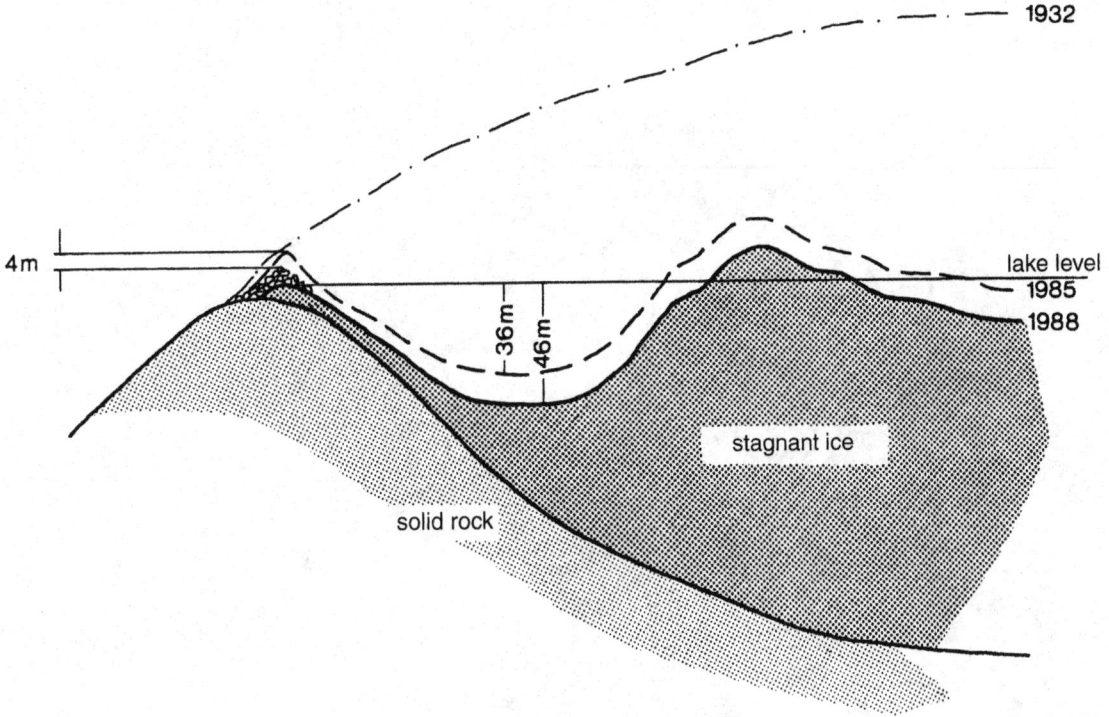

Fig. 0.3 The outer part of Lake 513 A in a longitudinal section. The outline shows the situations in 1932, 1985, and in 1988.

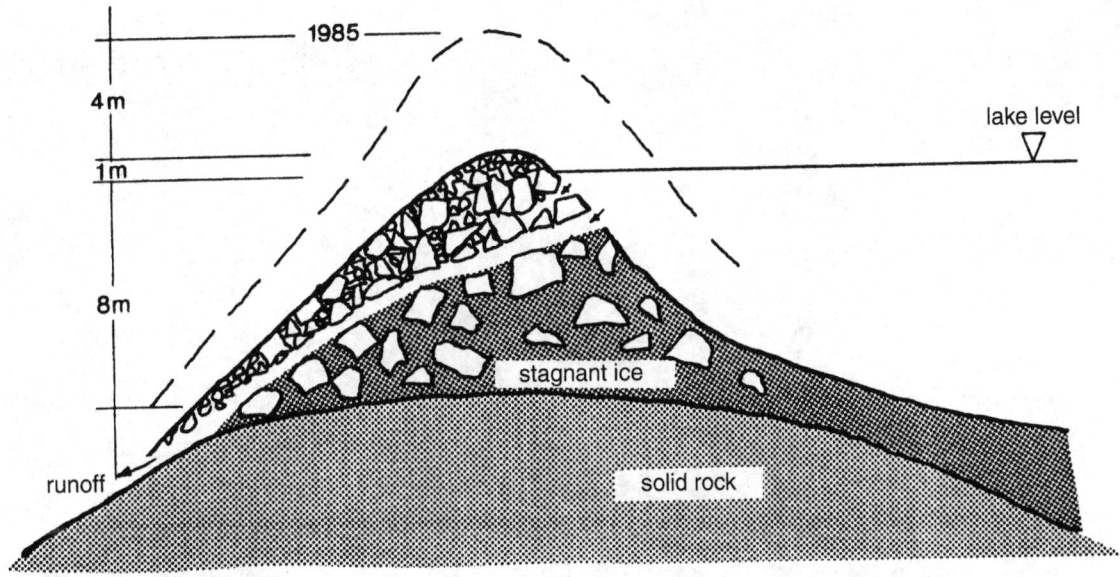

Fig. 0.4 The dam of 513 A drawn in section. The outline shows the situations in 1985 and in 1988.

Fig. 0.5 513 A. The frontal moraine in August 1988.
Photo: G. Kaser.

Fig. 0.6 513 A. The frontal moraine in August 1988.
Photo: G. Kaser.

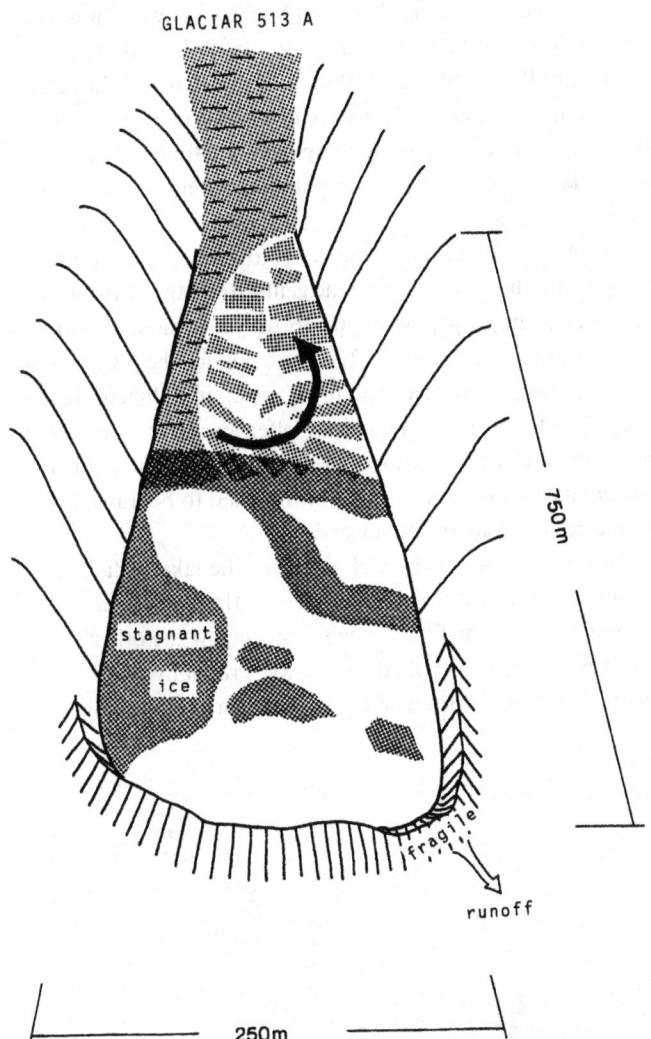

Fig. 0.7 Schematized picture of Lake 513 A in August 1988. The back part of the lake is covered with floating ice which is regularly supplied by the glacier of the Hualcán west face. Dead ice islands and open water surfaces lie at the front.

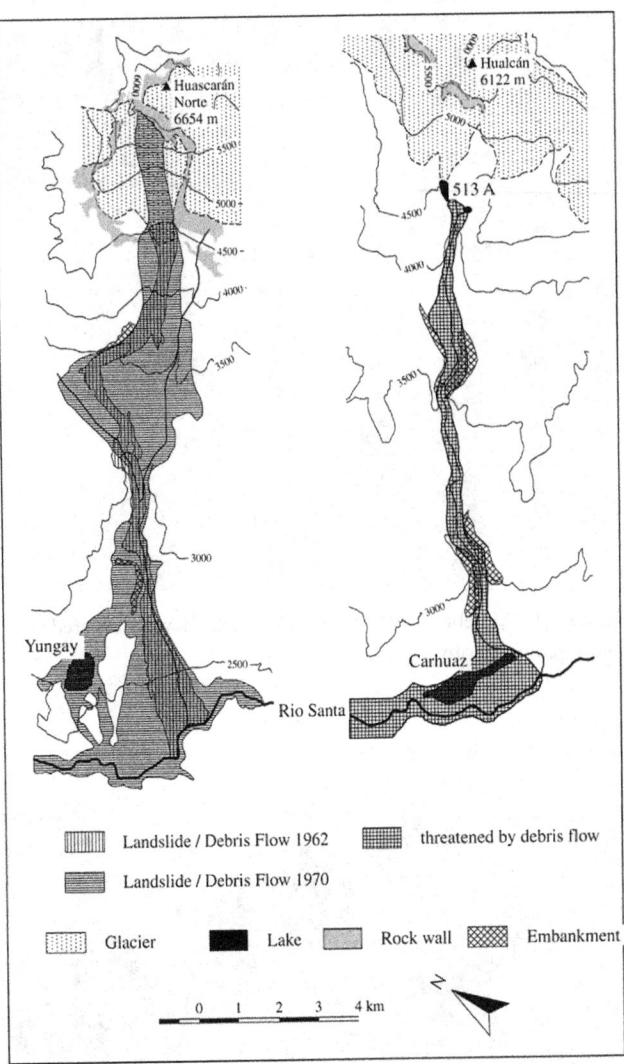

Fig. 0.8 (Left) The ice fall avalanche from Huascarán Norte on 30 May 1997 (Körner, 1983). (Right) The estimated extent of a mudflow from Hualcán (G. Kaser).

The analysis of the catastrophic mudslide on May 30, 1970 was useful; the mudslide was triggered by an earthquake and the following ice- and landslide from the western face of the Nevado Huascarán Norte subsequently buried the town Yungay and its 18 000 inhabitants (Patzelt, 1983). A comparison with the account of the events from the Nevado Huascarán Norte from the years 1962 and 1970 shows how dangerous 513 A was (Fig. 0.8 left).

The engineers M. Zamorra, C. Portocarrero, A. Huamán and M. Zapata developed the following solution to defuse the acute danger. Using the method of siphons for drainage, they estimated that two pipes, each about 20 cm wide, should within about three to four months lower the lake surface to such a level that digging through the frontal moraine should be

possible. Then, by excavating a tunnel for a permanent outflow, the level of the lake in its rock basin could be so lowered that even dead ice floating to the surface or large ice avalanches could not make the lake burst out in future.

Despite the difficult ascent to the 4600 m high lake, the first pipe was installed only a few weeks later, but this was not enough to lower the lake surface level in the necessary time. The 2000% currency inflation made the installation of the second pipe, which was only available by using foreign currency, impossible. However, private donations from Austria and England made it possible to install the second pipe in **January 1989**. By the end of March, the frontal moraine could be dug through (Figs. 0.9 and 0.10).

In fact, in **1991**, even before the tunnel had been started, an ice avalanche fell into the lake. The resulting wave enlarged the

Fig. 0.9 513 A, February 1989. The lake surface level is lowered using siphons. Photo: UGH.

new channel and caused a mudflow which was, however, slowed down on the Pampa (a former proglacial lake which is sediment filled), situated 1100 m further down and thus only caused minor damage to lower lying irrigation channels (Fig. 0.11). The extent of the mudflow would definitely have been much larger before the deliberate breaching of the loose frontal moraine.

In **1993** the necessary money was provided to blast a tunnel system into the rock to enable the gentle drainage of the water. In order to make it possible to do the construction work, a path had to be built (Fig. 0.12) and a camp for the porters and the construction workers had to be set up in a difficult rocky area. The following example may offer an impression of the extreme working conditions. Despite the laid out path, the pneumatic compressor for rock drilling had to be transported to the building site by 60 porters!

To simply blast a tunnel 20 m below the lake surface level would have caused a flood because of the huge hydrostatic pressure. Therefore, four tunnels – one below the other – had to be excavated, which made it possible gradually to lower the lake surface level through a connecting shaft (Fig. 0.13).

Fig. 0.10 513 A, April 1989. The breached frontal moraine.
Photo: UGH.

Fig. 0.11 513 A. The mudflow in 1991. Photo: G. Kaser, 1995.

Fig. 0.12 The 'path' to the lake 513 A. Photo: G. Kaser, 1995.

On **April 30, 1994,** the entrance to the lowest tunnel was blasted. Lake 513A now has 6 to 7 million m³ less water volume and the inhabitants of Hualcán and Carhuaz have since been safe (Fig. 0.14).

Lake 513 A is one of many lakes in the Cordillera Blanca that have developed in such a way and that are often unstable and dangerous. In the course of the retreat of the glaciers a high number developed in the twentieth century (Morales, 1969, 1979; Kinzl, 1970[1]; Lliboutry *et al.*, 1977[1]). Ames (1998) reports that there were about 30 moraine-dammed lakes in the 1930s, when the Alpenverein expeditions explored the Cordillera Blanca. By the end of the 1940s, there were already 68 lakes, at the beginning of the 1960s 99, and in 1970 an additional four more, which totalled 103. An incomplete list in the middle of the 1990s has brought the total to 138 proglacial lakes.

Ames (1998) reconstructed the retreat of a few selected glacier tongues and the origination of lakes in the Cordillera Blanca with the help of the extensive photo and map archives left by H. Kinzl in Innsbruck. Figs. 0.15 to 0.20 show different stages of development of these glacier tongues and lakes, and Fig. 0.21 shows their location (in the Quetchua language, *cocha* means lake and *raju* means snow and ice).

Some of the many lakes in front of the glaciers of the Cordillera Blanca have broken through, and some have caused damage. For example, the outbreak of the Palcacocha in the Quebrada (valley) Cojup on December 13, 1941 caused a mudflow which destroyed about a third of Huaraz and killed

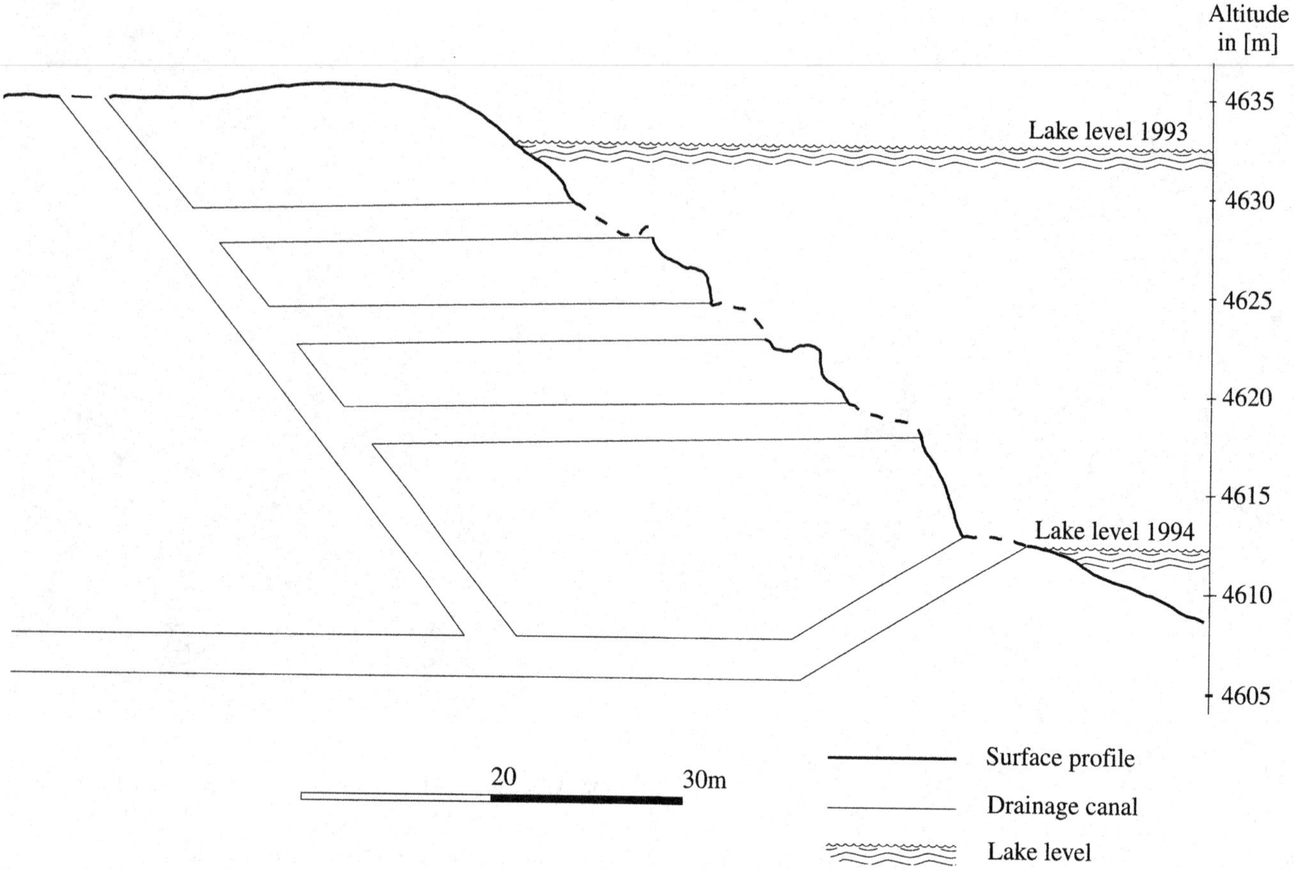

Altitude
in [m]

4635

Lake level 1993

4630

4625

4620

4615

Lake level 1994

4610

4605

20   30m

▬▬▬▬	Surface profile
────	Drainage canal
≈≈≈≈	Lake level

Fig. 0.13 A tunnel was excavated in steps into the rock barrier.
(Source: UGRH – Huaraz.)

Fig. 0.14 Lake 513 A after making it safe. Photo: G. Kaser, 1995.

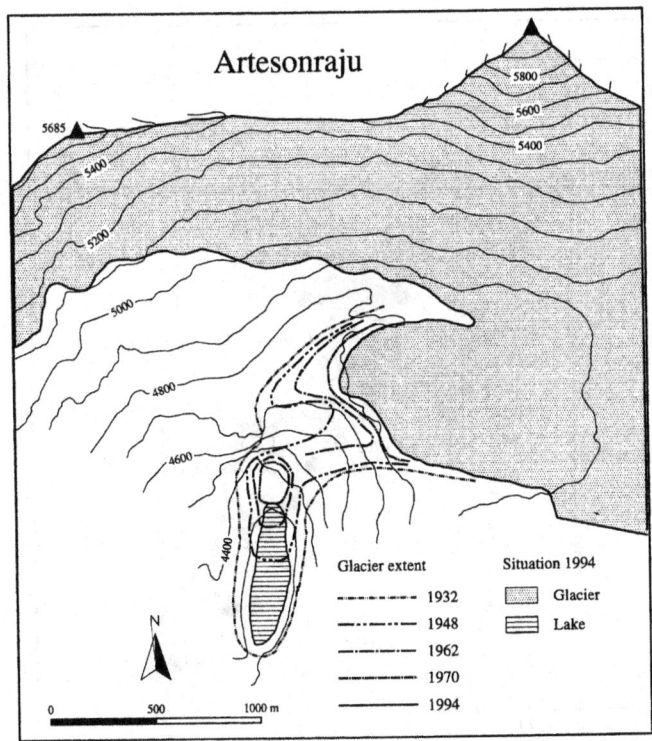

Fig. 0.15 The retreat of the Glaciar Artesonraju and the origin of the Artesoncocha (Ames, 1998).

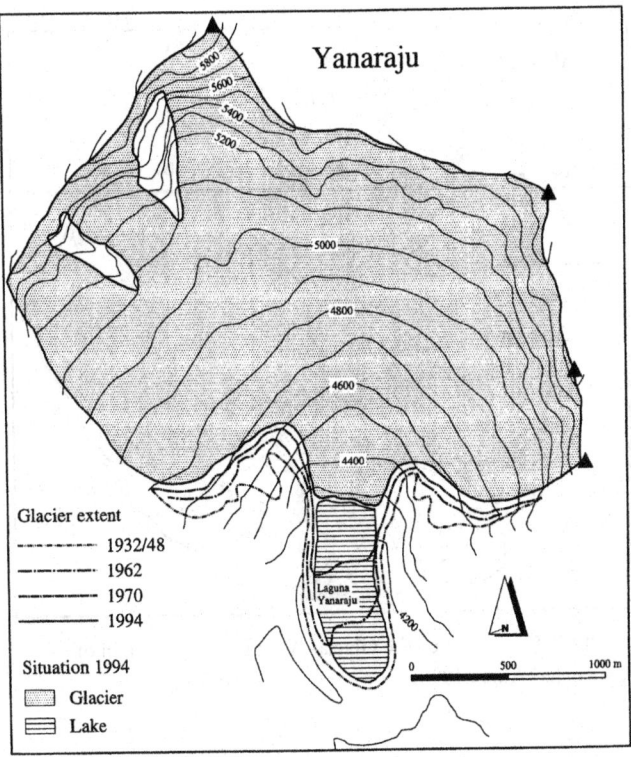

Fig. 0.17 The retreat of the Glaciar Yanaraju and the origin of the Yanacocha (Ames, 1998).

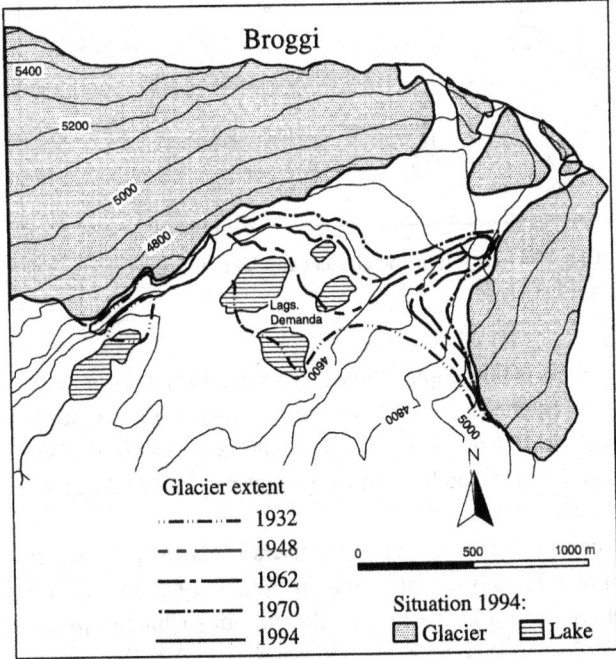

Fig. 0.16 The retreat of the Glaciar Broggi and the origin of the Lagunas Demanda (Ames, 1998).

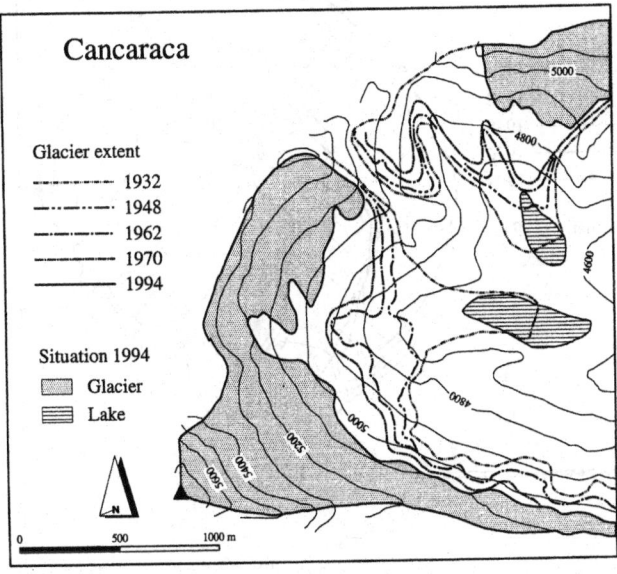

Fig. 0.18 The retreat of the Glaciares Cancaraca and the origin of the lakes in front of the face (Ames, 1998).

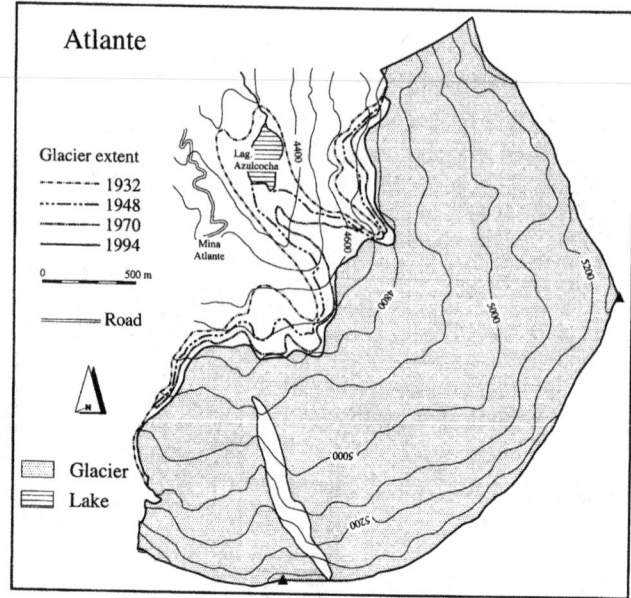

Fig. 0.19 The retreat of the Glaciar Atlante and the origin of Azulcocha (Ames, 1998).

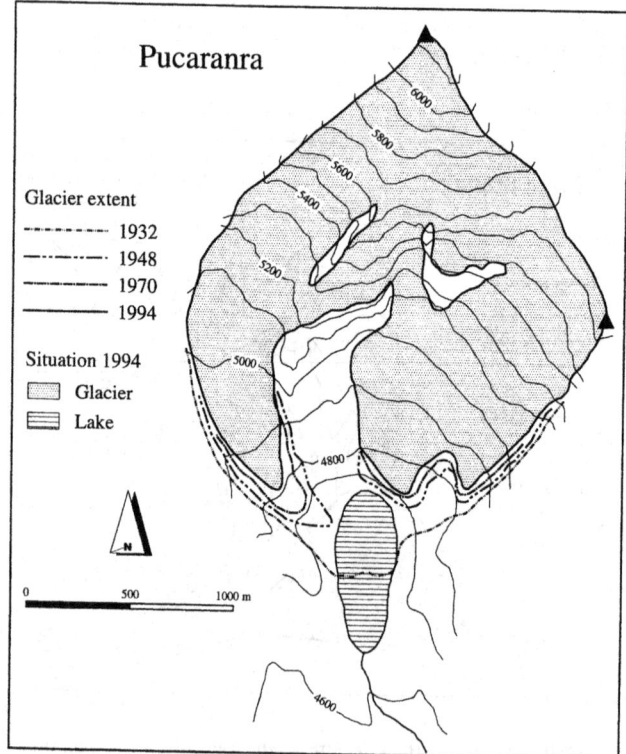

Fig. 0.20 The retreat of the Glaciar Pucaranra and the origin of the Pucaranracocha (Ames, 1998).

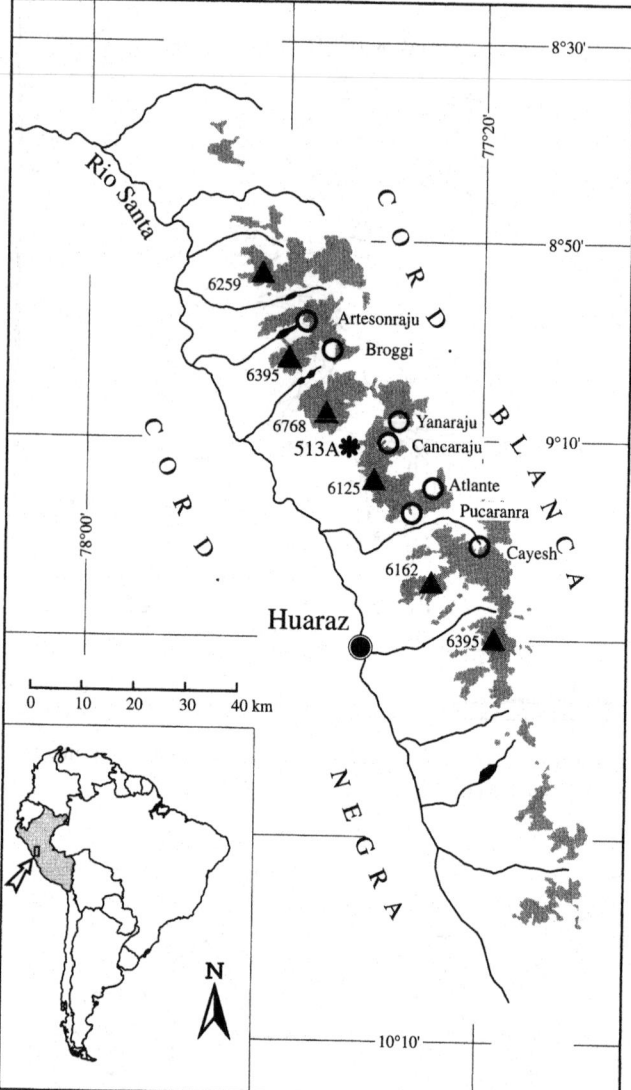

Fig. 0.21 The location of the glaciers and lakes in Figs. 0.15 to 0.20 (Ames, 1998).

more than 6000 people (Lliboutry *et al.*, 1977[1]; the authors give an extensive list of catastrophic lake ruptures in the Cordillera Blanca). A list of lakes that have ruptured and/or have undergone stabilization work is given in Table 0.1 according to Ames (1998).

A continuing retreat of the glaciers leads to the development of extremely unstable lakes. In August 1995, I visited the Quebrada Paria at the foot of the Nevados Chacraraju and Pirámide together with Jesus Gómez (UGHR–Huaraz) to look for suitable places for climatic stations on the eastern face of the Cordillera Blanca. The glacier coming from the face of the Nevado Chacraraju detached itself from its tongue, which is covered with debris, some time ago. A dead ice mass, about 1.5 km long and 400 m wide, is melting continuously. The first

Table 0.1. *Proglacial lakes in the Cordillera Blanca which ruptured (*) and/or where safety precautions were made (+).*

Name	φ (S)	λ (W)	Area (10^4 m^2)	Volume (10^6 m^3)
Safuna alta+	8°50.13'	77°37.06'	6.7	2.1
Pucacocha+	8°51.17'	77°37.78'	25.9	8.6
Jankarurish*	8°51.03'	77°40.35'	15.6	
Yuraccocha+	8°52.85'	77°44.01'	38.1	7.4
Artesoncocha*	8°58.40'	77°38.55'	7.3	
Laguna 69+	9°00.52'	77°36.48'	11.0	31.1
Artesa*+	9°06.64'	77°30.84'	2.5	
Hualcacocha+	9°09.45'	77°33.75'	17.1	3.8
Cochca+	9°12.75'	77°32.65'	7.5	0.9
Laguna 513*+	9°12.52'	77°33.95'	2.6	0.7
Lejiacocha+	9°16.99'	77°30.34'	18.8	1.5
Paccharuri+	9°16.95'	77°26.99'	28.7	7.6
Pucaranracocha+	9°20.07'	77°20.68'	17.8	2.9
Aquillpo+	9°20.13'	77°25.18'	38.6	
Ishinka+	9°23.01'	77°24.98'	9.8	
Mullaca+	9°25.75'	77°28.36'	10.8	
Llaca+	9°26.02'	77°26.57'	3.7	0.3
Palcacocha*+	9°23.69'	77°22.74'	6.3	0.5
Cuchillacocha+	9°24.44'	77°21.10'	14.1	2.2
Tullparaju*+	9°25.07'	77°20.47'	15.1	1.6
Shallap+	9°29.36'	77°21.29'	15.7	3.5
Rajucolta*	9°31.43'	77°20.76'	57.7	23.3
Tumarina*	9°26.18'	77°17.49'	10.6	
Allicocha+	9°14.59'	77°27.21'	26.5	5.3
Yanaraju+	9°07.95'	77°28.85'	20.9	7.3

Notes:
According to Ames (1998).
φ and λ are the geographic latitude and longitude respectively.

small lakes have already developed on its surface (Fig. 0.22). If similar climatic conditions continue and if the frontal moraine can withstand the pressure long enough, a similarly unstable lake like 513A could develop within a few years. This is not the only one in the Cordillera Blanca. Five to ten similar situations could develop within the next few years.

Cultivated land is found at these high altitudes particularly because of the possibility of irrigating with glacier runoff (Kinzl, 1943, 1944, 1965), and lies in the immediate area of the glaciers' influence. The high tectonic activity in the Cordilleras, which are at the edge of the subduction zone of the Pacific plate under the Nazca plate, the steep relief of the young and growing mountain range and its high relative differences in elevation, the large amounts of loose erosion material, the continuous retreat of the glacier and the vast settlements on the alluvial fan in the main valley, going back to the colonial settlement plans of the Spanish, very often create dangerous situations and have again and again been the causes of huge catastrophes. The Cordillera Blanca has been of immense interest for glaciologists, not only because of these facts but also because of the special location in the tropical climate.

The research project, planned and prepared in the summer of 1988, was accepted in July 1989 by the Austrian Science Foundation (Österreichischer Fonds zur Förderung wissenschaftlicher Forschung). At that time the terrorist activities by Sendero Luminoso and Tupac Amaru had also reached their climax in the Cordillera Blanca and my colleagues in Huaraz strongly advised me against fieldwork. The project had to be broken off and delayed even before work had begun.

It had already become obvious during the preparation for the project application that, despite increasing research on tropical glaciers, the differences between tropical glaciers and glaciers of the temperate and high zones have only been examined at a basic level (Kinzl, 1942; Lliboutry et al., 1977[2]; Jordan, 1979). A large part of this book is devoted to these theory-oriented considerations. The results will form the basis for future fieldwork on the various tropical glacier areas, especially the Cordillera Blanca, the area in the tropics with by far the most glaciers.

Thanks to Kinzl's work there is a considerable base of glaciological research and, thanks to the Unidad de Glaciologia y Recursos Hidricos, an abundance of climatological, hydrological and glaciological data exists. An important contribution to the present knowledge about glaciers in the Cordillera Blanca has been made by Alcides Ames who has, even after his retirement from the UGRH, dedicated himself to the glaciers of his homeland and has provided many glaciologists with his invaluable knowledge (Ames, 1985; Ames, 1994; Ames et al., 1989; Ames et al., 1994; Alean & Ames, 1994; Ames & Francou, 1995; Hastenrath & Ames, 1995[1,2]; Kaser, Ames & Zamora, 1990; Ames & Hastenrath, 1996; Kaser, Hastenrath & Ames, 1996; Kaser, Georges & Ames, 1996).

Fig. 0.22 The disintegrating tongue of the Glaciar Paria on the eastern face of the Cordillera Blanca. Photo: G. Kaser, 1995.

I

The nature of tropical glaciers

1   A brief introduction on the matter

'The basic knowledge about glaciers has been gained in the Alps. Even today glaciology has a certain alpine-centred attitude. The alpine valley glacier is for us the ideal type of a glacier with its structure consisting of firn field and tongue. We talk about glaciers without firn fields or without tongues when we see other glacier forms in outer-alpine mountains and, instinctively, we feel they are imperfect.

. . . Our knowledge of the regimen of the glaciers is mainly alpine as well. However, ideas won in the Alps cannot be easily transferred to the high mountains of other climatic zones. Glaciers are almost like living things and they almost react like them to the complex forms of high mountain climates. How more varied is the life of an alpine glacier with its seasonal change of accumulation and ablation compared to the one of a tropical ice field which does not show great contrasts during the year!

. . . only part of its [the comparative glacier research] questions can already be solved today. Some of them still lack observations in the outer European glaciated areas. This is especially true for the glaciers of the tropical zone despite the highly commendable work by H. Meyer, W. Sievers, Fr. Klute and others. The reason is easily understandable. The tropical glaciers are situated far from bigger settlements and even more so from the centre of scientific research. Their position in high altitudes makes travelling and working difficult due to the thin air and also due to the shortness of the tropical day.' (Kinzl, 1942)

The situation has not changed much up to the present. The knowledge of mountain glaciers is still distinctly *alpine-centred,* despite the highly commendable work by H. Kinzl, C. Troll, J. Whittow, H. Osmaston, J.A. Peterson, I. Allison, L. Lliboutry, S. Hastenrath, P. Kruss, E. Jordan, A. Ames, C. Schubert and others (see literature). *Travelling and working* has become incomparably more comfortable; the logistics in the area of the tropical high mountains, however, is still complicated (see above) and the air is still *thin.* The questions and problems arising for glaciology in the tropical high mountains, however, are just as exciting as in other areas of the world.

Glaciers react with more or less delay and in a complex manner to climatic variations and changes. Their advances leave behind signs in moraines which can be reconstructed and interpreted. They store information about past climates in their highest areas where no melting takes place. In order to be able to interpret this information it is necessary to examine the glaciers in their present behaviour, to understand them as 'climate meters' and 'climate graphs' and to 'calibrate' them.

The methods and theories of traditional and modern glaciology of mountain glaciers are still almost exclusively based on observations and research of mountains in the temperate zones. The following questions are yet to be investigated:

- How do glaciers in mountains of the tropics behave?
- To what extent is it justifiable to apply the traditional methods of research?
- Can the theories be transferred?
- How do tropical glaciers react to climatic variations and changes?

Thus, consequently:

- To which climatic variations do glaciers within the tropics display a reaction?

While the climate of the temperate zones is characterized by more or less a succession of quickly moving and changing synoptic patterns of different air masses, the atmosphere of lower latitudes is mainly homogeneous. This is true especially for the thermal character of the atmosphere. Tropical precipitation is, on the one hand, characterized by extensive advective processes and, on the other hand, embedded convective cells dominate the processes locally (among others Rudloff, 1981; Hastenrath, 1991[1]; Asnani, 1993). These circumstances require a different interpretation of the connection between glaciers and climate than is the case in the temperate and higher latitudes.

Of course, the following question is more immediate and goes beyond the characteristics of glaciers as climate meters:

How do tropical glaciers influence the human environment?

- as a (seasonal) water reservoir?
- as a source of danger?

- or, though only very locally and to a small extent, even as a source of ice for cooling purposes and as ice cream (Alean & Ames, 1994)?

Some of these questions will be dealt with in this book. The main focus is on the investigation of modern glacier variations in the tropics and their climatological interpretation. The period in question spans from the end of the maximum extent of glaciers in the middle of the last century up to today.

There are not many data on tropical glaciers (Kaser, 1995[1,2]; Kaser et al., 1996[2]) compared with those on temperate zones. Costly field work such as mass and heat budget studies are often faced with logistical problems. The reconstruction of glacier extents within the last 100 to 150 years can far less depend on earlier measurement results and historical documents, which is the case in many non-tropical mountains. Therefore, this work will first develop model ideas under defined boundary conditions in a deductive process and will then compare them with the data available. These data arise from investigations of other authors as well as from our research. If possible, data from many glacierized tropical mountains will be used but, very often, single examples must be sufficient. The glaciers of the Rwenzori mountains in Uganda play an important role as an example for the inner tropics and those of the Cordillera Blanca in Peru as an example for the outer tropics. A comparison with the best researched glaciers of the Alps, in most cases those of the Inner Ötz Valley and especially the Hintereisferner, will be repeatedly made.

After a delimitation of the tropics (chapter 2) and a description of the distribution of its glaciers (chapter 3), an ideal tropical climate will be established in a working hypothesis as the boundary condition (chapter 4). From these data, two different tropical glacier regimes will be deduced and their glaciological key variables will consequently be examined (chapter 5).

One of the most important key variables for a glacier is the vertical balance gradient and its variation with height. It is used to describe the variation of the mass budget with height. It considerably influences the reaction of other key variables such as the equilibrium line, volume, areal extent and position of the tongue towards possible climatic disturbances. Above all, the lack of an annual variation of air temperature leads one to expect fundamental differences with respect to the middle and high latitudes:

- The vertical profile of the specific mass balance will be examined in section 5.1.
- The consequent reaction of the equilibrium line EL to changes of single climatic variables will be discussed in section 5.2.
- The behaviour of the tongues is the aim of section 5.3.
- Consequently, the proportion of the tongue area compared with the whole area of the glacier (accumulation area ratio) AAR must be discussed with respect to tropical conditions (section 5.4).
- There are no winter-cold layers to be expected in the snow cover. This leads to corresponding consequences regarding the meltwater percolation, metamorphic processes and, therefore, the structure of glacier ice, and to the question about the dynamics of tropical glaciers (section 5.5).
- The lack of thermal seasons also has to have an influence on the role of glaciers in the water budget of tropical glacierized catchment areas (section 5.6).

These hypotheses are presented in detail, deduced theoretically and discussed on the basis of test results. During this process, a model of the well-tested Hintereisferner, as a representative of the conditions in the temperate zones, will be exposed to the defined tropical boundary conditions. The 'tropical Hintereisferner' is then compared with real tropical glaciers.

Observed and reconstructed modern glacier variations in the Rwenzori mountains will be discussed in chapter 6, those in the Cordillera Blanca in chapter 7, and those in tropical mountains as a whole in chapter 8. Their causes will be discussed on the basis of the knowledge gained.

A particular motivation for investigating the nature of tropical glaciers emerged from the latest paleoclimatological research results. Contrary to the common opinion until now that the tropics would only react slightly to global climatic changes, there are more and more research results that lead to the assumption that there has been an increase in temperature in the tropics, which is clearly higher than that of the temperate zones, since the Last Glacial Maximum. Latest analyses of two ice cores from the 'Garganta', the ice saddle between the two peaks of the Nevado Huascarán (Cordillera Blanca, Peru) give rise to the supposition of temperatures from the Last Glacial Maximum (Würm) which were 8°C to 12°C lower than today (Thompson et al., 1995; Thompson, 1995).

2 The delimitation of the tropics

The Greek term *tropics,* originally the name for the two latitudes at which the sun reaches the zenith once a year, is commonly used today for the region between those two latitudes. This term is also accepted as the name for the particular vegetational belt which one usually associates with 'typically tropical' vegetation. Similarly, 'typically tropical' soils, agricultural forms, life forms and the 'tropical climate' are also commonly associated with the tropics. Over the course of time, almost just as many delimitations as different approaches to the question have been suggested in addition to the astronomically clear and definite boundary conditions (among others Sapper, 1923; Fosberg *et al.,* 1961; Lauer, 1975). Many of them are climatological delimitations of the tropics.

From the glaciological point of view, radiation geometry as well as thermal and moisture conditions are of special interest.

The tropics are astronomically clearly defined latitudes. As mentioned above, the sun reaches the zenith at these two positions once a year, and within the area of the tropics twice a year. That is the main reason that, within the tropics, the annual variation of air temperature is smaller than its diurnal variation. Compared with other thermal delimitations this is also true for high altitude mountains without any limitations (Troll, 1943). Due to the differently delayed reaction of air temperature to the energy budget of different surfaces on the earth, and under the influence of advective processes in the oceans and in the atmosphere, the lines of equal diurnal and annual variation of air temperature ($\Delta T_{d} = \Delta T_{a}$) in the southern hemisphere are, on the one hand, mostly beyond the astronomical tropics, whereas they are close to the tropics in the 'more continental' northern hemisphere. This equilibrium line is a very useful delimitation for thermal-climatological investigation (Fig. 2.1).

A complex interaction between the energetic conditions and the dynamics of global circulation leads to a high degree of **thermal homogeneity** of the tropical atmosphere with regard to both **time** and **space**.

The moisture conditions in the tropics are also closely connected with the position of the sun. The meteorological equator is thermally and dynamically induced and thus part of global circulation. In the centre of this tropical circulation, convective processes dominate; these are characterized by the diurnal variation of the solar radiation, the available humidity, and the local relief. Delayed by a few weeks with regard to the solar oscillation, this meteorological equator, called Intertropical Convergence Zone (ITCZ) in the lower troposphere, reaches its points of return once a year and causes two more or less distinct rainy seasons between these turning points (Fig. 2.1).

The distribution of land and water prevents a spatially uniform oscillation of the ITCZ. The high heat capacity of water restrains the oscillation of the ITCZ, whereas it extends further over the continental areas during its annual path. Complex interactions between the moisture, the energetic and the biosystems, induced by the ITCZ, cause the permanently humid conditions of the inner tropics.

Some authors discuss the existence of permanent areas of low pressure above the extensive continental areas of South America and Africa and, consequently, a discontinuity of the ITCZ (e.g. Weischet, 1988). Latest analyses of isotope ratios in the precipitation, however, show that a clearly limited ITCZ oscillates over the land surfaces in South America (Rosanski, 1995; Rosanski & Araguás, 1995).

Beyond this simplified view there are, of course, processes in the tropics such as influxes of cold air from the temperate zones, monsoon-type appearances, low level jets, easterly waves, or the complex El Niño – Southern Oscillation (ENSO) phenomena (among others Fletcher, 1945; Riehl, 1954; Bonner, 1968; Lettau, 1976; Hastenrath, 1991[1]; Asnani, 1993). A simplified model may, however, be enough to define the limits of the tropics and, consequently, a definition of a 'tropical glacier':

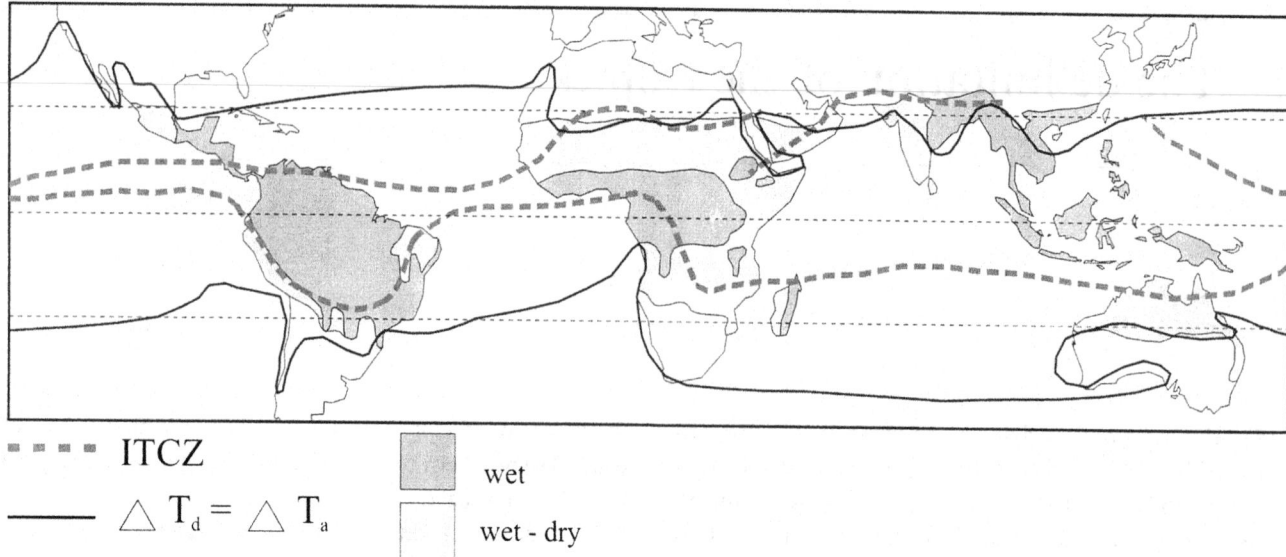

Fig. 2.1 The tropics. (Sources: Lauer, 1975; Liljequist and Cehak, 1984; Paffen, 1967.) ΔT_d, diurnal temperature range; ΔT_a, annual temperature range.

- The geometry of the incoming solar energy is distinctly different from that outside the tropics.
- The tropical atmosphere at each altitude is, from the thermal point of view, to a large extent continuously homogeneous within the lines $\Delta T_a = \Delta T_d$.
- The meteorological equator oscillating seasonally causes one to two rainy periods in which an oceanically narrow oscillation and/or interaction with extensive rain forests causes areas that are always humid.

These conditions make it possible to describe 'regimes of glaciers' which are clearly delimited from the temperate zones. For this reason, glaciers will only be called tropical which are clearly within all three defined delimitations. For all those glaciers that are within the astronomical tropics but outside the ITCZ and, therefore, in the extremely arid climate of the subtropics, different considerations of the energy and mass budget will have to be made. In that region, evaporation/sublimation, among other variables, plays a dominant role in the ablation process. That is the case with glaciers in northern Chile and southern Bolivia.

The Mexican volcanoes are outside the ITCZ, even though there is a humid and a dry season similar to some regions in the tropics. The reason for this is that the circulation of the northern hemisphere, reaching far south during the northern winter, causes arid conditions from December to April, and northeasterly trade winds bring humidity from the Carribean between May and October (Mosiño & García, 1974). Therefore, the Mexican glaciers are not 'tropical' under the above defined restrictions.

Although the ITCZ extends far to the north over South Asia during the monsoon season, the glaciers of the Himalayas lack both the astronomical and the thermal classification for the tropics.

The application of the line marking equal diurnal and annual temperature variations was further developed by Paffen (1967), prompted by Troll's depiction of thermoisopleths (Troll, 1943). The extreme positions of the ITCZ cannot always be clearly defined and they are not the same every year (among others Gruber, 1972). Therefore, its middle position varies with different authors depending on the period analysed. The middle positions of the ITCZ at the two solstices, as shown in Fig. 2.1, are from the textbook on meteorology by Liljequist & Cehak (1984).

3 The distribution of glaciers in the tropics

On the basis of the boundary conditions defined in the previous chapter, tropical glaciers are distributed over the three highest East African mountains, the Indonesian Puncak Jaya (Carstensz Mountains) in Irian Jaya and the South American Andes between Bolivia and Venezuela. Table 3.1 shows the distribution of the actual surface areas of tropical glaciers according to countries and Fig. 3.1 depicts their relative proportions of area. Relevance and accuracy of the data are different in the individual countries. Therefore, the data can only give a rough overview. As there has been a general retreat in all tropical mountains in the twentieth century, which has also led to considerable area losses in the past twenty years, some of the individual values given in Table 3.1 and, consequently, the total area also must be assumed to be too large.

With a total glacier area which is definitely below 2500 km^2 today, tropical glaciers cover about 4% of the area of all mountain glaciers and about 0.15% of the global glacier area (WGMS, 1989).

More than 99% of the tropical glacier areas are distributed over the South American Andes, more than 70% alone over the Cordilleras of Peru. The Cordillera Blanca, with its 723 km^2 (state of 1970; Ames *et al.*, 1989), covers 36.7% of the Peruvian glacier area and 26.1% of the total tropical glacier area. It is, consequently, by far the most glacierized tropical high mountain region.

Table 3.1. *The areas of the tropical glaciers according to Jordan (1991) brought up to date*

	Area			
	(km^2)	(%)	Year	Sources
Rwenzori	1.7	0.06	1990	chapter 6
Mt. Kenya	0.4	0.01	1993	Hastenrath (1995)
Kilimanjaro	4.9	0.18	*c.* 1970	Hastenrath (1984)
Africa	7.0	0.25		
Irian Jaya	3.0	0.11	1988	Peterson & Peterson (1994)
Colombia	108.5	3.92	*c.* 1950	
Venezuela	2.7	0.10	*c.* 1950	
Equador	112.8	4.08	*c.* 1970	
Peru	1972.0	71.24	1970	
Bolivia	562.0	20.30	*c.* 1980	
South America	2758.0	99.64	*c.* 1950–*c.* 1980	
Total	2768.0	100	1950s – 1990s	

Estimated total surface area for 1990: $< 2.5 \times 10^3$ km^2

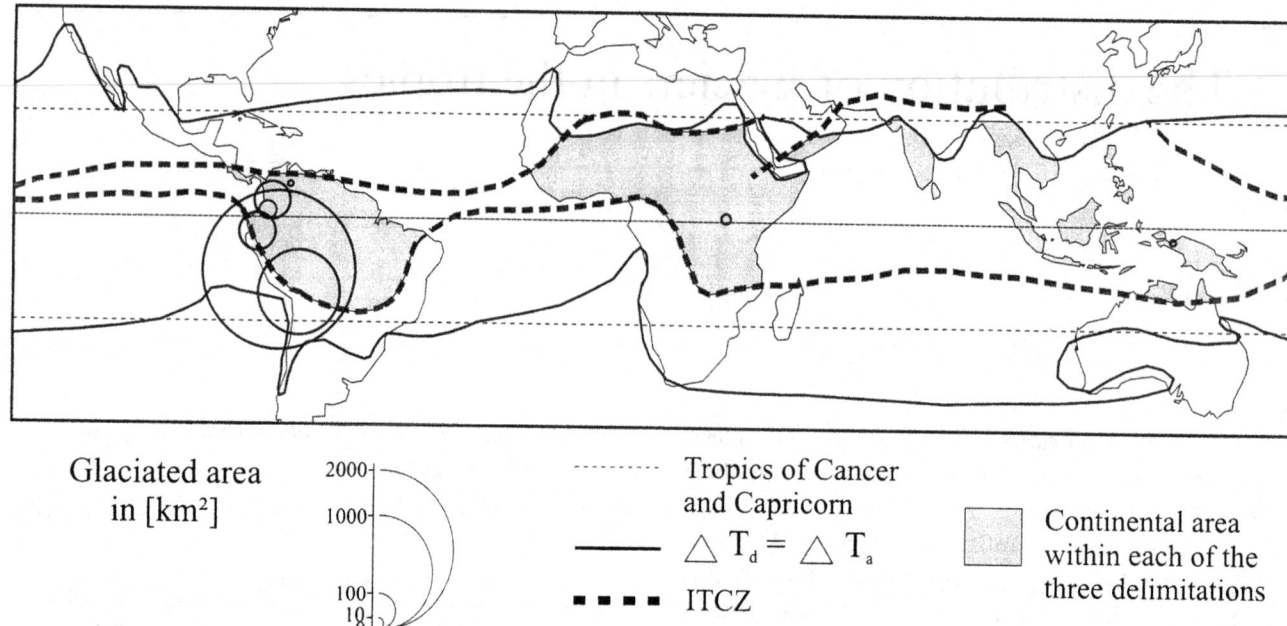

Glaciated area in [km²]

2000
1000
100
10
0

········· Tropics of Cancer and Capricorn

——— $\triangle T_d = \triangle T_a$

▪▪▪▪ ITCZ

Continental area within each of the three delimitations

Fig. 3.1 The distribution of the glacier areas in the tropics by country.

4 Working hypotheses

4.1 A SIMPLIFIED ASSUMPTION

The criteria suggested and discussed in chapter 2 describe a tropical atmosphere which is homogeneous, compared with one of the mid- or the high-latitudes, and seen from the temporal as well as from the spatial point of view. Under this circumstance, a glacier regime has to function differently from that in a climate that is characterized by a succession of winter (accumulation) and summer (ablation). Additionally, the climatological interpretation of glacier fluctuations can be made in a more direct manner than would be the case with the climate in the west wind zone of the mid-latitudes, which is characterized by a very dynamic and complex interplay of different air masses.

In order to examine the thermal homogeneity as the most significant difference between the tropics and the mid-latitudes, the following formula uses a simplified but clear boundary condition:

At any given altitude in the tropical atmosphere the temperature is constant both spatially (φ,λ) and temporally ($t>1$ day): $\delta T(t, \varphi,\lambda)=0$.

Apart from the complete lack of a thermal seasonality, this assumption also includes a constant altitude of the 0 °C level. The result of this is a constant altitude of the lower snowfall limit.

4.2 HOW VALID IS THIS ASSUMPTION?

The real conditions more or less deviate from this strict assumption. The dependence of the distribution of the annual variation of air temperature on the geographical latitude was first examined by Lautensach (1952) and, independently, by Iwanow (1959). Paffen (1967) discussed the two results and compared them with an analysis of his own from which Fig.

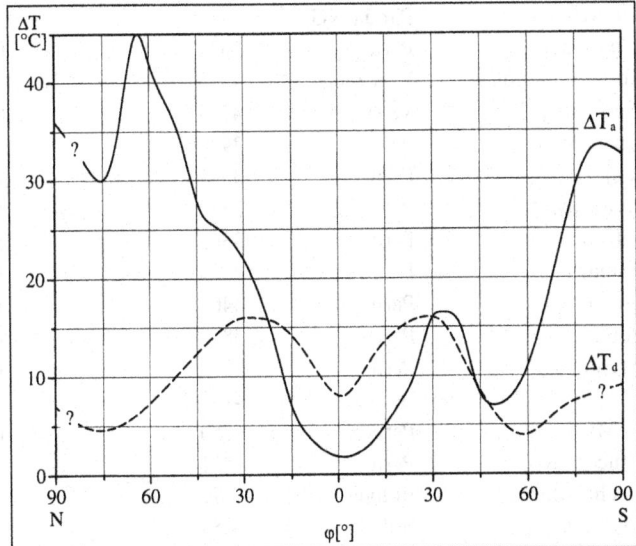

Fig. 4.2.1 Annual and diurnal variations of the air temperature (ΔT_a and ΔT_d) and its dependency on the geographical latitude φ. (According to Paffen, 1967.)

4.2.1 has been drawn. Paffen examined not only the annual but also the diurnal variation of air temperature and its dependency on the geographical latitude (Fig. 4.2.1). While the annual variation in the inner tropics comes to about 2 °C and the diurnal variation to about 8 °C, the annual variation increases towards the equilibrium line ($\Delta T_a = \Delta T_d$) to about 15 °C. The diurnal variation decreases slightly towards the mid-latitudes. The annual variation, however, greatly increases, especially in the northern hemisphere. The conditions in the southern hemisphere, which are not so distinct, can be attributed to the dominating water surfaces in these latitudes.

It can only be estimated with difficulty to what extent these models are also valid for the conditions in the tropical high mountain regions. Climatic data are, like all measured data from the tropical high mountain regions, extremely rare.

Table 4.2.1. *Annual variations of the maximum (ΔT_{ax}), mean (ΔT_a) and minimum (ΔT_{an}) air temperatures at tropical stations above 2000 m a.s.l., calculated from monthly means*

Station	Country	Altitude [m a.s.l.]	φ	ΔT_{ax}(°C)	ΔT_a(°C)	ΔT_{an}(°C)	Source
Equator Station	Kenya	2762	00°S	5	2	2	1
Quito	Equador	2812	00°S	3	0	3	1
Cotopaxi	Equador	3560	01°S		0.3		6
Rio Pita	Equador	3860	01°S		0.2		6
Kisozi – Collina	Burundi	2155	03°S		2		1
Bogotá	Colombia	2547	04°N	2	1	2	1
Ertsberg	Irian Jaya	3600	04°S		0.5		3
Mucubaji	Venezuela	3550	05°S		1.2		2
Mt. Wilhelm	Papua NG	3480	06°S		1.1		2
Cajamarca	Peru	2642	07°S	2	2	4	1
Loma Redonda	Venezuela	4065	08°N	4.9	2.9	3.4	7
Pico Espejo	Venezuela	4765	08°N	6.5	2.7	8.1	7
Querococha	Peru	3980	09°S	1.7	0.8	2.3	5
Jauja	Peru	3389	11°S	4	2	5	1
Huancayo	Peru	3350	12°S		2		1
Cuzco	Peru	3365	13°S	3	4	9	1
Asmara	Eritrea	2321	15°N	4	5	6	1
Imata	Peru	4405	15°S		6		1
Puno	Peru	3852	15°S		4		1
Arequipa	Peru	2332	16°S		3		1
El Alto	Bolivia	4103	16°S	4	3	6	1
El Misti	Peru	5850	16°S		5		4
Paucarany	Peru	4451	17°S		6		1
Cochabamba	Bolivia	2570	17°S		7		1
Charana	Bolivia	4059	17°S		7		1
Oruro	Bolivia	3706	18°S		9		1

Sources: 1, Rudloff (1981); 2, Barry (1992); 3, Allison & Bennett (1976); 4, Troll (1959); 5, Kaser *et al.*, (1990); 6, Jordan (1983); 7, Schubert (1972).

Lauscher (1966) examined the diurnal variations of air temperature at mountain stations. According to his findings, higher diurnal variations can be expected in tropical mountains than in lowlands, and a location in a valley shows higher results than a location on a free slope.

In Table 4.2.1 annual variations of the air at stations above 2000 m a.s.l. have been compiled and Fig. 4.2.2 shows these variations (ΔT_a) in relation to their geographical latitude φ.

The annual variation of the mean temperature remains below $\Delta T_a = 3\,°C$ up to a geographical latitude of 12°. However, there is no station among those mentioned which would describe the conditions near glaciers or even less near a mean equilibrium line of a glacier. Most of the stations are situated on plateaus and a few of them in towns. It can only be assumed that the annual temperature variations are low in the tropical high mountain regions. This is only indicated by the

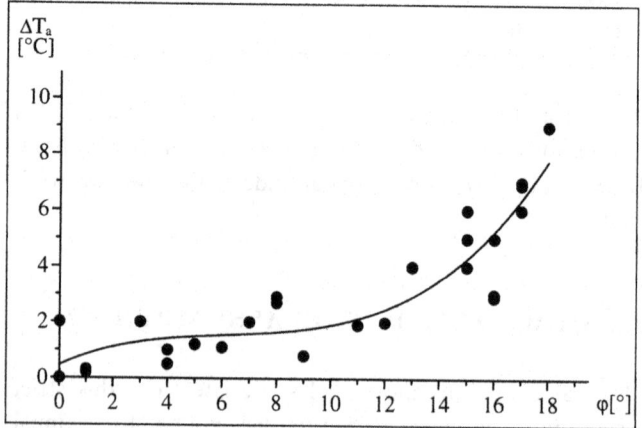

Fig. 4.2.2 The annual variation of the air temperature ΔT_a dependent on geographical latitude. (Sources: Rudloff, 1981; Barry, 1992; Allison & Bennett, 1976; Troll, 1959; Kaser *et al.*, 1990.) The regression curve was calculated as a second-degree polynomial.

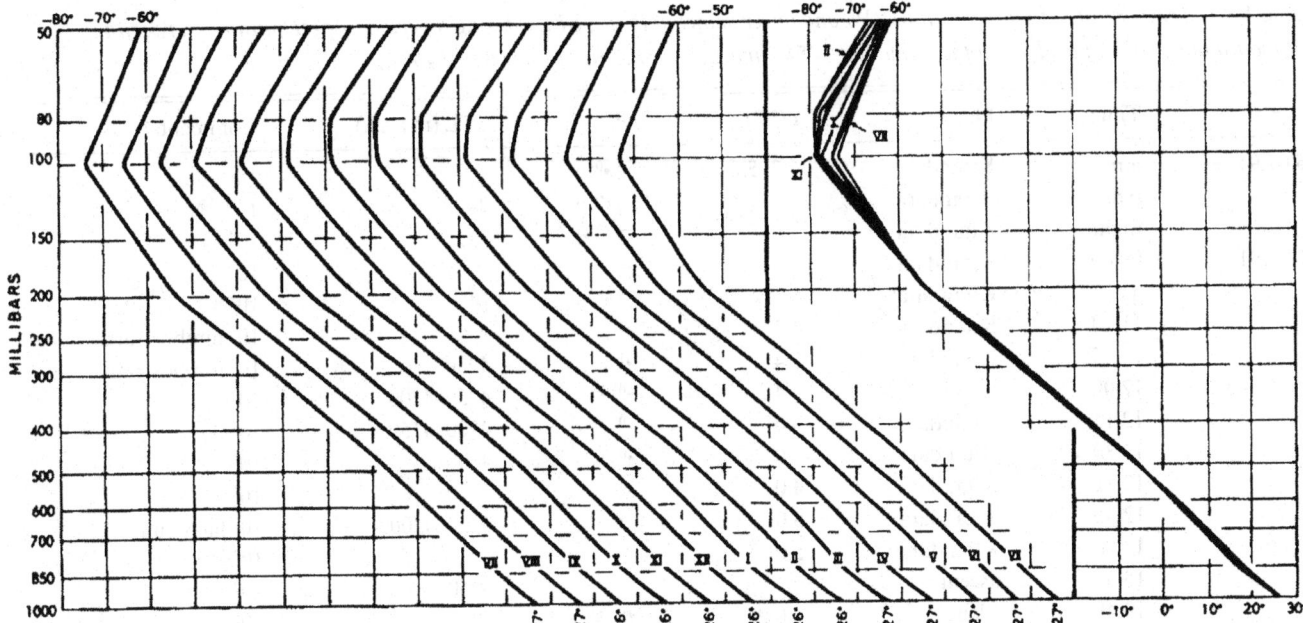

Fig. 4.2.3 Mean monthly temperature profiles above Balboa station on the Panama Canal. (Source: Crutcher, 1969.)

scarce available information and cannot be proved. An indication that the annual temperature variations in the tropical high mountains are extremely low is shown by the vertical temperature profiles in the free atmosphere taken from ascents with radiosondes (Crutcher, 1969; Prohaska, 1973; Asnani, 1993), which show generally low annual variations with their minimum just below the 500 hPa level (Fig. 4.2.3).

Results from a short period record in front of the tongue of the Carstensz Glacier in Irian Jaya, where carefully ventilated temperatures were measured in the 'base camp' at 4251 m altitude, show that the diurnal temperature variation close to the tropical glacier is very small (Allison & Bennett, 1976). With $\Delta T_d = 2.7\,°C$, the variation was clearly lower here than at the closely situated Ertsberg station (3600 m) with $\Delta T_d = 3.4\,°C$.

4.3 THE VARIATION OF THE LOWER SNOWFALL LIMIT

The consequence of the extremely small annual temperature variations is that the altitude of the 0 °C level, thus also the lower snowfall limit, varies only slightly. Such variations, however, can be effective under certain circumstances. Alcides Ames (Ames, 1995, personal communication), who has examined and regularly visited, among others, the Glaciar Yanamarey in the Cordillera Blanca for many years, reports that there is almost exclusively rain on the glacier tongue

during the wet period. During the dry period, single occasions of precipitation also create thin snow covers just below the glacier tongues. While the ice-free area is soon free of snow again, the snow can cover the whole glacier for a few days under the dry conditions which normally follow, and keep the shortwave albedo at high values. This shows that areas at the periphery of the tropics are exposed to 'subtropical' dry conditions which are temporally limited. The conditions of these areas will be discussed in more detail below.

Variations of the lower snowfall limit can also appear in the areas of the inner tropics with very small annual temperature variations. On the one hand, this can depend on the time of the day of precipitation and, on the other hand, it can also take place during and because of the precipitation itself. Table 4.3.1 compiles results that have been observed and measured which show the drop of the lower snowfall limit in the course of an occurrence of precipitation.

The observations were made during cartographical work in the Rwenzori mountains where the temperature was measured with a whirling thermometer. Precipitation had already started on the night of July 5. When it had grown in intensity in the morning of July 7, the temperature and the lower snowfall limit had fallen, only to rise again when the sun had come through. When, on July 8, the cloud cover over the Elena hut got thinner and the diffuse shortwave radiation increased, the temperature increased again and reached results of slightly above 0 °C, the snowfall changed into sleet and drizzle and the fallen snow started to melt again.

The seasonal variations of the lower snowfall limit compared with the ones in the temperate zones are so small under

Table 4.3.1. *Temperature (t), precipitation (P) and lower snowfall limit (SFL) during a period of precipitation between the Kitandara hut (4027 m a.s.l.) and the Elena hut (4540 m a.s.l.) on the Rwenzori (6–8.7.1991)*

Date	Time	Place	t (°C)	P	SFL (m a.s.l.)	Comments[a]
06.07.91	a.m.	Kitandara	4.7–5.0	□ •[0] •[1]		10[2]
	p.m.	Kitandara		□ •[0] •[1]	4400	10[2]
	night	Kitandara		□ •[0] •[1]		10[2]
07.07.91	09.00	Kitandara	4.0	•[0]		10[2]
	a.m.	Kitandara	5	□ •[0] •[1]		10[2]
	11.20	Kitandara	4.5	•[1]		10[2] nimbostratus
	11.45	Kitandara	3.4	•[1]	4250	10[2] nimbostratus
	12.00	Kitandara	3.4	•[0-1]	4250–4200	10[2]
	12.15	Kitandara	3.0	•[0]	4200–4150	10[2]
	12.30	Kitandara	3.0	•[0]	4200	10[2]
	12.38	Kitandara	4.0	•[0]	4200	10[2]
	12.45	Kitandara	5.0	□	4250–4300	10[2] hazy sun
	14.00	Kitandara	5.0	□		10[2]
	15.30	ascent			4250	10[2]
	19.40	Elena	−0.2	*[0-1]		10[2]
	20.00	Elena	−0.5	*		fog
08.07.91	08.30	Elena	−0.5	sleet		fog
	09.05	Elena	−0.1	*_•		fog
	09.30	Elena	−0.2	sleet		fog
	12.20	Elena	0.2	□	snow melts	fog, strong diff. radiation
	12.40	Elena	0.2	□		fog, strong diff. radiation
	13.30	Elena	0.1	sleet		fog

Notes:

Precipitation: • rain; □ drizzle; * snowfall. Intensity: [0] light, [1] moderate, [2] heavy.
[a]Cloud cover: 10 = 10/10 = overcast. Density: [0] light, [1] moderate, [2] heavy. According to Gutmann (1948).
Source: observations and measurements made by G. Kaser.

inner tropical conditions that they can be neglected. Platt (1966) reports from the Lewis Glacier (Mount Kenya) that daily melting processes were observed during the 'long rainy season' and that the meteorological conditions above the glacier are very similar to the ones in the dry months. For a short time, the lower snowfall limit may be exposed to comparatively small variations during occurrences of precipitation. However, in lower regions, the snow does not stay for very long. The degree of influence these variations have depends much on the size and above all on the vertical extent of a glacier. The assumption of a constant height of the 0°C isotherm and the lower snowfall limit in the inner tropics may well be true for bigger glaciers and glacial-climatological considerations.

The assumptions stated can also be applied, despite rare data, to the humid period of those regions which are reached by the ITCZ only once per year. The temperature variation is very small during that time (Fig. 7.2.2) and, when precipitation leads to cooling just as in other geographical latitudes (Steinacker, 1980), there are no lasting effects. This is also true for precipitation which occurs at a lower snow level due to lower temperatures by night. The snow melts just a few hours later (Jordan, 1979). However, the processes during the dry periods have to lead to a more or less clear deviation from the 'ideal-tropical' conditions and thus, from a glaciological point of view, to the definition of an 'outer tropical' regime as an intermediate regime between the tropical and the subtropical ones.

4.4 HOW TROPICAL ARE THE OUTER TROPICS? A SECOND ASSUMPTION

The Peruvian Cordillera Blanca, where the largest part of the tropical glaciers are concentrated, can be clearly assigned to the outer tropics (this will be discussed in detail in section 7.2). There, the annual temperature variations are still small, being only $\Delta T_a = 0.8$ °C (Table 4.2.1) at the Querococha station in a high valley in the south of the mountain range, although each variable of the heat balance could be expected to lead to a different result during the distinct dry period (Kaser *et al.*, 1990).

However, going further south and closer to the Tropic of Capricorn, the annual temperature variations increase, but still remain far below those that denote winter and summer in the temperate zones.

Therefore, the temperature variation to a small extent, and mainly the annual moisture variation (air humidity, cloud cover, precipitation) and the consequent variations of individual variables of the heat budget (global radiation, atmospheric longwave radiation, latent heat flux) lead us to a second working hypothesis:

In the outer tropics, the climate and glaciers are only 'tropical' within time limits. During the dry period, subtropical conditions are predominant.

4.5 TWO TROPICAL GLACIER REGIMES

Following the two hypotheses, two completely different types of tropical glacier regimes can be depicted which can be clearly differentiated from those of the mid- and high latitudes:

- on the one hand, the type of regime in the permanently humid inner tropics, and
- on the other hand, the type in the outer tropics, characterized by a period of precipitation and a dry period.

In Fig. 4.5.1 the two types are compared with a schematic regime of the temperate zones. The conditions at the Rwenzori and in the Cordillera Blanca are examples of these ideal models. The Rwenzori represents the inner tropics where there is precipitation all year round, increasing with a relatively slight variation of intensity during the twice yearly passage of the ITCZ. Constant temperature conditions, but also conditions of the heat budget that change little compared with the alternating humid outer tropics, lead to an ablation that appears throughout the whole year. Similar or even more distinct processes can be assumed for the glaciers of the Irian Jaya. Deviations from this idealized ablation process and their possible causes on the Rwenzori were discussed by Whittow (1960). However, neither Whittow's work nor any of the works quoted there show whether they deal with gross ablation or with the net sums of accumulation and ablation, possibly in the firn region.

In the Cordillera Blanca, the annual change from the period of precipitation to the dry period is distinct (Niedertscheider, 1990; Kaser *et al.*, 1990; see also Fig. 7.2.2). It can be used as an example for the conditions in the outer tropics. The ablation,

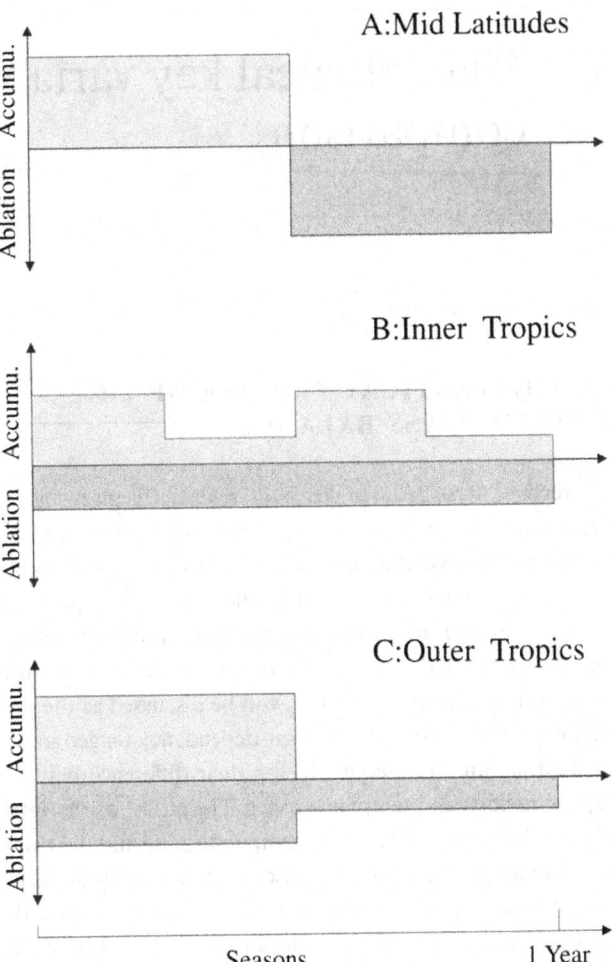

Fig. 4.5.1 A schematic comparison of tropical and outer tropical glacier regimes with those of the temperate zones (after Kaser *et al.*, 1996[2]).

however, is not constant, despite extremely low temperature variations. The lower ablation measured during the dry period (Kaser *et al.*, 1990 and section 5.1, Table 5.1.4) has to be ascribed, above all, to higher evaporation/sublimation, lower atmospheric longwave radiation and occasionally higher albedo.

While the seasonal variation of accumulation and ablation is dominated by the seasonal variation of the temperature in the mid and also in the higher latitudes, the glacier regimes are controlled by the moisture seasonality in the tropics. The spatial distribution of accumulation and ablation on the glaciers is determined by the variation of the temperature. While the whole glacier area stays far below freezing point during the winter in temperate and higher zones, there is ablation on the tongues of the tropical glaciers throughout the whole year. Accumulation, however, is limited to periods of precipitation and to the high areas of the glaciers. This has effects on the hydrological and glaciological records and key variables.

5 Glaciological key variables: tropics and mid-latitudes in comparison

5.1 THE VERTICAL PROFILE OF THE SPECIFIC MASS BALANCE

The vertical mass balance profile of a glacier is given by the variation of the vertical mass balance gradient with altitude. It describes the essential relation between a glacier and its climatic setting. Thus, the vertical gradient of the specific mass budget db/dz and the deriving mass balance profile, as well as the activity index $d\dot{b}/dz$ as the variation of the specific mass budget rate $\dot{b}$ with the altitude z, will be discussed as the first glaciological key variables for their dependency on geographical latitude and, consequently, for their difference between tropical regions and temperate zones. The model of the vertical mass balance profile for the temperate zone and the tropical regions makes it possible to discuss the essential causes for the differences in the behaviour of glaciers depending on the latitude. First results of these ideas were presented by Kaser (1995[1,2]) and Kaser et al. (1996[2]).

5.1.1 The activity index and the vertical mass balance profile of a glacier

Kuhn & Herrmann (1990) describe the **activity index** of a **steady-state** glacier as follows. In the steady-state case, the vertical changes due to the mass balance (positive in the accumulation area and negative in the ablation area) have to be replaced by the vertical component (ρv_z) of the ice flux at every point on the surface:

$$b = \rho v_z \qquad (5.1.1)$$

The two sides of the equation have the dimension of mass per unit area and time. The mass budget can also be linked, via the continuity of the mass flow which is reduced to

$$\rho \vec{\nabla} \vec{v} = \rho \frac{\partial v_x}{\partial x} + \rho \frac{\partial v_z}{\partial z} \qquad (5.1.2)$$

in an incompressible medium and on the assumption of a negligible lateral divergence ($\partial v_y/\partial y = 0$) on the tongue of a glacier, to the longitudinal flow velocity v_x:

$$\frac{d\dot{b}}{dz} = \rho \frac{\partial v_z}{\partial z} = -\rho \frac{\partial v_x}{\partial x} \qquad (5.1.3)$$

On a steady-state glacier, the full net accumulation per unit time $\dot{C}$ has to be transported through the cross-section Q under the equilibrium line with the mean longitudinal velocity $\bar{v}_{x\,max}$ there, the position of maximum ice discharge:

$$\dot{C} = Q\bar{v}_{x\,max} \cdot \rho \text{ [mass per time]} \qquad (5.1.4)$$

The integration of equation 5.1.3 over the length X from the end of the tongue to the equilibrium line, where v_x reaches its maximum, establishes a relationship with equation 5.1.4 between the temporal derivative of the balance gradient and the net accumulation $\dot{C}$:

$$\frac{d\dot{b}}{dz} = -\rho \frac{v_{x\,max}}{X} = \frac{\dot{C}}{XQ} \qquad (5.1.5)$$

The vertical gradient of the specific mass budget rate is, therefore, proportional to the net accumulation rate divided by the volume of the tongue ($\approx XQ$) and certainly warrants the term 'activity index' which has been applied by other authors (e.g. Kuhn, 1980[2]). With a constant accumulation rate and a constant cross-section of the glacier at its equilibrium line, the length of the tongue is only dependent inversely on the vertical variation of the budget rate. However, the activity index is not dependent on the geographical latitude, under the assumption that the vertical variation of the mass balance parameters per time unit is independent of it. Still, if one rigorously follows the assumption that at any given altitude in the tropical atmosphere the temperature is constant both spatially and temporally, then the lower snowfall limit has to be assumed to be constant as well. If this lies above the glacier terminus, then the accumulation rate is markedly influenced. In this way, the geographical latitude influences the activity index.

If the activity index is integrated over the time of a mass balance year $\Delta t = 1a$, the **vertical mass balance gradient** of a glacier results:

$$\frac{db}{dz} = \int_{t=0}^{t=1a} \frac{\partial \dot{b}}{\partial z} \, dt \text{ [mass per surface unit and altitude]} \qquad (5.1.6)$$

($t = 0$ is the beginning of the ablation period). This vertical mass balance gradient is composed of a vertical accumulation gradient and a vertical ablation gradient, in which the accumulation rate $\dot{c}$ can be effective throughout the year ($\Delta t = 1a$). But the ablation rate $\dot{a}$ is only effective during the period of ablation ($\Delta t = \tau$):

$$\frac{db}{dz} = \frac{\partial c}{\partial z} + \frac{\partial a}{\partial z} = \int_{t=0}^{t=1a} \frac{\partial \dot{c}}{\partial z} \cdot dt + \int_{t=0}^{t=\tau} \frac{\partial \dot{a}}{\partial z} \cdot dt \quad (5.1.7)$$

As the period of ablation depends on the geographical position, especially on the geographical latitude φ, the vertical mass balance gradient also changes with it ($db/dz(\varphi)$).

Thus, the influence of the geographical latitude on the vertical mass balance gradient and the consequent mass balance profile is twofold: (a) due to the duration of the period of ablation and (b) due to the consistency of the altitude band where snow changes into rain in the tropical high mountain areas. Although the activity index $d\dot{b}/dz$ is independent of (a), the constant lower snowfall limit has an influence on it.

A comprehensive summary of the discussion about the vertical mass balance gradient was compiled by Kuhn (1980[2]).

5.1.2 The modelling of the vertical mass balance profile

Just as the specific mass budget of a glacier over a specific period is composed of the sum of specific accumulation (positive) and specific ablation (negative)

$$b(z) = c(z) + a(z) \quad [\text{kg m}^{-2}] \quad (5.1.8)$$

this is also true for the vertical mass balance gradient

$$\frac{db}{dz} = \frac{\partial c}{\partial z} + \frac{\partial a}{\partial z} \quad [\text{kg m}^{-2}\,\text{m}^{-1}] \quad (5.1.9)$$

The differential of the specific balance has the following form:

$$db = \frac{\partial c}{\partial z}\,dz + \frac{\partial a}{\partial z}\,dz \quad [\text{kg m}^{-2}] \quad (5.1.10)$$

Ablation usually consists of the meltwater runoff ($m(z)$) and the sublimation processes ($s(z)$) directed into the atmosphere and is, with the respective latent heat fluxes $Q_M(z)$ for melting and $Q_L(z)$ for sublimation, consequently part of the heat balance on the surface of a glacier:

$$Q_R(z) + Q_S(z) + Q_M(z) + Q_L(z) = 0 \quad [\text{MJ m}^{-2}\,\text{d}^{-1}] \quad (5.1.11)$$

$Q_R(z)$ is the heat flux resulting from the radiation balance and $Q_S(z)$ the sensible heat flux. Minor other fluxes are neglected. However, the specific ablation is

$$a(z) = m(z) + s(z) = \tau(z)\left(\frac{1}{L_M} Q_M(z) + \frac{1}{L_S} Q_L(z)\right) \quad (5.1.12)$$

with the heat of fusion $L_M = 0.334$ MJ kg^{-1}, the heat of sublimation $L_S = 2.835$ MJ kg^{-1}, and the duration of the ablation season τ, counted in days [d].

If $Q_M(z)$ is replaced with the help of equation 5.1.11, the result is

$$a(z) = -\tau(z)\left[\frac{1}{L_M}(Q_R(z) + Q_S(z)) + \left(\frac{1}{L_S} - \frac{1}{L_M}\right)Q_L(z)\right] \quad (5.1.13)$$

The sensible heat flux $Q_S(z)$ can be derived with the help of the heat transfer coefficient for turbulent exchange α_S in [MJ m^{-2} d^{-1}°C^{-1}] from the difference in temperature between the atmosphere and the surface of the glacier ($T_a(z) - T_s(z)$) in [K]. The radiation balance is composed of the absorbed portion of global radiation $G(z)(1 - r(z))$, the atmospheric incoming longwave radiation $A(z)$ and the outgoing longwave radiation $E(z)$. Both the outgoing as well as the incoming longwave radiation can be calculated from climatic data with the help of the Stefan–Boltzmann equation. Thus, the specific ablation is

$$a(z) = -\tau(z)\left\{\frac{1}{L_M}\left[G(z)(1 - r(z)) + \varepsilon_a \sigma T_a(z)^4 - \varepsilon_s \sigma T_s(z)^4 + \alpha_S(T_a(z) - T_s(z))\right] + \left(\frac{1}{L_S} - \frac{1}{L_M}\right)Q_L(z)\right\} \quad (5.1.14)$$

ε_a and ε_s are emission coefficients of the atmosphere near the surface and the surface of the glacier. The Stefan–Boltzmann constant is $\sigma = 4.9 \times 10^{-9}$ MJ m^{-2} d^{-1} K^{-4}. If based on a **reference level z_0 where $T_a = 273.15$ K $= 0$°C, which is equal to the 0°C level during the ablation period**, and taking into account the following assumptions:

- the surface temperature $T_s = 273.15$ K $= 0$°C over the entire glacier,
- the vertical gradient of the effective global radiation is $\partial G(1 - r)/\partial z = 0$, and
- the vertical gradient of the latent heat flux is $\partial Q_L/\partial z = 0$,
- $4\varepsilon_a \sigma\, 273.15^3 = \alpha_R$,

then the vertical ablation gradient at z_0 is

$$\left.\frac{\partial a}{\partial z}\right|_{z_0} = -\frac{\partial \tau}{\partial z}\left\{\frac{1}{L_M}[G(1-r) + \varepsilon_a \sigma T_a^4 - \varepsilon_s \sigma T_s^4 + \alpha(T_a - T_s)] + \left(\frac{1}{L_S} - \frac{1}{L_M}\right)Q_L\right\} - \tau\frac{1}{L_M}\left[\alpha_R\frac{\partial T_a}{\partial z} + \alpha_S\frac{\partial T_a}{\partial z}\right] \quad (5.1.15)$$

Considering that the term within the curly brackets is the ablation a_0 at z_0 divided by the respective number of days, and assuming that τ changes linearly with altitude starting from the value τ_0 at z_0, the equation 5.1.15 simplifies to

$$\left.\frac{\partial a}{\partial z}\right|_{z_0} = -\frac{\partial \tau}{\partial z}\left\{\frac{a_0}{\tau_0}\right\} - \left(\tau_0 + \frac{\partial \tau}{\partial z}dz\right)\frac{1}{L_M}\left[\alpha_R\frac{\partial T_a}{\partial z} + \alpha_S\frac{\partial T_a}{\partial z}\right] \quad (5.1.16)$$

According to equation 5.1.10, the differential of the specific mass balance at z_0 is

$$db = \frac{\partial c}{\partial z}dz - \left\{\frac{\partial \tau}{\partial z}\frac{a_0}{\tau_0} + \left(\tau_0 + \frac{\partial \tau}{\partial z}dz\right)\frac{1}{L_M}\left[\alpha_R\frac{\partial T_a}{\partial z} + \alpha_S\frac{\partial T_a}{\partial z}\right]\right\}dz$$

$$(5.1.17)$$

This equation describes the change of the specific balance with altitude, and thus the mass balance profile, under the assumption that it exclusively depends on the vertical gradients of the accumulation and the air temperature as well as the duration of the period of ablation and its variation with altitude. Possible influences of the remaining heat balance key variables, as well as their dependency on the vertical variation of the duration of the period of ablation, are disregarded here.

The specific mass balance, changing along discrete altitude steps Δz above and below z_0, is

$$\Delta b = \frac{\partial c}{\partial z}\Delta z - \left\{\frac{\partial \tau}{\partial z}\frac{a_0}{\tau_0} + \left(\tau_0 + \frac{\partial \tau}{\partial z}\Delta z\right)\frac{1}{L_M}\left[\alpha_R\frac{\partial T_a}{\partial z} + \alpha_S\frac{\partial T_a}{\partial z}\right]\right\}\Delta z$$

$$(5.1.18)$$

5.1.3 The vertical mass balance profile in the mid-latitudes

As a first step, a vertical mass balance profile for the Hintereisferner is modelled from climatic data under the assumption of equilibrium conditions and is then compared with a measured profile. This, consequently, forms an model for the mid-latitudes. The aim is not to find the best model for the Hintereisferner, but to find a simple formulation which can be transferred to the postulated climatic differences essential in the tropical regions. The position of the 0 °C level and the duration of the ablation period, as well as its influence on the mass balance profile, are at the centre of investigation.

The following values for the model key variables in the equation 5.1.18 are from publications by Kuhn et al. (1979) as well as Kuhn (1979, 1980[1,2]):

- the mean duration of the annual period of ablation at the altitude of the summer 0 °C level (June to September) is $\tau_0 = 100$ d,
- the duration of the period of ablation changes with the altitude with $\partial \tau / \partial z = -0.1$ d m^{-1},
- the accumulation changes with the altitude with the value of $\partial c/\partial z = 1$ kg m^{-2} m^{-1},
- the air temperature (about 2 m above the surface of the glacier) changes with the altitude with $\partial T_a/\partial z = -0.0065$ K m^{-1}.

Additionally:

- It is assumed that the surface of the glacier below the 0 °C level is continuously melting. The surface temperature is therefore $T_s = 273.15$ K = 0 °C and the sensible heat flux is directed towards the surface. It is further assumed, for simplification, that no ablation appears above the 0 °C level, although it has been proved possible (Kuhn, 1987).
- All those key variables of the energy balance that do not depend on the vertical gradient of the air temperature are assumed not to change with the altitude and, therefore, not to have an influence on the vertical mass balance gradients and, consequently, on the form of the mass balance profile.
- The annual ablation at the altitude of the summer 0 °C level (June–September) a_0 is, as an approximation, set equal to the ablation at the mean equilibrium line. This, in turn, is equal to the accumulation at the equilibrium line, which is 1600 kg m^{-2} at the Hintereisferner (Kuhn et al., 1979).

The constants needed for the solution of the equation 5.1.18 are chosen as follows:

- The mean heat transfer coefficient as an integrative quantity to parameterize heat conductivity and turbulent exchange in the boundary layer between the surface of the glacier and the atmosphere is given by Kuhn (1979) as $\alpha_S = 1.7 \pm 0.2$ MJ m^{-2} d^{-1} K^{-1}. Kuhn (1989) calculates a value of $\alpha_S = 0.5 - 2.7$ MJ m^{-2} d^{-1} K^{-1} from wind profiles measured at different glaciers. Tanzer (1986) gives the value of $\alpha_S = 0.5$ MJ m^{-2} d^{-1} K^{-1} as a long-term mean at the equilibrium line of the Hintereisferner, Funk (1985) a weekly mean of $\alpha_S = 0.9 - 1.5$ MJ m^{-2} d^{-1} K^{-1} for the Rhone Glacier and Schug (1987) $\alpha_S = 0.8$ MJ m^{-2} d^{-1} K^{-1} for a three-day period at the Schwarzmilzferner. Using the empirical formula introduced by Escher-Vetter (1980) according to Hofmann (1965),

$$\alpha_s = 0.49 \text{ MJ m}^{-2}\text{K}^{-1} \quad \sqrt{v/\text{m s}^{-1}} \qquad (5.1.19)$$

where v is the wind speed in [m s^{-1}] measured at a height of 4 m above the surface, then the values of $\alpha_S = 1.1$ MJ m^{-2} d^{-1} K^{-1} at $v = 5$ m s^{-1} and $\alpha_S = 1.56$ MJ m^{-2} d^{-1} K^{-1} at $v = 10$ m s^{-1} can be calculated for the heat transfer coefficient. It can be assumed that, because of the dependency of the heat transfer coefficient (Kuhn, 1989) on the wind speed and the roughness of the surface, this coefficient is lower in the area of the equilibrium line than above the bare ice surface of the tongue. The tongue is rougher during the period of ablation and, as a rule, the glacial wind is better developed there. In the forefield of the Vernagtferner (Austria), there are wind speeds mainly between 2 m s^{-1} and 4 m s^{-1} (Escher-Vetter, 1980). During a period of sunny weather in summer (9 days), a mean diurnal middle of $v = 5.0$ m s^{-1} was measured in front of the tongue of the Hintereisferner and $v = 3.3$ m s^{-1} at the Hintereis station at 3030 m (12 days) (Kaser, 1983[1,2]). All in all, a heat transfer coefficient that lies between 1 and 1.5 MJ m^{-2} d^{-1} K^{-1} has to be presumed on snow-free tongues. For the calculation of the mass balance profile of the Hintereisferner, a heat transfer coefficient of $\alpha_S = 1.5$ MJ m^{-2} d^{-1} K^{-1} shows the best correspondence with the measured distribution in altitude of the specific mass budget (Fig. 5.1.1). However, it goes without saying that α does not only parameterize the complex procedures of the sensible heat exchange but also compensates for all the neglected vertical changes.

- It is assumed, and again heavily simplified, that the emission coefficients of atmosphere (ε_a) and glacier surface (ε_s) are each close to 1 (for the model $\varepsilon_a = \varepsilon_s = 1$ is used). As, presumably $\varepsilon_a < 1 \approx \varepsilon_s$ (Kuhn, 1980[2]), this simplification also leads to an underestimation of the vertical change of the ablation.
- The initial level of the calculations is the 0 °C level ($z = 0$) in summer from which the mass balance profile is calculated upward ($z = 0 \Uparrow$) and downward ($z = 0 \Downarrow$).

The amounts of the variables and constants used in the model calculation are compiled in Table 5.1.1. Fig. 5.1.1 compares the modelled mass balance profile with the measured values from the balance year 1966/67 which had a balanced mass budget with the mean specific mass budget of $\bar{b} = + 20$ kg m^{-2} and can, therefore, be adopted as a steady-state situation.

Table 5.1.1. *Variables and constants for the calculation of the vertical mass balance gradient in the mid-latitudes*

	Mid-latitudes
τ_0	100 d
a_0	1600 kg m^{-2} a^{-1}
$\partial\tau/\partial z$	-0.1 d m^{-1}
$\partial c/\partial z (z = 0\Uparrow)$	1 kg m^{-2} m^{-1}
$\partial c/\partial z (z = 0\Downarrow)$	1 kg m^{-2} m^{-1}
$\partial T_a/\partial z$	-0.0065 K m^{-1}
T_s (abl.)	273.15K
T_s (acc.)	T_a
α_S	1.5 MJ m^{-2} d^{-1} K^{-1}
ε_a	1
ε_s	1
σ	4.9×10^{-9} MJ m^{-2} d^{-1} K^{-4}
L_M	0.334 MJ kg^{-1}

Notes:
See Fig. 5.1.1. Meaning of the symbols and discussion of the amounts: see text.

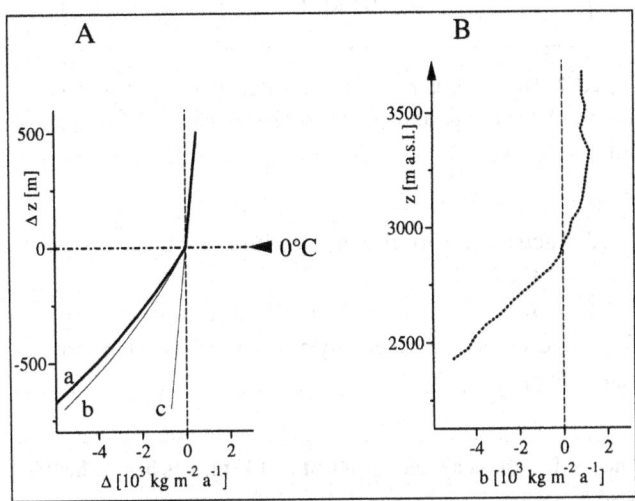

Fig. 5.1.1 The variations (Δ) of specific accumulation (c), ablation (a) and mass budget (b) with altitude were calculated from climatic data for the Hintereisferner (A), using the 0 °C line in summer as reference level. These are compared with the measured balance altitude distribution from the balanced budget year 1966/67 ($\bar{b} = + 20$ kg m^{-2}) (B).

In order to reach an even better comparison, the measured curve is best fitted by a parallel displacement along the coordinate axes (Fig. 5.1.2).

The curves correspond to a high degree and are characterized by a continuous increase of the vertical mass balance gradient toward the tongue which depends on $\partial\tau/\partial z < 0$ and on an

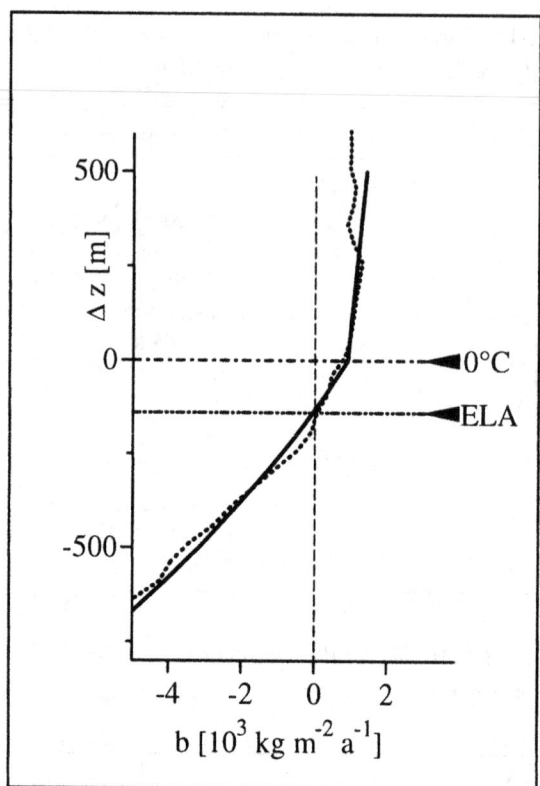

Fig. 5.1.2 The model curve of the Hintereisferner is best fitted into the measured mass balance profile (dotted line) through parallel displacement.

albedo increasing with decreasing altitude. The last is disregarded in the model.

The vertical mass balance gradient at the altitude in question is decisive for the sensitivity of the equilibrium line on climatic changes.

The vertical mass balance gradient at 400 meters below the 0 °C level is 7.8 kg m^{-2} m^{-1} (Fig. 5.1.2). It is adopted as a representative for the mid-latitudes.

It is especially remarkable that the equilibrium line is about 130 m below the 0 °C level. This corresponds well with the inner-Alpine conditions on the Hintereisferner. In a long-standing mean (1969 to 1978), summer temperatures (May to September) of $t_a = 0.4$ °C were measured at the Hintereis station (3030 m). Along the mean vertical temperature gradient of -0.0065 K m^{-1}, the altitude of the 0 °C level is calculated as 3090 m. The mean equilibrium line was found at 2930 m during the same period, which is 120 m below the 0 °C level in summer (data from Kuhn *et al.*, 1979).

The deviations in the accumulation area can be explained

without hesitation by concluding that the accumulation decreases towards the high and wind-exposed positions mostly as a result of eroding drift.

The deviations along the curve in the ablation area have to be mainly attributed to the vertical change of the incoming radiation which is neglected in this model formulation. A different duration of snow-free ice surface and snow-free surroundings will cause, on average, an increased gain of energy from the shortwave radiation budget as well as from the longwave radiation from the increasingly snow-free morainic slopes towards the tongue.

5.1.4 The vertical mass balance profile in the tropics

For the modelling of the vertical mass balance profile, it is assumed that the ablation period in the tropical regions is $\tau = 365$ days per year. This implies that $\partial \tau / \partial z = 0$ and leads to a marked simplification of equation 5.1.18. It is also assumed that there is mixed precipitation below the constant height of the mean annual 0 °C level. The proportion of rain increases with the moist adiabatic gradient downwards and is 100% at 400 m below the 0 °C level. With a transition zone of 400 m, it is roughly presumed that there is a mixed zone as well as that the snowfall limit decreases during an event of precipitation (see observations on the Rwenzori, Table 4.3.1). The energy gathered from rain is omitted because of the minimal differences in temperature from that of the melting surface. Only these assumptions will be applied to the model of the Hintereisferner.

A mid-latitude alpine glacier is exposed to ideal tropical conditions in a model. How is the mass balance profile going to change and what consequences is this going to have on the glaciological key variables?

The assumption of a constant ablation also implies that there is no change of the duration of the ablation period $\partial \tau / \partial z = 0$ d m^{-1}. Thus, the vertical ablation gradient is linear. The calculation of the change of the accumulation with the change in altitude occurs in two steps. For the altitude zones above the 0 °C level, as at the Hintereisferner, a linear increase with height $\partial c / \partial z = 1$ kg m^{-2} m^{-1} is assumed. Below that, the value increases to $\partial c / \partial z = 4$ kg m^{-2} m^{-1}. This value is calculated from the assumption that the annual accumulation is 1600 kg m^{-2} at the 0 °C level (mean value at the mean equilibrium line at the Hintereisferner; Kuhn *et al.*, 1979) and is zero 400 m below that level. The values used in the model calculations are compiled in Table 5.1.2.

The mass balance profile modelled for tropical conditions

Table 5.1.2. *Variables and constants for the calculation of the vertical mass balance gradient in the inner tropics*

	Tropics
$\tau_{z=0}$	365 d
$\partial\tau/\partial z$	0 d m^{-1}
$\partial c/\partial z(z=0\Uparrow)$	1 kg m^{-2} m^{-1}
$\partial c/\partial z(z=0\Downarrow)$	4 kg m^{-2} m^{-1}
$\partial T_a/\partial z$	−0.0065 K m^{-1}
T_s (abl.)	273.15 K
T_s (acc.)	T_a
α_S	1.5 MJ m^{-2} d^{-1} K^{-1}
ε_a	1
ε_s	1
$*\Rightarrow\bullet$	400 m
σ	4.9×10^{-9} MJ m^{-2} d^{-1} K^{-4}
L_M	0.344 MJ kg^{-1}

Notes:
See Fig. 5.1.3. Meaning of the symbols and discussion of the amounts: see text.

has been compared with measured curves from the lower latitudes in Fig. 5.1.3. For this purpose, only two measured and one calculated mass balance profiles are available for comparison.

Mass balances have been available for the Lewis Glacier on Mount Kenya since 1971 (Hastenrath, 1984, 1991[2]; Hastenrath *et al.*, 1989). However, there is no year in the whole series that had a balanced mass budget, as was the case on the Hintereisferner in 1966/67, although there are two years in which, with reversed signs, the same mass balance was measured. As the mass balance profiles in the two balance years 1986/87 with $\bar{b}=-770$ kg m^{-2} and 1988/89 with $\bar{b}=+770$ kg m^{-2} have very similar forms, a curve is determined for a balanced mass budget as an arithmetical mean between the two measured ones and the corresponding equilibrium line is established (Fig. 5.1.4).

During the Australian Universities Expeditions from 1971 to 1973, the mass budget and, consequently, the mass balance profile were determined for 1972 with the direct glaciological method for the two largest glaciers of the Punjak Jaya, the Meren and the Carstensz Glaciers (Allison, 1976). Hastenrath

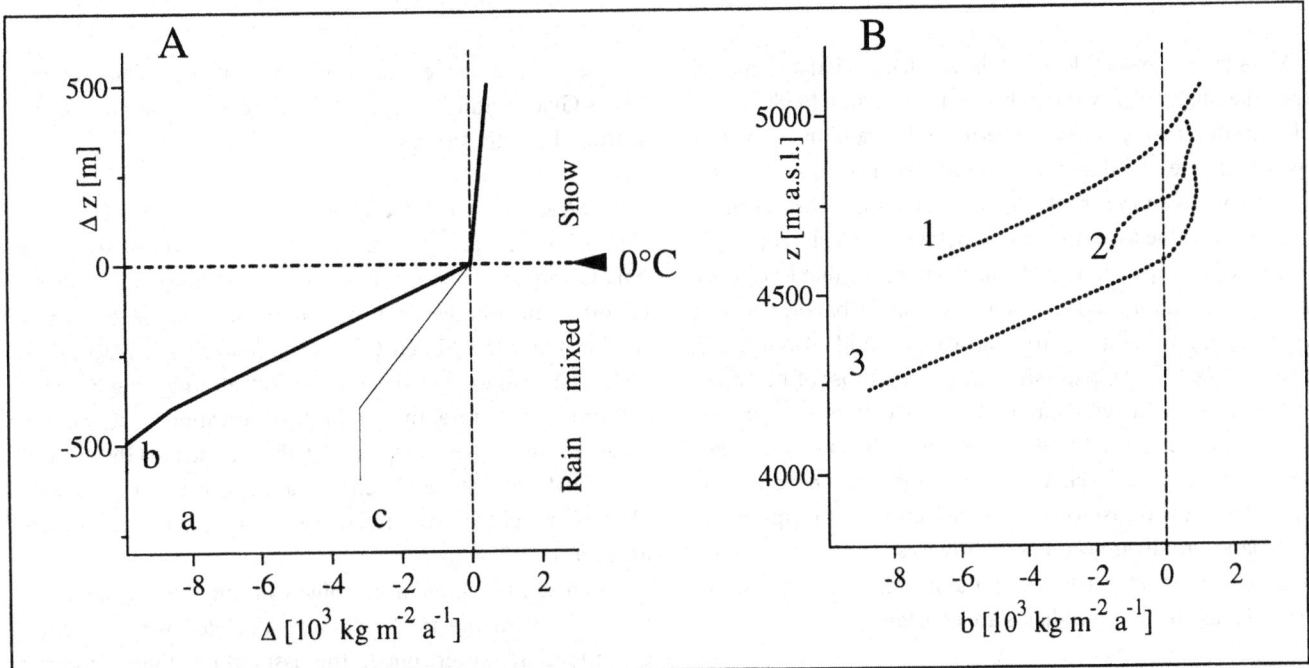

Fig. 5.1.3 The variations (Δ) of the specific accumulation (c), ablation (a) and mass budget (b) with the altitude were modelled (A), taking the 0 °C level as a reference level and under the assumption of ideal-tropical climatic conditions. This is compared with the measured mass balance profile of tropical glaciers (B). (1) Glaciar Yanamarey, Cordillera Blanca (Hastenrath & Ames, 1995[2]); (2) Lewis Glacier, Mt. Kenya (Fig. 5.1.4); (3) Meren and Carstensz Glacier, Irian Jaya (Allison, 1976).

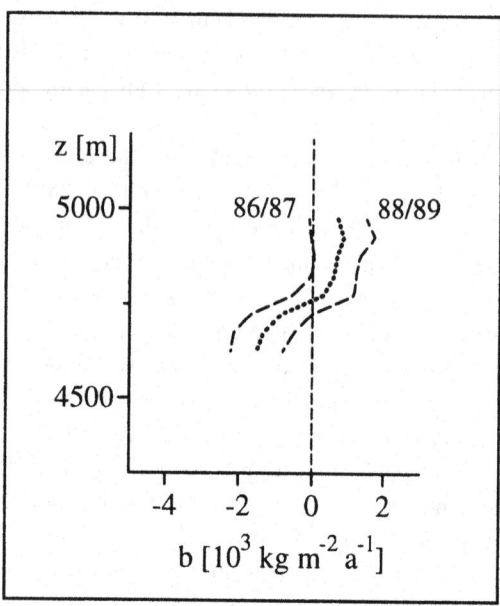

Fig. 5.1.4 Lewis Glacier, Mount Kenya. A mass balance profile for a balanced budget year (dotted line) was determined from the two measured mass balance profiles for 1986/87 with $\bar{b} = -770$ kg m^{-2} and for 1988/89 with $\bar{b} = +770$ kg m^{-2}.

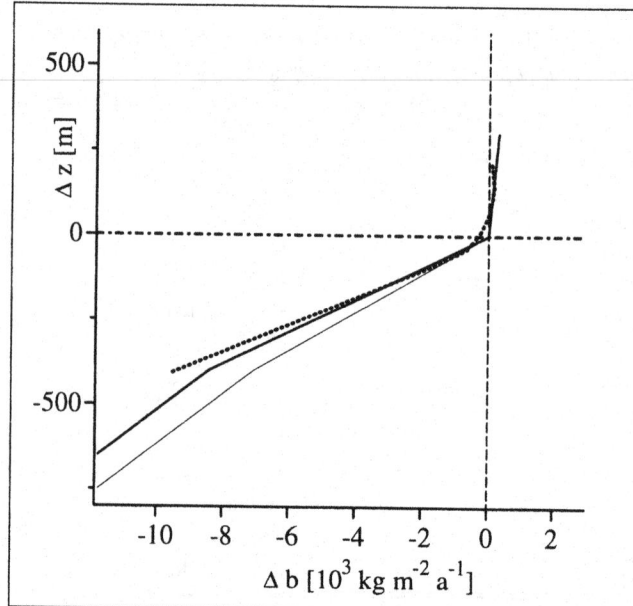

Fig. 5.1.5 The measured mass balance profile (dotted line) of the Meren and Carstensz Glaciers is optimally fitted into the model curves by a translatory shift (thick line: $c_{z_0} = 3000$ kg m^{-2}, thin line: $c_{z_0} = 1600$ kg m^{-2}).

& Ames (1995[2]) have calculated the mean mass balance profile from the changes in volume between 1977 and 1988 for the Glaciar Yanamarey in the southern Cordillera Blanca. In both cases, there are no values available for a balanced mass budget.

The sole assumption of temporal and spatial isothermia, together with the assumption that below the 0 °C level there is first mixed precipitation and then, 400 m below that level, only rainy precipitation, leads to a vertical mass balance profile which distinguishes itself from those of the mid-latitudes and corresponds, in its first approximation, with those of the lower latitudes. While the gradient above the 0 °C level is similar to the one in the mid-latitudes, it becomes distinctly stronger below the reference level, as shown by the sharp bend in Fig. 5.1.5. Below the transition into exclusively rainy precipitation, the balance gradient becomes slightly weaker. Apart from the discrete changes at places set by the model assumptions, the vertical mass balance gradients are constant.

For the tropics, a balance gradient can be calculated that is 17.5 kg m^{-2} m^{-1} within the snow-rain mixed area and 13.5 kg m^{-2} m^{-1} below that area (Fig. 5.1.3).

How well does this model describe mass balance profiles measured at individual tropical glaciers in detail? How can possible deviations be explained? These questions are first investigated

for the glaciers in the wet-tropical Punjak Jaya, then for the Lewis Glacier and, finally, for the Glaciar Yanamarey representing the outer tropics.

PUNJAK JAYA (IRIAN JAYA)

Fig. 5.1.5 compares the calculated with the measured mass balance profile. The mass balance profile reported by Allison (1976) is an average from measurements on the Carstensz (0.9 km^2) and the Meren Glaciers (1.9 km^2) and is valid for 1972. The values for the lowest 200 altitude meters were extrapolated from a three-month observation. The specific balances were $\bar{b} = -60$ kg m^{-2} at the Carstensz Glacier and $\bar{b} = -510$ kg m^{-2} at the Meren Glacier. The mean value of both glaciers, weighted with reference to the respective surface areas, is $\bar{b} = -374$ kg m^{-2}.

The mean mass balance profile can only be conditionally compared with the model curves calculated with balanced conditions. However, under the assumption that the curve moves horizontally parallel in different balances (e.g. Hoinkes, 1970), the result may still be valid. The strong change of the balance gradient at the 0 °C level is clearly visible on the glaciers of the Punjak Jaya. Below that the gradient, as the one calculated in the model, is constant. Yet, in reality it is stronger than the one calculated with the data from Table 5.1.2 (thin line in Fig. 5.1.5). A better fit of the model is obtained if the annual accumulation at the 0 °C level is formulated higher.

This, in turn, is more realistic as Allison & Bennett (1976) report a mean annual precipitation of about 3000 mm from stations in the surroundings of the mountains. Hence, the thick line in Fig. 5.1.5 was calculated with $c_{z_0} = 3000$ kg m^{-2}, resulting in an accumulation gradient in the snow-rain transition zone of $\partial c / \partial z$ ($z = 0 \Downarrow$) = 7.5 kg m^{-2} m^{-1}. The difference from the measured values is only slight and can be related to the simplified assumptions.

A statement about the steady-state position of the equilibrium line cannot be given from the data available. But, if the mass balance profile is moved to positive balance values under the assumption that it maintains a constant form (Hoinkes, 1970), then the equilibrium line has to be below the 0 °C level. Schwerdtfeger (1976), Kuhn (1980[2]) and Ohmura et al. (1992) have analysed the position of the mean equilibrium line in its relation to the position of the 0 °C level during the period of ablation (see Fig. 5.2.1 below). They show the equilibrium line of the glaciers within the wet tropics of South America to be uniformly slightly below the 0 °C level. Due to the modest amount of data available, a discussion in further detail is not possible.

The characteristic of wet tropical mass balance profiles mainly depends on the facts that ablation takes place all year round and that the 0 °C level is almost constant. Not only the strong balance gradients along the tongue, but also the abrupt transition to weak gradients in the accumulation area can be described clearly with this model.

LEWIS GLACIER (MOUNT KENYA)

The mass balance profile (Fig. 5.1.4) established for the Lewis glacier for a balanced mass budget is, as for the Hintereisferner (Fig. 5.1.2), best fitted along the coordinate axes through a translatory shift (Fig. 5.1.6). The scales on the axes of the model results with their relative values and those of the measured curves with their absolute values are again equal. In consideration of the local conditions, the real accumulation at the 0 °C level was estimated. From details given by Hastenrath (1984), a mean annual sum of 717 mm precipitation at the Austrian Hut (4800 m) on the Lewis Glacier between 1974 and 1972 can be estimated. For the calculation of the thick line in Fig. 5.1.6 it was assumed that $c_{z_0} = 800$ kg m^{-2} and, consequently, that $\partial c / \partial z$ ($z = 0 \Downarrow$) = 2 kg m^{-2} m^{-1}.

Although the Lewis Glacier, with an area of about 0.3 km^2, is much smaller than the Hintereisferner (about 9 km^2) and, therefore, strongly shaped by local influences, the curve derived from measurements corresponds well with the modelled one, except for the area of the lowest tongue. The drastic change at the 0 °C level, as well as the strong gradient below

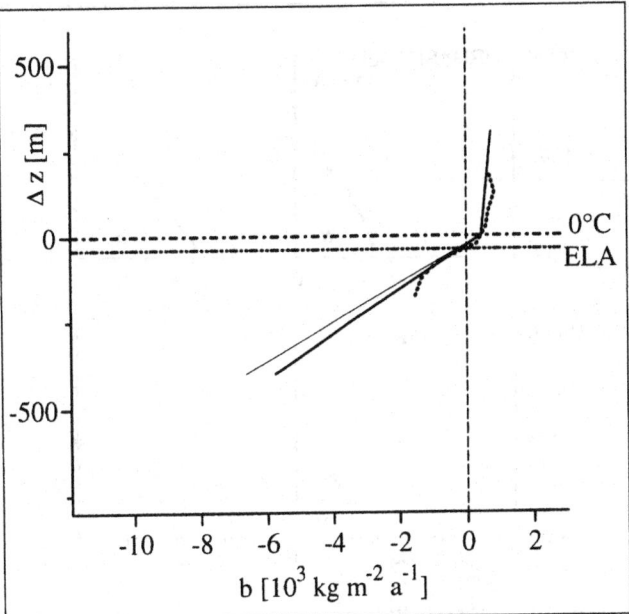

Fig. 5.1.6 The model curves are best fitted into the mass balance profile from measurements on the Lewis Glacier (dotted line) through a translatory shift (thick line: $c_{z_0} = 800$ kg m^{-2}, thin line: $c_{z_0} = 1600$ kg m^{-2}).

that level, are distinct. This suggests that the assumption of the snow–rain transition zone at 400 m below the 0 °C level is within the correct order of magnitude. The lower snowfall limit during the course of precipitation sometimes reaches Mackinder's Camp, which is situated 300 m below the tongue of the Lewis Glacier. Local influences (shading, possible drifting of snow) can lead to weak balance gradients in the area of the tongue. The equilibrium line is just below the 0 °C level and thus corresponds with the conditions analysed by Schwerdtfeger (1976), Kuhn (1980[2]) and Ohmura et al. (1992) for the glaciers within the wet tropics of South America.

Also in this case, the balance gradient in the ablation zone as well as its transition to the one in the accumulation zone is most satisfactorily calculated with the simple assumptions.

GLACIAR YANAMAREY (CORDILLERA BLANCA)

Although, at first glance, the altitude distribution of the specific mass balance in the Cordillera Blanca has generally quite tropical characteristics (Fig. 5.1.3), with closer observation the following differences from the analyses by Kaser et al. (1990), Niedertscheider (1990) and Kaser et al. (1996[1]) have to be considered:

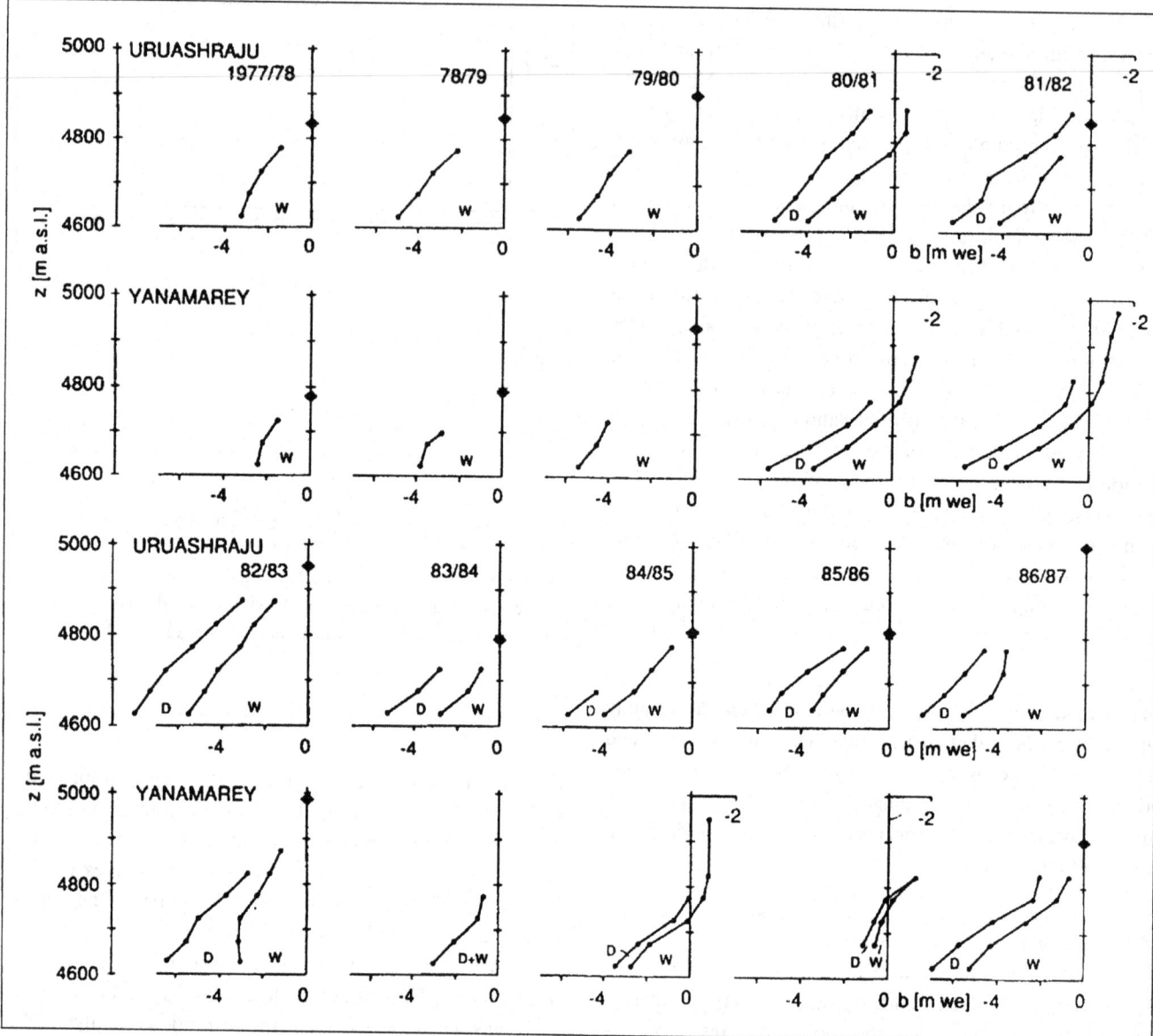

Fig. 5.1.7 The variation of the specific mass budgets *b* with the altitude *z* at the tongues of the Yanamarey and Uruashraju Glaciers (Cordillera Blanca) during the wet (W) and the dry period (D). (From Kaser *et al.*, 1990.)

A. Fig. 5.1.7 shows that ablation takes place during the wet as well as the dry period on the tongues of the Yanamarey and Uruashraju Glaciers. Vertical changes of the specific balance, however, only develop during the wet period. During the dry period the same amount ablates at all altitudes.

The reasons for missing gradients during the dry period can only be assumed:

• A considerable part of the ablation during the dry period takes place through sublimation of snow and ice

as well as evaporation of meltwater. This is assumed to be the reason that the mean diurnal amount of ablation at the Glaciar Yanamarey, and even more so at the nearby Glaciar Uruashraju, is lower during the dry period in some years despite higher radiation than during the wet period (Table 5.1.3). The impression of heavy evaporation and sublimation processes was received during field work, when no remarkable diurnal increase of glacial streams (Yanamarey, Uruashraju, Gajap, Artesonraju) was observed on cloudless days during the dry period.

The atmospheric vapour pressure, as a mainly conservative term in the glacier wind system, and eventually stronger turbulent exchange above the tongue, as well as small changes of the surface temperature with

Table 5.1.3. *Values of the mean diurnal net ablation during the wet (b_w) and the dry period (b_d) at an altitude of 4600–4650 m at the Glaciar Uruashraju and at the Glaciar Yanamarey (Cordillera Blanca)*

	Uruashraju			Yanamarey		
	b_w (mm we)	b_d (mm we)	$100\, b_d/b_w$ (%)	b_w (mm we)	b_d (mm we)	$100\, b_d/b_w$ (%)
1980/81	−17.36	−11.11	64	−14.48	−15.75	109
1981/82	−18.87	−15.88	84	−15.38	−17.12	111
1982/83	−24.01	−18.63	78	−14.34	−22.61	158
1984/85	−21.79	−10.39	48	−12.74	−5.88	46
1985/86	−16.70	−14.64	88			
1986/87	−24.20	−15.03	62	−22.64	−13.73	61

Source: According to Kaser *et al.* (1990).

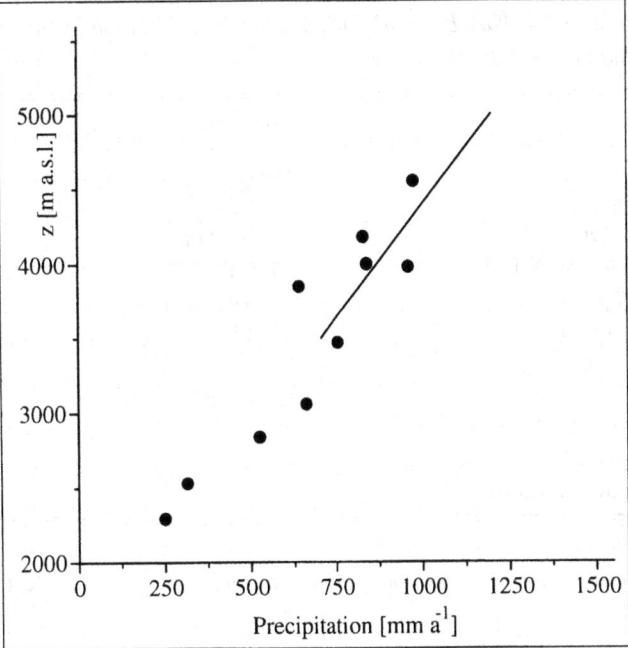

Fig. 5.1.8 Precipitation distribution with increased altitude in the Cordillera Blanca (according to Niedertscheider, 1990).

changes in altitude could be the reasons that sublimation and evaporation show small differences with increased altitude. But the fact that the mass balance gradients are also missing for those years, where ablation is higher in the dry period than in the wet period, shows that this effect can only offer a partial explanation at best.

• Compared with the wet period, in which precipitation appears almost always as rain on the tongues, occasional snowfall during the dry period also covers the lower parts of the glacier tongue of the Glaciar Yanamarey (Ames, personal information; own observation). This can also limit its vertical gradient, in addition to a general reduction of the shortwave radiation balance.

These approaches, however, do not offer a plausible complete explanation. It has to remain unresolved due to missing observations and measurements. This, however, does not change the fact that a vertical mass balance gradient is only clearly effective during the wet period (October–April and, therefore, $\tau = 212$ d).

B. From precipitation data originating only from the catchment area of the Rio Santa, Fliri (1968) calculated an increase of precipitation with altitude of $\partial P/\partial z\,(z = 0\Downarrow)$ $= 0.46$ kg m^{-2} m^{-1}. Niedertscheider (1990), who had a temporally more extended collection of data available (1953–1986), calculated a gradient of $\partial P/\partial z\,(z = 0\Downarrow) = 0.26$ kg m^{-2} m^{-1} from all data available between the Hidroelectra (1386 m) and Cahuish (4550 m) stations. This dependency on altitude, which also includes stations in the Cordillera Negra situated in the west (Fig. 0.1), is

superimposed on a decrease of precipitation along the Rio Santa from north to south. A profile, reaching from the Rio Santa into the Cordillera Blanca between Caraz (2286 m) and Parón (4185 m), comes to $\partial P/\partial z\,(z = 0\Downarrow) =$ 0.34 kg m^{-2} m^{-1}. This profile is, however, situated in the northern Cordillera Blanca which is wetter than the area in which the Yanamarey and Uruashraju glaciers are situated. Fig. 5.1.8 displays all Niedertscheider's precipitation stations which are situated in the Cordillera Blanca, except the Safuna station (4275 m a.s.l.) whose strong deviation is still unexplained.

Values from the Cordillera Negra and from the main valley were omitted. The curve in Fig. 5.1.8 was plotted by hand. From this, a mean gradient of $\partial P/\partial z\,(z = 0\Downarrow) =$ 0.30 kg m^{-2} m^{-1} can be assumed. An extrapolation along this gradient up to the mean equilibrium line, which is at about 5000 m in the Cordillera Blanca (Kaser *et al.*, 1996[1]), and the assumption that precipitation equals accumulation, results in a value of $c_{z_0} = 1200$ kg m^{-2}. There are, however, no measurement stations at the altitude of the mean equilibrium line or even at that of the glaciers in general.

Under these conditions (A and B), a mass balance profile for the outer tropical conditions can be calculated. The values in Table 5.1.4 are compared with those of the mid-latitudes and the inner tropics.

Table 5.1.4. Variables and constants for the calculation of the vertical mass balance gradients in the mid-latitudes, the inner tropics and the outer tropics

	Mid-latitudes	Inner tropics	Outer tropics
$\tau_{z=0}$	100 d	365 d	212 d
$\partial \tau / \partial z$	-0.1 d m^{-1}	0 d m^{-1}	0 d m^{-1}
$\partial c / \partial z (z = 0 \Uparrow)$	1 kg m^{-2} m^{-1}	1 kg m^{-2} m^{-1}	0.3 kg m^{-2} m^{-1}
$\partial c / \partial z (z = 0 \Downarrow)$	1 kg m^{-2} m^{-1}	4 kg m^{-2} m^{-1}	3.0 kg m^{-2} m^{-1}
$\partial T_a / \partial z$	-0.0065 K m^{-1}	-0.0065 K m^{-1}	-0.0065 K m^{-1}
T_s (abl.)	273.15 K	273.15 K	273.15 K
T_s (acc.)	T_a	T_a	T_a
α_S	1.5 MJ m^{-2} d^{-1} K^{-1}	1.5 MJ m^{-2} d^{-1} K^{-1}	1.5 MJ m^{-2} d^{-1} K^{-1}
ε_a	1	1	1
ε_s	1	1	1
snow–rain zone		400 m	400 m

Notes:

For meaning of the symbols and discussion of the values: see text.

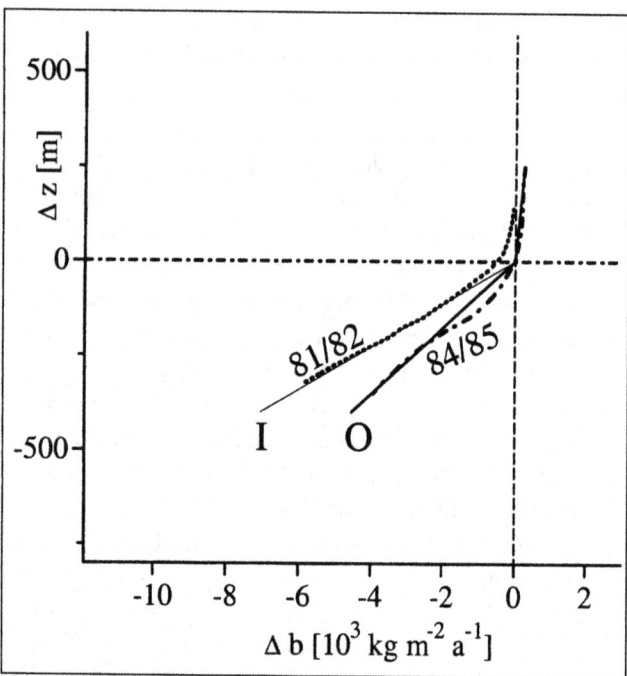

Fig. 5.1.9 Fitting of reconstructed balance altitude distributions (broken lines) measured at the Glaciar Yanamarey to two modelled curves (solid lines). (I) Inner Tropics, (O) Outer Tropics.

In Fig. 5.1.9, the mass balance profiles for the inner tropics (I) and for the outer tropics (O) are displayed. Curve B1 in Fig. 5.1.3 determined by Hastenrath & Ames (1995[2]) has similar gradients in the ablation area as the tropical model, but it clearly shows different values in the accumulation area. The calculation from long-term changes of volume naturally results in a smoothed distribution and suppresses sharp transitions to different gradients. Therefore, two curves from individual years, which were determined from level measurements in the accumulation area of the Glaciar Yanamarey, have been inserted into Fig. 5.1.7. Independently from each other, the measured curves have again been best fitted through a translatory shift. A discussion of the position of the mean equilibrium line in relation to the 0 °C level is also not possible for the Cordillera Blanca.

The curve from 1984/85 (Fig. 5.1.9, broken and dotted line) correlates well with the modelled curve for the outer tropics. The curve from 1981/82 (broken line) correlates well with the curve for the inner tropics. Most of the measured curves for the other years not pictured in Fig. 5.1.7 would lie between the shown curves. However, the reasons for the generally strong gradients remain unclear so far. There are insufficient data to formulate any assumptions and to evaluate whether conditions are in principle different during an ENSO (El Niño – Southern Oscillation) occurrence. The query of the extent to which the conditions at the Glaciar Yanamarey can be transferred to the many larger glaciers in the Cordillera Blanca remains unanswered. Furthermore, the conditions in the north of the Cordillera Blanca seem to be slightly wetter (Niedertscheider, 1990).

However, similarly to the case for inner tropical glaciers, the two measured examples show a relatively sharp transition from strong gradients in the ablation area to weak ones in the accumulation area. This shows that the assumption of a mainly constant 0 °C level is correct and of considerable influence.

Thus a constant 0°C level has a characteristic influence in the outer tropics too, but so far there are no data on the processes during the dry period or for the annual variation of the mass balance gradients on the tongues.

5.1.5 Summary of section 5.1

The influence of the geographical latitude on the vertical mass balance profile is twofold: (a) the duration of the ablation period and (b) the assumption of an altitude of transition from snow to rain which is constant in tropical high mountains. While the activity index $\partial b/\partial z$ is independent of (a), the constant snowfall limit has an influence on it.

The change of the specific mass balance with altitude can be described and calculated under the assumption that it only depends on the following points:

- the duration of the ablation period,
- its decrease with altitude,
- the vertical gradient of the accumulation, and
- the vertical gradient of air temperature.

The form of the tropical mass balance profile, which is different from the one of the mid-latitudes, is based to a thermally homogeneous atmosphere.

The vertical mass balance gradients, calculated under the model assumptions, are as follows:

- close to the equilibrium line on a glacier tongue in the mid-latitudes on average 7.8 kg m^{-2} m^{-1},
- constant in the area of a snow–rain mixture zone in the tropics at 17.5 kg m^{-2} m^{-1} and
- constant below the snow–rain mixture zone in the tropics at 13.5 kg m^{-2} m^{-1}.

The tropical balance gradients along the tongue of the outer tropical Glaciar Yanamarey (Cordillera Blanca), however, leave unanswered questions.

5.2 THE REACTION OF THE EQUILIBRIUM LINE ALTITUDE TO CLIMATIC VARIATIONS

The reaction of the equilibrium line altitude (ELA) to changed climatic conditions depends on the vertical mass balance gradient at the respective altitude. In the tropical regions it undergoes a sharp change in the area of the 0°C

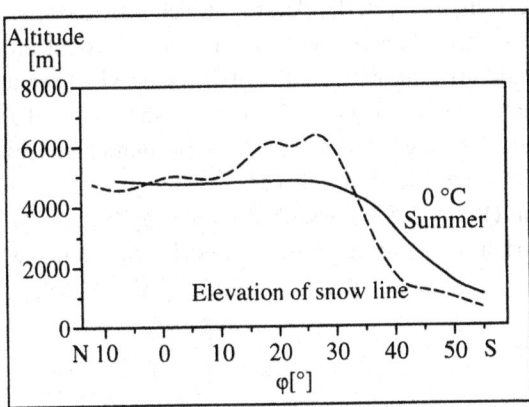

Fig. 5.2.1 The position of the snow line (mean equilibrium line) and the (summer) 0°C level in the South American Andes (according to Kuhn, 1980[2]).

level. Therefore, for questions regarding the sensitivity of the equilibrium line altitude to climatic changes, it is important to know its relative position to the 0°C level. In wet areas, the high accumulation depresses the altitude of the equilibrium line (ELA < 0°C level), and in dry areas the ablation processes (sublimation) can compensate for low annual accumulation even above the 0°C level (ELA > 0°C level).

According to Schwerdtfeger (1976), Kuhn (1980[2]) and Ohmura *et al.* (1992), the ELA is situated just below the 0°C level in the wet inner tropics of South America and, consequently, in the area of strong changes of the specific mass balance with height (Fig. 5.2.1).

The ELA approaches the 0°C level with decreasing humidity towards the outer tropics, only to rise far above it in the extremely dry subtropics of southern Bolivia and northern Chile. Fig. 5.2.1 shows the ELA slightly above the 0°C level for the Cordillera Blanca. It can be assumed that these conditions can be transferred to other tropical mountains.

Climatic changes result in a change of the specific mass balance δb, which is caused by changes of one or more of the following variables: the accumulation with the respective change δc, the air temperature (δT_a), the shortwave radiation balance ($\delta[G(1-r)]$); G is the global radiation, r is the albedo on the glacier surface), and the latent heat flux (δQ_L). The last one corresponds to changes of evaporation, condensation, sublimation, and/or resublimation δs. The different effects of these changes on the vertical shift of the equilibrium line (ΔELA) will now be discussed individually for each variable. Each change will be calculated as if it is solely responsible for a given ΔELA. The starting point is the assumption that the equilibrium line is elevated by ΔELA$_m$ = 100 m in the mid-latitudes.

This corresponds with the elevation which was reconstructed for the Eastern Alps between their maximum extent around 1850 and the situation around 1970 (Gross, 1987). Then it will be attempted to calculate a ΔELA_t below the 0°C level along tropical balance gradients with the corresponding necessary amounts of the individual changes.

Kuhn (1980[1,2], 1989) described the energy balance on the equilibrium line and developed a model formulation, which allows for the investigation of the connection between the individual climatic changes and the reaction of the equilibrium line. Kaser et al. (1996[1]), similar to Kuhn (1989), have added the latent heat flux as a variable to this model.

Applied to the equilibrium line, where the mass balance is by definition $b = 0$, it follows according to equations 5.1.8 and 5.1.14, that

$$c = -a = \tau\left\{\frac{1}{L_M}[G(1-r) + \varepsilon_a\sigma T_a^4 - \varepsilon_s\sigma T_s^4 + \alpha_S(T_a - T_s)] + \left(\frac{1}{L_S} - \frac{1}{L_M}\right)Q_L\right\} \qquad (5.2.1)$$

If δc and δQ are climatic variations of accumulation and heat balance, and if it is assumed for simplification that the latent heat flux does not change with altitude ($\partial Q_L/\partial z = 0$), then these variations can be compensated through adjustment of the altitude Δh of the ELA along the vertical mass balance gradient (Fig. 5.2.2):

$$\frac{\partial c}{\partial z}\Delta h + \delta c = \frac{\tau}{L_M}\left(\frac{\partial Q}{\partial z}\Delta h + \delta Q\right) \qquad (5.2.2)$$

For the next step, the vertical gradients $\partial/\partial z$ of the global radiation G, the albedo r, the incoming longwave radiation and the latent heat flux will also be omitted close to the equilibrium line. The assumption of a melting surface $T_s = 273.15$ K during the whole ablation period causes a constant longwave emission. After a linearization of the incoming longwave radiation and under the simplifying assumption that the air temperature is $T_a = 273.15$ K on the ELA, the variation of the incoming longwave radiation is

$$\delta A = \varepsilon 4\sigma \cdot 273.15^3\,\delta T_a = \alpha_R\delta T_a \qquad (5.2.3)$$

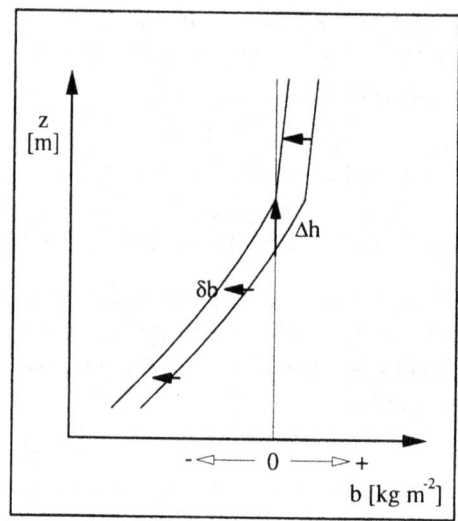

Fig. 5.2.2 Adaptation of the altitude Δh of the ELA to a climatic disturbance δb.

Under these assumptions, the conditions for an adjustment to a change δc can be calculated as

$$\delta c = \tau\left\{\frac{1}{L_M}\left[\delta[G(1-r)] + \alpha_S\left(\frac{\partial T_a}{\partial z}\Delta h + \delta T_a\right) + \alpha_R\delta T_a\right] + \left(\frac{1}{L_S} - \frac{1}{L_M}\right)\delta Q_L\right\} - \frac{\partial c}{\partial z}\Delta h \qquad (5.2.4)$$

If only one cause is assumed to be effective, then its value is calculated for a given shift of the equilibrium line Δh as

$$\delta T_a = \frac{\dfrac{\partial c}{\partial z}\Delta h - \tau\dfrac{1}{L_M}\alpha_S\dfrac{\partial T_a}{\partial z}\Delta h}{\tau\dfrac{1}{L_M}(\alpha_S + \alpha_R)} \qquad (5.2.5)$$

$$\delta c = \tau\frac{1}{L_M}\alpha_S\frac{\partial T_a}{\partial z}\Delta h - \frac{\partial c}{\partial z}\Delta h \qquad (5.2.6)$$

$$\delta[G(1-r)] = L_M\frac{1}{\tau}\frac{\partial c}{\partial z}\Delta h - \alpha_S\frac{\partial T_a}{\partial z}\Delta h \qquad (5.2.7)$$

$$\delta Q_L = \frac{\dfrac{\partial c}{\partial z}\Delta h - \tau\dfrac{1}{L_M}\alpha_S\dfrac{\partial T_a}{\partial z}\Delta h}{\tau\left(\dfrac{1}{L_S} - \dfrac{1}{L_M}\right)}\ [\text{MJ m}^{-2}\,\text{d}^{-1}] \qquad (5.2.8)$$

The corresponding values of sublimation are

$$\delta s = \tau\frac{1}{L_S}\delta Q_L\ [\text{kg m}^{-2}\,\text{a}^{-1}] \qquad (5.2.9)$$

The adaptation of the altitude of the equilibrium line can be calculated from one or the sum of more climatic changes:

$$\Delta h = \frac{\frac{\tau}{L_{\mathrm{M}}}\{\delta[G(1-r)] + \alpha_{\mathrm{R}}\delta T_{\mathrm{a}} - \alpha_{\mathrm{S}}\delta T_{\mathrm{a}}\} + \left(\frac{\tau}{L_{\mathrm{S}}} - \frac{\tau}{L_{\mathrm{M}}}\right)\delta Q_{\mathrm{L}} - \delta c}{\frac{\partial c}{\partial z} - \frac{\tau}{L_{\mathrm{M}}}\alpha_{\mathrm{S}}\frac{\partial T_{\mathrm{a}}}{\partial z}} \qquad (5.2.10)$$

While it can be assumed for glaciers in the mid-latitudes that, in the area of the equilibrium line, $\partial\tau/\partial z = 0$, and that there are no vertical gradients of the changes $(\partial(\delta c)/\partial z = 0,$ $\partial(\delta T_{\mathrm{a}})/\partial z = 0,$ $\partial(\delta[G(1-r)])/\partial z = 0$ and $\partial(\delta Q_{\mathrm{L}})/\partial z = 0)$, the accumulation gradient in the tropics changes according to $c_0 + \delta c$ at the 0 °C level, and with constant altitude of the snow–rain transition zone of 400 m, with

$$\frac{\partial c}{\partial z} = \frac{c_0 + \delta c}{400\ \mathrm{m}} \qquad (5.2.11)$$

(see Fig. 5.2.3). Equation 5.2.6 develops on a tropical glacier to

$$\delta c = \frac{\tau\frac{1}{L}\alpha_{\mathrm{S}}\frac{\partial T_{\mathrm{a}}}{\partial z}\Delta h - \frac{c_0}{400\ \mathrm{m}}\Delta h}{1 + \frac{\Delta h}{400\ \mathrm{m}}} \qquad (5.2.12)$$

With a given change of the accumulation, and hypothetically this being the only effective variable, the altitude of the equilibrium line moves by

$$\Delta h_{\mathrm{T}}(\delta c) = \frac{\delta c}{\tau\frac{1}{L_{\mathrm{M}}}\alpha_{\mathrm{S}}\frac{\partial T_{\mathrm{a}}}{\partial z} - \frac{c_0 + \delta c}{400\ \mathrm{m}}} \qquad (5.2.13)$$

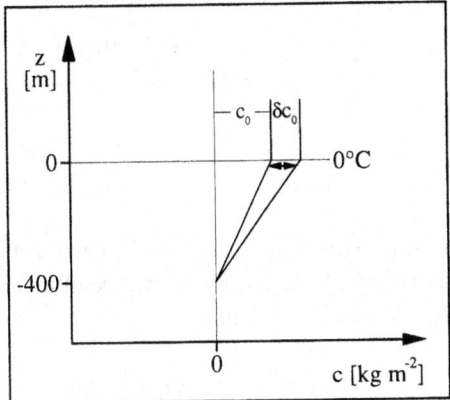

Fig. 5.2.3 The variation of the vertical balance gradient $\partial c/\partial z$ due to a climatic disturbance δc on a tropical glacier.

For those tropical glaciers where ELA < 0 °C line, the effect of a shift of the 0 °C level on the equilibrium line due to a possible change of the temperature δT_{a} must be calculated (Fig. 5.2.4). The 0 °C level is moved by

$$\Delta h_0 = \delta T_{\mathrm{a}}\left(\frac{\partial T_{\mathrm{a}}}{\partial z}\right)^{-1} \qquad (5.2.14)$$

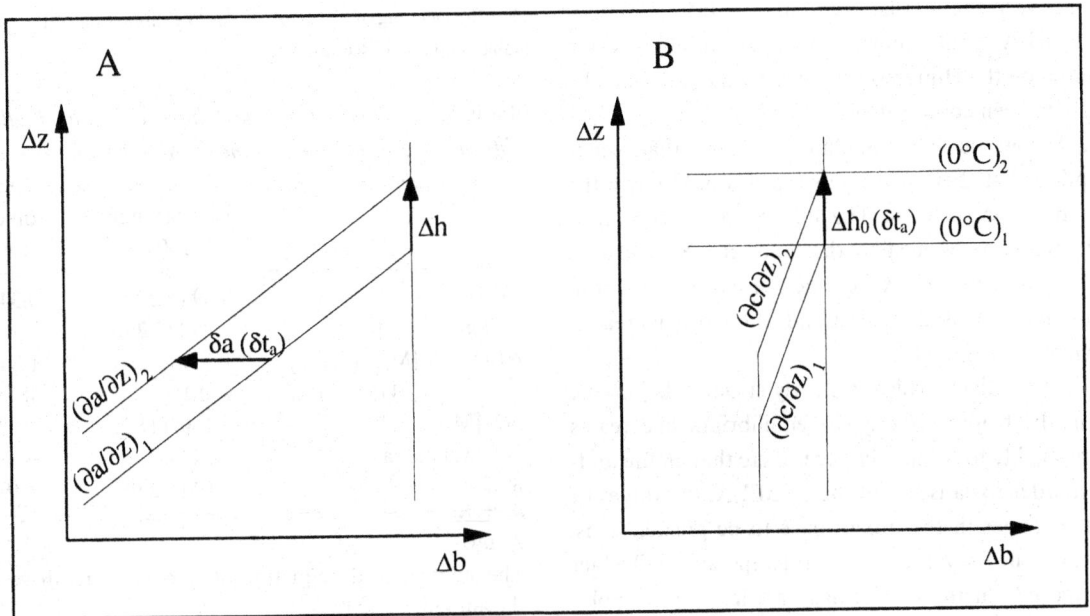

Fig. 5.2.4 On tropical glaciers the ELA moves due to a disturbance δT_{a} not only because of higher melting rates for Δh (A), but additionally because of a shift of the 0 °C level for Δh_0 (B).

along the vertical gradient of the air temperature. This leads to the fact that, on tropical glaciers, the ELA moves due to a change δT_a, not only because of higher melting rates for Δh, but altogether by

$$\Delta h_T(\delta T_a) = \Delta h + \Delta h_0 \qquad (5.2.15)$$

Equation 5.2.10 develops to

$$\Delta h_T(\delta T_a) = \frac{\tau \frac{1}{L_M}\{\delta[G(1-r)] + \alpha_R \delta T_a - \alpha_S \delta T_a\} + \left(\frac{\tau}{L_S} - \frac{\tau}{L_M}\right)\delta Q_L - \delta c}{\frac{\partial c}{\partial z} - \tau \frac{1}{L_M}\alpha_S \frac{\partial T_a}{\partial z}} + \delta T_a \frac{\partial z}{\partial T_a} \qquad (5.2.16)$$

For a given variation of altitude of the tropical equilibrium line, the hypothetical change in the temperature, which is solely responsible for the shift Δh, is:

$$(\delta T_a)_T = \frac{\tau \frac{1}{L_M}\alpha_S \frac{\partial T_a}{\partial z}\Delta h_T - \frac{\partial c}{\partial z}\Delta h_T}{2\tau \frac{1}{L_M}\alpha_S - \frac{\partial c}{\partial T_a} - \tau \frac{1}{L_M}\alpha_R} \qquad (5.2.17)$$

Possible causes for an assumed shift of the equilibrium line $\Delta h = 100$ m for the mid-latitudes, the inner and the outer tropics, should now be calculated based on these equations. According to Kuhn (1980[2]) and Ohmura et al. (1992), the equilibrium line for the inner tropics is assumed to be below the 0°C level (ELA < 0°C level).

As in the discussion regarding the mass balance profile in section 5.1, the Hintereisferner is again numerically exposed to idealized tropical conditions. For the calculation of the basic differences between the mid-latitudes and the tropics in connection with ELA–climate interactions, key variables which were measured on the Hintereisferner are used as a basis. The data have again been compiled in Table 5.2.1.

Table 5.2.2 shows the values of climatic changes that, under the assumption that they each occur alone, would move the equilibrium line by $\Delta h = 100$ m. The numbers of the equations used are in parentheses next to the amounts. Note that a decrease of the latent heat flux, i.e. a decrease of sublimation, is favourable for increased melting and, therefore, increased ablation, which makes the ELA rise.

The balance gradient, which is clearly stronger below the 0°C level in the tropics, is why the equilibrium line reacts clearly more weakly to climatic changes there than in the mid-latitudes. In order to cause a shifting of ΔELA, the values of the climatic changes in the tropics have to be greater. This, however, is not true for variations of air temperature. The fact that a change δT_a in the tropics has an effect on the ablation–altitude distribution, as well as on the accumulation–altitude distribution via the shift of the 0°C level, makes the

Table 5.2.1. *Variables and constants for the calculation of the ELA–climate interaction in the mid-latitudes and in the inner tropics (ELA < 0 °C level)*

	Mid-latitudes	Inner tropics
$\tau_{z=0}$	100 d	365 d
c_0	1600 kg m^{-2}	1600 kg m^{-2}
$\partial c/\partial z(z=0\Uparrow)$	1 kg m^{-2} m^{-1}	1 kg m^{-2} m^{-1}
$\partial c/\partial z(z=0\Downarrow)$	1 kg m^{-2} m^{-1}	4 kg m^{-2} m^{-1}
$\partial T_a/\partial z$	−0.0065 K m^{-1}	−0.0065 K m^{-1}
α_S	1.5 MJ m^{-2} d^{-1} K^{-1}	1.5 MJ m^{-2} d^{-1} K^{-1}
ε_a	1	1
$*\Rightarrow\bullet$		400 m
σ	4.9 × 10^{-9} MJ m^{-2} d^{-1} K^{-4}	4.9 × 10^{-9} MJ m^{-2} d^{-1} K^{-4}
L_M	0.334 MJ kg^{-1}	0.334 MJ kg^{-1}
L_S	2.835 MJ kg^{-1}	2.835 MJ kg^{-1}

Notes:

For the meaning of the symbols and the discussion of the values, see tables and text in section 5.1.

Table 5.2.2. *Climatic changes which each individually cause a shifting of the equilibrium line of Δh = 100 m*

	Mid-latitudes ($\tau = 100$ d)	Inner tropics ($\tau = 365$ d)
δT_a [K]	0.69 (5.2.5)	0.34 (5.2.18)
δc [kg m^{-2} a^{-1}]	−392 (5.2.6)	−1172 (5.2.13)
$\delta[G(1-r)]$ [MJ m^{-2} d^{-1}]	1.301 (5.2.7)	1.341 (5.2.7)
$\delta[G(1-r)]$ [MJ m^{-2} a^{-1}]	130.1	489.5
δQ_L [MJ m^{-2} d^{-1}]	−1.484 (5.2.8)	−1.520 (5.2.8)
δQ_L [MJ m^{-2} a^{-1}]	−148.4	−554.8
δs [kg m^{-2} a^{-1}]	−52 (5.2.9)	−196 (5.2.9)

Notes:

The numbers of the equations used are in parentheses next to the amounts.

equilibrium line of inner tropical glaciers react more sensitively to changes of temperature than in the mid-latitudes.

In the wet inner tropics (this only becomes evident there, as ELA < 0 °C level), the absolute amounts of accumulation are definitely higher than in the model assumption. For this reason, the vertical accumulation gradient, as well as the balance gradient, are stronger in the snow–rain mixture zone and, consequently, the sensitivity of the tropical ELA to climatic variations can be assumed to be even lower than depicted here.

In the outer tropics, where the equilibrium line is situated above the 0 °C level, a shifting of the snow–rain transition zone has no effect on the position of the equilibrium line.

The equilibrium lines of glaciers in wet tropical high mountains generally react more weakly to climatic changes than those in the mid-latitudes. However, they clearly react more strongly to variations of air temperature.

5.3 THE REACTION OF GLACIER TONGUES TO CLIMATIC VARIATIONS

The change of the tongue of a glacier is the easiest glaciological key variable to be measured. Especially in the tropical high mountains, such data are often the only reference points for glaciological–climatological examination. The question about climatic reasons for the observed changes of tongue positions and, consequently, the question about the expected reaction to a given climatic change are, however, even more difficult to answer.

A first possibility of examining the reaction of glacier tongues to climatic changes is the analysis of the spatial extent of a tongue under certain mean climatic conditions. Kuhn *et al.* (1985) show that different vertical mass balance profiles influence the extent of a tongue considerably on the basis of the analysis of neighbouring tongues of the Hintereisferner and the Kesselwandferner. Contrary to the macroclimatic reasons for different vertical balance profiles as discussed here, the reasons for the contrasts in the case of the two neighbouring glaciers in the Ötztal are, however, differences in relief and in exposure.

In a first step, the length of the glacier tongue in the steady-state ideal case is described in the two climatic zones. In a second step, new climatic conditions are assumed after a change and the corresponding new length of the tongue is calculated.

5.3.1 The length of the tongue in the steady-state case

In section 5.1.1, the connection between the activity index of a glacier $\partial \dot{b}/\partial z$, on the one hand, and on the other hand the whole net accumulation of one year $\dot{C}$, the cross-section Q under the equilibrium line, and the length X of the tongue from its end up to the equilibrium line are described. The mass flux can also be calculated via the density of ice ρ from the maximal horizontal speed $v_{x\,max}$ which prevails under the equilibrium line (according to equation 5.1.5):

$$\frac{\partial \dot{b}}{\partial z} = \frac{\dot{C}}{X \cdot Q} = -\rho \cdot \frac{v_{x\,max}}{X} \qquad (5.3.1)$$

After the integration over the period of one year ($\Delta t = 1a$), and under the assumption of a constant vertical balance gradient $\partial b/\partial z$, the length of the tongue is

$$X = -\rho v_{x\,max} \Delta t \frac{1}{\partial b/\partial z} = \frac{C}{Q} \frac{1}{\partial b/\partial z} \qquad (5.3.2)$$

As the vertical balance gradient exclusively refers to the tongue in this equation, it is assumed that this gradient is constant for the glaciers in the tropics as well as approximately constant for those in the mid-latitudes (Figs. 5.1.3 and 5.1.1). Therefore, under the assumption of equal accumulation gradients and sums and equally dynamic behaviour, the length of the tongue in the steady-state case depends solely and inversely proportionally on the vertical mass balance gradient. There is no reason for tropical glaciers to have a different dynamic behaviour from those in the mid-latitudes (see section 5.5.3).

In other words, the length of a glacier tongue depends on the linearity of the vertical profile of the specific mass balance. The marked change of the mass balance gradient from small values in the accumulation zone to large values in the ablation zone, as given for tropical conditions (section 5.1.1), is the cause of shorter tongues.

In case of the same mass flux through the equilibrium line cross-section, the glaciers in the tropical regions have shorter tongues than those in the mid-latitudes.

Alessandro Roccati says, in his 'Zusammenfassende Übersicht über die geologischen, petrographischen und mineralogischen Beobachtungen' (De Filippi, 1909),

'The glaciers of the Rwenzori belong to the so-called equatorial type; i.e. these are ice caps which can have a considerable thickness and more or less cover the peaks of the mountains. Branchings which force their way into the valleys and rarely, and only over a small distance, cross the lower limit of perpetual

Fig. 5.3.1 The debris-covered tongue of the Glaciar Kinzl in the Cordillera Blanca which emerges from the northern face of the Nevado Huascarán. Photo: G. Kaser, July 1995.

snow, which is between 4450 and 4500 m, reach from the caps downwards.'

Exceptions can be expected where tongues either heavily covered with debris (Fig. 5.3.1) or lying in shaded ravines (Fig. 5.3.2) have particular heat budgets, which are locally completely different. Glaciers with very different area–altitude distributions will behave differently, as shown in the example of the Hintereisferner and the Kesselwandferner (Kuhn *et al.*, 1985). Fig. 5.3.3, however, shows slopes covered with ice, which seem to be cut off horizontally, in the Quebrada Paria on the eastern face of the Cordillera Blanca.

> 'En casi toda la longitud de la cordillera nevada en el Perú, y principalmente en la parte que recorre el Departamento de Ancash, los nevados no se extienden por las quebradas como en la cadena de los Alpes en Europa. Aqui comunmente forman picos elevados con mucha inclinación, en los que se ve **la nieve en bancos cortados casi a pico** y al pie de los quales se observa casi siempre una laguna.' (Raimondi, 1873)

'Almost everywhere along the snow-covered Cordilleras of Peru, and principally in the part which stretches through the Departamento de Ancash, the glaciers do not reach the valleys like in the Alps of Europe. Here, commonly, the peaks reach high elevations with steep inclinations, where the snow fields seem to be abruptly cut and, where one almost always observes a lake at the foot of these fields.' (Raimondi, 1873)

A possibility for comparing the lengths of tongues in the mid-latitudes and the tropics is offered by the graphical depiction of the vertical extent between the highest points and the ends of the tongues of glaciers of different sizes (Kaser, 1995[1,2]). In Fig. 5.3.4, data from the Cordillera Blanca, Mount Kenya and the Irian Jaya is compared with that from the Ötztal Alps. In this particular analysis, the Cordillera Blanca is denoted by the glaciers of the Huascarán–Chopicalqui Massif, which represent the situation in 1970 (unpublished analysis). The data from Mount Kenya in 1963 are from Hastenrath *et al.* (1989). For the Irian Jaya, the extents for 1972 of the two biggest glaciers, Meren and Carstensz (Allison, 1976), are depicted. For the Ötztal Alps, the glaciers in the catchment area of Niedertalache and Rofenache (Austrian glacier inventory, 1969, unpublished) are given. The individual glaciers are arranged according to each mountain group, with their

Fig. 5.3.2 The Moore Glacier of Mount Baker in the Rwenzori mountains. Photo: K. Gabl; December 1991.

Fig. 5.3.3 There are only slight differences in altitude of the ends of the tongues in the Quebrada Paria (Cordillera Blanca). Photo: G. Kaser; July 1995.

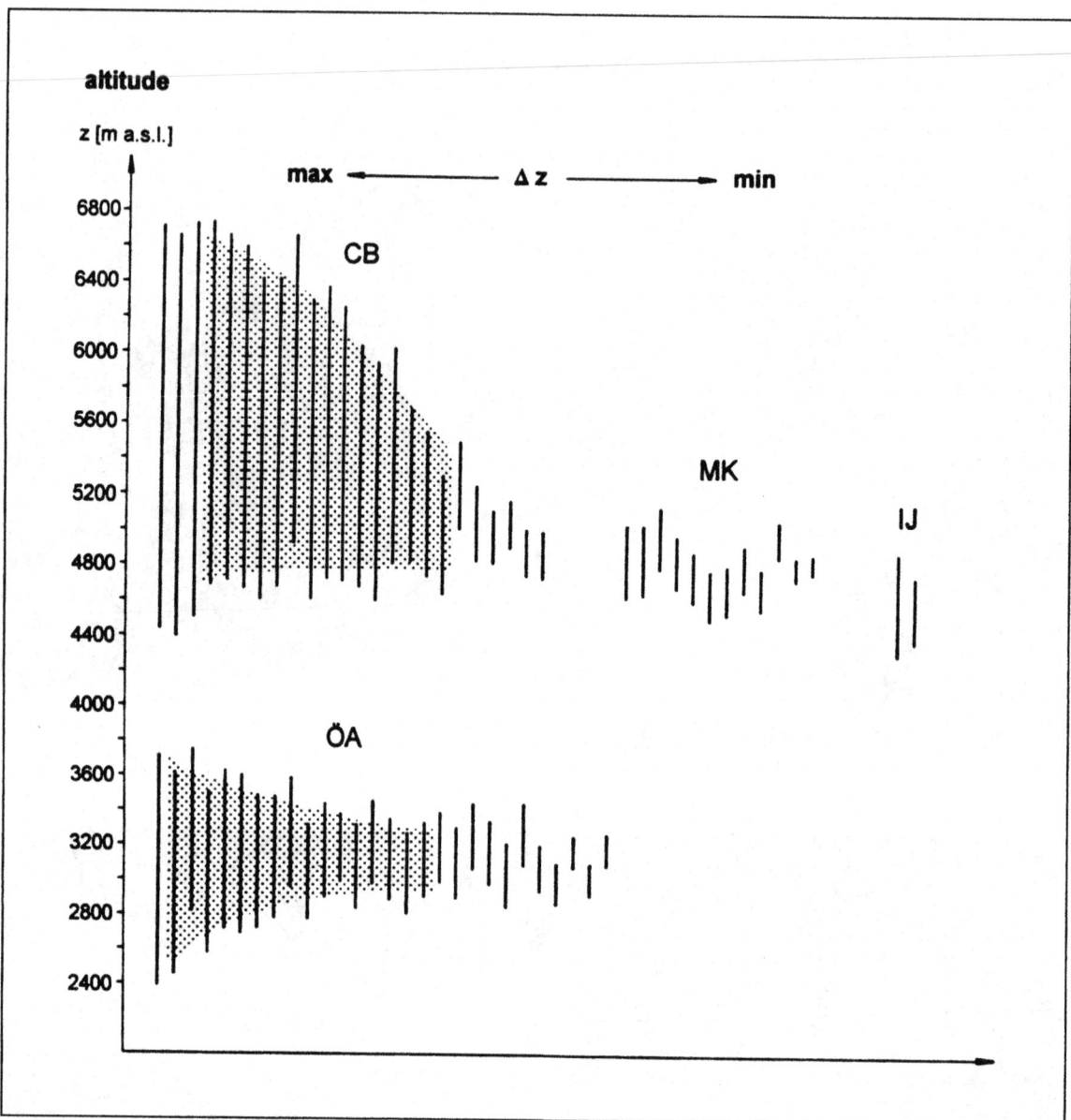

Fig. 5.3.4 The altitudinal extents of glaciers between their highest point and their tongues in three tropical and one Alpine mountain groups. CB: Huascarán–Chopicalqui Massif, Cordillera Blanca, 1970 (own analysis from aerial pictures); MK: Mt. Kenya, 1963 (Hastenrath *et al.*, 1989); IJ: Meren and Carstensz Glacier, Irian Jaya, 1972 (Allison, 1976); ÖA: catchment area of Niedertalache and Rofenache, Ötztal Alps, 1969 (Austrian glacier inventory, 1969, unpublished). The glaciers of each mountain group are arranged according to their altitude differences Δz decreasing from left to right. (According to Kaser, 1995².)

vertical extent decreasing from left to right. It is assumed for all glaciers within each particular area that they have neither distinct differences in accumulation nor differences in relief and, therefore, the altitudes of the tongues mainly depend on the mass balance profile.

While the alpine glaciers show a clear symmetry between highest and lowest points, the different altitudes of origin of the glaciers in the Huascarán–Chopicalqui Massif have hardly any influence on the altitude of the tongues. They all end at more or less the same altitude. Only the three largest glaciers have tongues situated lower than the others. The first and the third glaciers from the left in Fig. 5.3.4 are the glaciers with the survey numbers I1 and K1 (Ames *et al.*, 1989), which both originate on the peak of the Huascarán Sur and which are both part of the

Table 5.3.1. *The length* X *of a glacier tongue with different speeds in the horizontal longitudinal direction at the equilibrium line of the glacier* $v_{x\,max}$ *and with different vertical balance gradients* $\partial b/\partial z$

$v_{x\,max}$ (m a^{-1})	X_m (m) ($\partial b/\partial z = 7.8$ kg m^{-2} m^{-1})	X_{t_1} (m) ($\partial b/\partial z = 17.5$ kg m^{-2} m^{-1})	X_{t_2} (m) ($\partial b/\partial z = 13.5$ kg m^{-2} m^{-1})
50	5769	2571	3333
45	5192	2314	3000
40	4615	2057	2667
35	4038	1800	2333
30	3462	1543	2000

Notes:

X_m, mid-latitudes; X_{t_1}, tropical regions, snow–rain zone; X_{t_2}, tropical regions, below the snow–rain zone $X_m:X_{t_1} = 2.24$; $X_m:X_{t_2} = 1.73$ below the snow–rain zone. Mean $X_m:X_{t_1} = 2.24$:1.

glacierized western face (Glaciar Raimondi) of the mountain (see Figs. 7.1.2, 7.4.2). They have about the same width from the firn field to the tongue and, therefore, an extraordinary area–altitude distribution. The second glacier is the Glaciar Kinzl (Fig. 5.3.1), the tongue of which is heavily covered with debris and offers special heat budget conditions. The difficulty of delimiting on aerial photographs such debris-covered tongues with dead ice masses may also be a reason for the low value of the tongue position. An exact mapping of the tongues of the Raimondi Glaciers, extending far over the undissected rock shelves, is also more difficult than in definite valleys.

There are also some glaciers in the Huascarán–Chopicalqui Massif whose firn basins are framed by high rock walls (Glaciar Schneider, A3 and C3 in the glacier inventory; Fig. 7.4.2, chapter 7). A definition of the highest point is difficult in these cases. Under the assumption that avalanches from these rock walls into the firn basins add considerably to the accumulation, the highest rocks of the catchment areas were taken as the highest points.

The glaciers with a small altitudinal extent, the small glaciers in all mountain groups as well as the very flat glaciers in the Irian Jaya in Fig. 5.3.4, vary in this respect in tropical areas as well as in the Alps without any clear tendency. Local effects of relief, exposition and climate dominate as discussed by Kruss & Hastenrath (1990), Kaser & Noggler (1991) and Kaser & Georges (1997), among others.

5.3.2 The reaction of the position of the tongue to climatic changes

If, under the assumption of equal accumulation conditions, the vertical balance gradient on the glacier tongues considerably influences their extent for steady-state condition, then a change of altitude of the end of a tongue, due to a continuing climatic change, must also be different in the tropical regions from the respective change in the Alps.

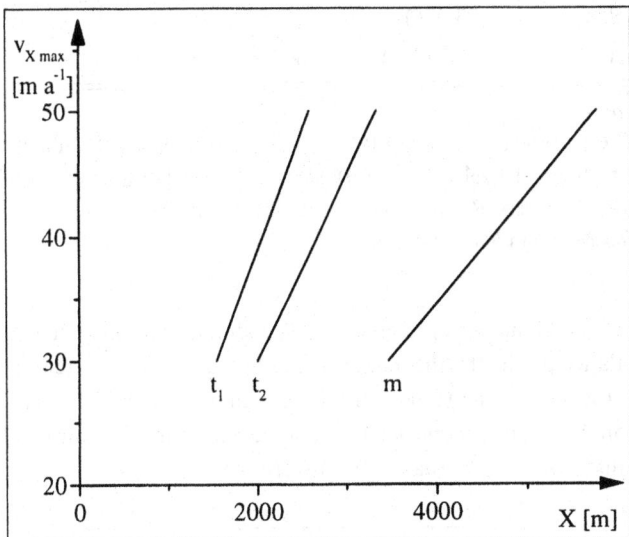

Fig. 5.3.5 The length of the glacier tongues and their changes under different climatic conditions. *m*, mid-latitudes; t_1, tropical regions within the snow–rain transition zone; t_2, tropical regions below the snow–rain zone.

The 'dynamic (rheological)' reaction can be shown with the equation 5.3.2 in the form of

$$X = -\rho v_{x\,max}\Delta t\,\frac{1}{\partial b/\partial z} \qquad (5.3.3)$$

by changing the value of the mass flow $|\rho v_{x\,max}|$ as an indirect consequence of a continuing climatic change and by calculating the length of the tongue for the new steady-state case. The change of the length of the tongue (X_m) under mid-latitude conditions, under tropical conditions within the snow–rain transition zone (X_{t_1}), and under those conditions below that zone (X_{t_2}) are given in Table 5.3.1 and are graphically depicted in Fig. 5.3.5. A speed of $v_{x\,max} = 50$ m a^{-1}, which corresponds to a mean value at the Hintereisferner (Kuhn *et al.*, 1979), is assumed for the initial situation. In the

Table 5.3.2. *Relative area changes of glaciers in the tropical regions and in the Austrian Alps*

Classes [1] (km²)	ÖA (n)[a] 1850–1969 (%)	IJ (n)[b] 1850–1972 (%)	Lewis (n)[c] 1850–1974 (%)	ÖA[a] 1920–1969 (%)	CB (n)[d] 1920–1970 (%)	Lewis (n)[c] 1920–1974 (%)	RU (n)[e] 1906–1955 (%)
− 0.15	67.8 (85)	78.2 (2)		45.3	73.4 (4)		58.0 (1)
− 0.30	60.3 (70)	75.3 (2)		41.5	50.2 (3)		20.6 (1)
− 0.50	55.5 (52)		53.3 (1)	36.8	35.7 (1)	47.9 (1)	14.8 (1)
− 1.00	52.4 (76)	64.0 (1)		35.1	35.0 (4)		
− 1.50	44.4 (25)			29.8	22.8 (2)		
− 2.00	42.6 (17)	62.7 (1)		28.1			
− 3.00	40.2 (14)			26.4	18.7 (6)		
− 5.00	33.9 (23)	60.4 (1)		22.4	20.8 (3)		
− 9.00	29.2 (6)			20.4	8.2 (4)		
>9.00	23.0 (6)			14.1			
Total	40.8 (374)	64.2 (7)		26.2	18.0 (27)		

Notes:

(1) Glacier extents around 1970; (n) number of glaciers; (a) Ötztal Alps (Gross, 1987); (b) Irian Jaya (Allison, 1976); (c) Lewis Glacier, Mt. Kenya (Patzelt *et al.*, 1984); (d) Cordillera Blanca, Huascarán–Chopicalqui Massif (own evaluations); (e) Moore, Elena and Speke Glaciers, Rwenzori (Kaser & Noggler, 1996).
Source: From Kaser (1995²).

various climates represented by the different vertical mass balance gradients, the length of the tongue reacts differently to the speed. The values for the remaining variables in equation 5.3.3 are $\rho = 900$ kg m^{-3} and $\Delta t = 1$ year. The various values for $\partial b / \partial z$ are taken from section 5.1.

The tongues of tropical glaciers are shorter than those in the mid-latitudes because of the continuous ablation. For the same reason, with a given change in the mass supply from the accumulation area, the lengths of tropical tongues vary less than those of the mid-latitudes.

Measured and reconstructed changes of glacier areas indicate such actual differences. Table 5.3.2 (from Kaser, 1995²) shows the relative area changes of glaciers in the tropics and in the Austrian Alps since the last maximum extents in the middle of the 1800s, and at the beginning of the twentieth century. The glaciers are divided into classes according to their area extents.

When comparing the Ötztal Alps (1920–1969) with the Cordillera Blanca (1920–1970), the small glaciers in the Cordillera Blanca show distinctly larger relative area losses, whereas all glaciers in the Irian Jaya were exposed to high losses. All these glaciers have small altitudinal extents. When raising the equilibrium line, possibly above the highest point of the glacier, those glaciers in the tropical regions are more affected than those in the mid-latitudes due to a continuous ablation at all altitudes.

The Lewis Glacier on Mount Kenya shows similar retreat values to comparably large glaciers in the Alps. The retreat of the glaciers at the Rwenzori refer to a different period but show tendencies similar to those in the Alps.

Compared with the Alpine glaciers, the largest glaciers in the Cordillera Blanca show smaller area losses. The reason for the smaller wastage of the large glaciers in the Cordillera Blanca can be found, on the one hand, in their strong and protecting debris cover (Glaciares Kinzl and Schneider are the two largest; they both have tongues which are covered with debris) and, on the other hand, as discussed in connection with the length of the tongue, in having extraordinary area–altitude distributions (Raimondi K1 (third largest) and I1 (fourth largest)).

The immediate effect of changed ablation conditions is greater in the tropical regions than that in the mid-latitudes due to the constant ablation period in the former.

At this point, the question regarding the differences in reaction and response times arises.

5.3.3 Reaction time and response time of glaciers to climatic changes

Firstly, how much time elapses before a climatically induced change reaches the tongue of a glacier via mass budget and ice

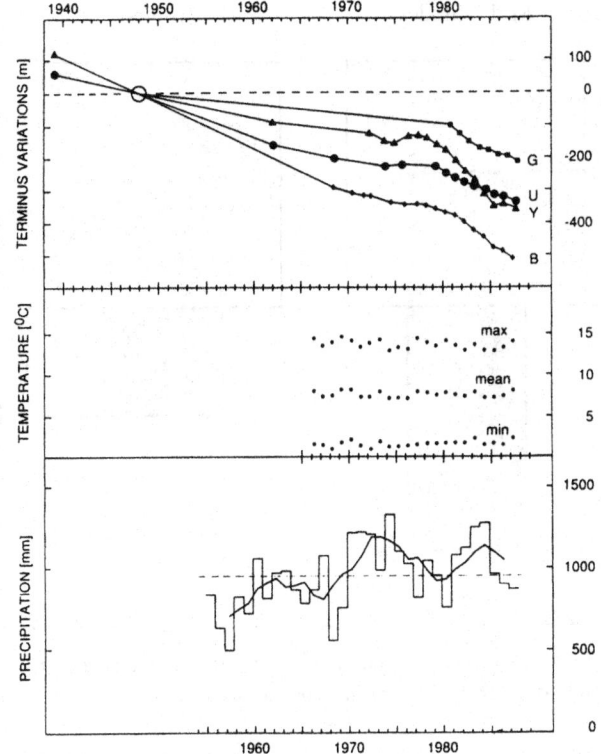

Fig. 5.3.6 Changes of lengths of glacier tongues of Gajap (G), Uruashraju (U), Yanamarey (Y) and Broggi (B), starting at a common reference point in 1948. For Uruashraju and Yanamarey, the positions have been reconstructed since 1939. In the middle of the figure, the monthly mean annual temperatures for the Querococha station (3955 m a.s.l.) are given. At the bottom, the annual precipitation amounts, as well as a five-year running mean, are shown. The broken line shows the mean amount of the precipitation amounts over the analysed period. (From: Kaser *et al.*, 1990.)

dynamics, and causes an advancing or retreating movement (**reaction time**)? The individual rheological peculiarities of a glacier make a reasoned answer to this question more difficult. The biggest problem when interpreting the changes of tongues is that some causal climatic fluctuations happen at time scales which are mostly shorter than the reaction times. Consequently, a large number of overlapping causes can be responsible for an observed signal.

A statistical approach is not satisfactory due to the few data available from tropical regions. Kaser *et al.* (1990) have analysed a 20-year-long series of tongue lengths and climatic data in the Cordillera Blanca (Fig. 5.3.6). The results indicate that the changes of the tongues of four relatively small glaciers have, on the one hand, a certain correlation with the variation of the temperature and, on the other hand, a delayed reaction to the influence of variation of precipitation of about four years.

A short advance of the Speke Glacier on the Rwenzori at the beginning of the 1960s even occurred simultaneously with higher precipitation, which Temple (1968) derives from an increased runoff of the Mubuku River, and which was confirmed by mountain climbers who had to deal with extraordinarily large amounts of snow during their ventures (Fantin, 1968; Osmaston & Pasteur, 1972). It is, however, in no way advisable to draw general conclusions from these few observations regarding the special behaviour of tropical glaciers.

Theoretical approaches allow the modelling of the **response time** that lies between one steady-state extent of a glacier and a second due to a continuous climatic change. These model approaches are derived from Finsterwalder's ideas of steady-state glaciers (Finsterwalder, 1897). Different approaches lead to different orders of magnitude of reaction times. The theory of kinematic waves, mainly developed by Nye (among others 1960, 1963[1,2], 1965), leads to extremely long memory functions. Accordingly, Alpine glaciers with a length of about 10 km would need about 10^2–10^3 years in order to adjust to a new balance. The results of Kruss (1984), who examined the reaction of the tongue of the Lewis Glacier for sine-shaped variation of the mass budget, show similarly high values if the asymptotic approach of a maximum length of the tongue is interpreted as the arrival at a new steady-state. Jóhannesson *et al.* (1989[1,2]) offer a simplified approach which calculates reaction times of 10^1–10^2 years for alpine valley glaciers, which correspond more closely with analyses of a series of observations (Porter, 1986; Reynaud, 1978, 1983). The discrepancies between the different approaches for solutions are discussed by Jóhannesson *et al.* (1989[1,2]) and by Paterson (1994). The approach proposed by Jóhannesson *et al.* (1989[1,2]) does not directly allow the quantification of different response times for glaciers with different mass balance profiles.

In an earlier work (Kaser, 1995[2]), the different behaviour was indicated, greatly simplified, in a graphic approach, resulting in the following observation:

It can be concluded that tropical glaciers react faster, and above all to changes of ablation, than those in the mid-latitudes.

Glaciers, which develop entirely into ablation areas due to a sufficient elevation of the equilibrium line, are consequently more affected in the tropics and disappear faster than comparable glaciers in the mid-latitudes. This seems to be the case at the moment with the glaciers on Irian Jaya and with most of the glaciers on the Rwenzori (see part III).

The occurrence of shorter tongues has, of course, a corresponding influence on the ratio of accumulation to ablation area and, therefore, on a further essential key variable which is used to estimate the altitude of the equilibrium line (ELA) of recent and past glacier extents (among others Kerschner, 1990).

5.4 THE ACCUMULATION AREA RATIO OF TROPICAL GLACIERS

Moraines are evidence of glaciers which mark an almost steady-state situation on the culmination of an advancing period. Under this assumption, a mean altitude of the equilibrium line ELA (or lower snowfall limit) can be assigned to such a reconstructed glacier and, therefore, a reconstruction of the prevailing climate can be estimated (Kerschner, 1990).

A method often used to determine the ELA of a glacier under equilibrium conditions is the application of a presumably constant accumulation area ratio (AAR) between the accumulation area and the total area of a glacier (AAR = $S_c:S_{tot}$). This is comparable to the ratio $k = S_c:S_a$ which describes the ratio between accumulation area and ablation area, and which is used mainly in German-speaking countries.

In the Alps, experience has shown that, in a balanced mass budget case, the surface of the accumulation area of a glacier is double that of the ablation area. Consequently, $k = 2:1$ and AAR = 0.67. This relation is well proved for the Alps through mass budget examinations (Gross *et al.*, 1977) and is confirmed by Kerschner (1990) for the quasi-steady case of the maximum glacier extent around the middle of the last century. The area ratio method is applied by determining the hypsographic curve from a large-scale map for the glaciers that are to be examined. By inserting the area ratio, the altitude of the equilibrium line is deduced.

In order to use this method, which has been developed and calibrated for the Alps, a few requirements have to be fulfilled:

- steady-state or near (quasi-) steady-state conditions,
- an area–altitude distribution with maximum area portions around the mean equilibrium line,
- a linear course of the hypsographic curve in the area of the mean equilibrium line.

It is often the case that single glaciers do not correspond to these conditions. Therefore, this method should only be used, except for well-examined individual cases, to determine the climatic lower snowfall limit as a mean over larger glacierized areas. The ratio calibrated for the Alps of $S_c:S_a = 2:1$ was often also used for glaciers in lower latitudes for lack of better knowledge (e.g. Finsterwalder, 1987).

The accumulation area ratio mainly depends on the linearity of the mass balance profile (see section 5.1.3). Because of the distinct change from weak mass balance gradients in the accumulation area to marked gradients along the tropical

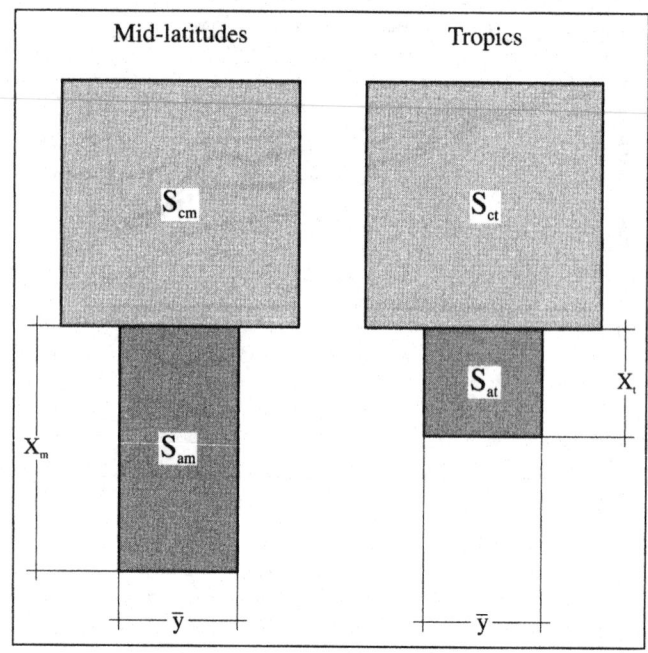

Fig. 5.4.1 A schematic depiction of the area ratios of a glacier in the mid-latitudes and under ideal tropical conditions.

glacier tongues, the AAR and the ratio k have to be higher in the tropics in principle than in the mid-latitudes. As in the previous chapters, a model alpine glacier is again exposed to ideal tropical conditions (Fig. 5.4.1).

On a steady-state alpine glacier the ratio of accumulation area to ablation area is

$$S_{cm}:S_{am} = 2:1 \qquad (5.4.1)$$

If the glacier is exposed to tropical conditions until it finds a new balance, then the area ratio is

$$S_{ct}:S_{at} = a:1 \qquad (5.4.2)$$

Under the assumption that the mean width of the tongue $\bar{y} = \bar{y}_m = \bar{y}_t$ is equally large under both conditions, the areas of the two ablation zones are only dependent on the respective lengths of the tongue X

$$S_{am} = X_m \bar{y} = \frac{S_{cm}}{2} \qquad (5.4.3)$$

and

$$S_{at} = X_t \bar{y} = \frac{S_{ct}}{a} \qquad (5.4.4)$$

The ratio of the accumulation to ablation areas is then

$$a = \frac{S_{ct}}{X_t \bar{y}} \qquad (5.4.5)$$

The accumulation area is assumed to remain constant in both cases

$$S_{cm} = S_{ct} \qquad (5.4.6)$$

This corresponds to the assumptions made in the previous chapters. Thus, equation 5.4.5 becomes

$$a = \frac{S_{cm}}{X_t \bar{y}} = 2 \cdot \frac{X_m \cdot \bar{y}}{X_t \bar{y}} = 2 \cdot \frac{X_m}{X_t} \qquad (5.4.7)$$

The values to be inserted can be found in Table 5.3.1.

The accumulation:ablation area ratio is double the ratio of the lengths of the tongues. If the whole tongue of a tropical glacier is in the snow-rain zone, then the tongue length ratio Xm:Xt1 = 2.24:1 (Table 5.3.1), the area ratio Sct:Sat1 = 4.48:1 and the Accumulation Area Ratio AAR = 4.48/(4.48 + 1) = 0.82.

If part of the tongue reaches into the pure rain zone, then the division values would be proportionally a little smaller in correspondence with the lower balance gradient of $\partial b/\partial z = 13.5$ kg m^{-3}.

Basically, it has to be assumed that the AAR and the $S_c{:}S_a$ ratio are clearly higher in the tropics than in the mid-latitudes. The values given, however, are only valid if there are very similar area–altitude distributions to those in the Alps, and if the climatic conditions in the tropical regions are close to ideal. Appropriate observations to check the calculated ratio values are missing. The more one moves away from ideal tropical conditions, the smaller the ratio values will be. As they depend heavily on the duration of the ablation period τ, it has to be assumed that they are clearly higher in the outer tropics, as well as in the subtropics, than in the mid-latitudes.

The questions regarding topographical conditions have to be examined for each individual case before the method is used. Similar considerations have to be made when using other cartometric methods (toe-to-headwall altitude ratio THAR; area–altitude balance ratio method AABR) which are used to determine lower snowfall limits.

5.5 THE TEMPERATURE DISTRIBUTION IN THE SNOW COVER AND ITS CONSEQUENCES

There are five processes that contribute to snow cover compression (e.g. Kuhn & Herrmann, 1990; Paterson, 1994):

- Breaking of snow flakes in the wind.
- Compression through the load of the snow cover (**pressure metamorphism**). It is dominant in cold snow.

- Diffusion of vapour from convex to concave surfaces of crystals (**destructive, isothermal metamorphism**). It leads to the decay of points and edges.
- Diffusion of vapour from warmer to colder layers (**constructive metamorphism**). Along a temperature profile in the snow cover, a related vapour pressure profile exists corresponding to the saturated vapour pressure at each temperature. The diffusion of vapour thus caused from higher to lower vapour pressure is responsible for changes which appear, in the course of time, in the grain structure and in the density profile of the snow cover. It leads to the formation of angular shapes.
- Infiltration and refreezing of meltwater (**regelation metamorphism**). The compression of the snow cover is accelerated at the melting temperature and the grain structure becomes more simple. The maximum compression of snow to firn and to ice is faster with regelation than in cold snow. With continual refreezing, waterproof layers develop. This can occur in cold layers which develop from cold seasons or when the surface freezes at the end of the ablation period.

Compared with other substances, snow has a high heat capacity, which, together with the low ability for thermal conductivity, is the reason for a high insulation effect. Thus, differences in temperature are temporally and spatially conserved in the snow cover and lead to the corresponding consequences:

- At the end of the ablation period the surface freezes to a waterproof layer.
- The snow cover, accumulating on top of this waterproof layer, becomes colder towards the top in the course of the cold season. The constructive metamorphism draws vapour from the autumn layer and leaves behind characteristic grain forms.

Both processes are fundamentally responsible for the development of annual layers which can be identified in the accumulation area of a glacier. In the mid- and the high latitudes, where such zones clearly develop in the course of thermal seasons, this fact is used as the main condition for the measurement of accumulation in snow pits when using the direct glaciological method for the determination of the mass budget (e.g. Hoinkes, 1970).

5.5.1 The temperature distribution in snow and firn under ideal tropical conditions and the expected consequences

In tropical regions, the lack of thermal seasons must inevitably have the consequence that the snow cover and the firn body are isothermal. In very high and well-ventilated locations, the constantly low temperatures ensure that the snow surface will not melt and, consequently, cold firn and ice bodies can develop. This fact made it possible, for example, for a core to be extracted at an altitude of 6000 m in the Garganta between the two peaks of Huascarán in the Cordillera Blanca and interpreted (Thompson *et al.*, 1995; Thompson, 1995).

In lower locations, snow cover, which is isothermal at and slightly below 0°C, develops above the equilibrium line. This is true for most tropical accumulation areas. Such snow cover displays several features:

- Annual layers, which are recognized quite clearly in mid-latitudes due to the dust collected in summer and the freezing over in late autumn, are hardly identifiable in tropical regions. Most dust, that might have been collected, is washed out depending on the size of the grains. Meltwater percolates unobstructedly through several annual layers (percolation zone, Paterson, 1994 according to Benson, 1961 and Müller, 1962). Consequently, the traditional determination of accumulation in snow pits or drill cores leads to values which lie below the actual accumulation.

- Because of the constant temperature and so lack of diffusion of vapour, decomposition and regelation dominate the metamorphism of the snow crystals. The firn grains become more rounded and more uniform there than in the mid-latitudes.

- The same causes lead to the more compact structure of the glacier ice. The question of whether and how this affects the flow behaviour of the ice has to be posed.

Before these assumptions are examined with the help of observations and measurements, some additional consequences can be anticipated. For example the evidence of any small temperature variations is not conserved in the snow cover but is broken down by constantly percolating meltwater.

In the outer tropics, it can be assumed that iced dirt layers develop during distinct dry periods which are similar to those that mark the end of the ablation period in the mid-latitudes. It is conceivable that strong nocturnal radiation and high evaporation allow for a freezing of the surface on which dust concentrates from the snow free surroundings. The question arises, however, whether these layers will ever be thick, closed and cold enough in order to be waterproof, or whether they will be soaked and softened by penetrating meltwater during the wet season.

Under the given assumptions the following is true:

In the accumulation areas of tropical glaciers, snow and firn layers are isothermal. Consequently, waterproof layers are non-existent at the beginning of a budget year and thus the annual accumulation cannot be determined with the help of traditional methods. The grain shapes of firn are simple and uniform.

5.5.2 Accumulation measurements on tropical glaciers

The consequences of isothermal composition of snow and firn layers, which are close to the melting point, have an influence, above all, on accumulation measurements.

The only accumulation measurements from the Rwenzori mountains were carried out by Bergstrøm (1955). In the snow pits that had been dug into the firn area of the Elena Glacier in August 1952, Bergstrøm found dust horizons which he attributed to the respective dry periods. There are no reports about the problems mentioned above.

Allison (1976) reports that, following investigation at the Meren and the Carstensz Glaciers in the Irian Jaya in 1972, no stratigraphy worth mentioning was seen in 10 m deep drill holes in the firn areas of the two glaciers. He attributes the isothermal conditions measured in the firn bodies to the extremely low annual fluctuation of air temperature which he assumes is even lower over the glacier than that measured at the Ertsberg Mine (see section 4.2). There are no reports about considering these observations when determining the accumulation at the gauging poles.

Platt (1966) reports from the Lewis Glacier on Mount Kenya, where the annual mass balance has been determined since 1978 (Hastenrath, 1984, 1991[2]), that melting occurred daily during a measurement period in the long rainy season, despite predominant accumulation on the glacier surface. However, Platt has omitted a discussion regarding the possible consequences for the determination of the mass balance.

Measurement tests and observations in the Cordillera Blanca indicate that the assumptions derived from the isothermal conditions of the snow cover are correct. For many years, Alcides Ames and his colleagues from the Unidad de Glaciologia y Recursos Hidricos have tried to examine the mass balance on the Yanamarey and Uruashraju glaciers using this method. It was mostly impossible to identify the surface of the previous dry period when measuring the accumulation in the pits. Attempts to mark it failed as soot layers marked by poles were gone the following year (Ames, 1995; personal information). Only the ablation measurements have been carried out successfully.

In July 1995, field work on the Artesonraju and Yanamarey glaciers in the Cordillera Blanca brought similar results. B. Francou, P. Wagnon (ORSTOM, La Paz), J. Gómez, W. Tamayo, A. Tinoco, A. Valverde (UGRH, Huaraz) as well as B. Noggler and G. Kaser (Inst. f. Geographie, Univ. Innsbruck), took part in this field work (Francou *et al.*, 1995[3]; Kaser, 1996).

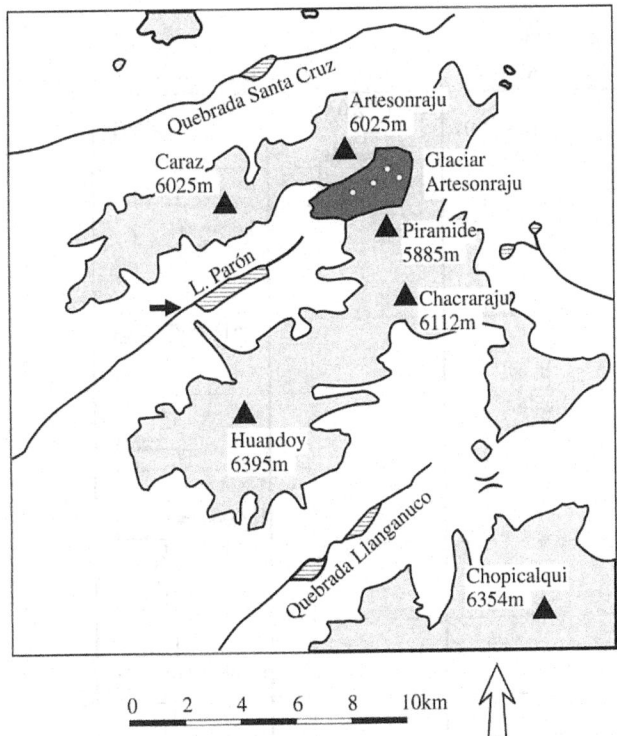

Fig. 5.5.1 Cordillera Blanca, June 1995. The position of the drill sites on the Glaciar Artesonraju are indicated by white dots. The glacierized areas are shaded, and the Parón runoff gauging station is indicated by the arrow.

On the Glaciar Artesonraju, drill cores up to 11 m deep were drawn with a hand drill (SIPRE) at four locations between 5350 m and 4950 m a.s.l.. (Figs. 5.5.1 and 5.5.2). The only recognizable changes in the stratigraphy were a few differences in the size of the firn grain, which were difficult to determine, a few thin ice lamellas, and a density which increases with depth and which registered several higher undefinable values (Fig. 5.5.3). Furthermore, the drill cores were consistently wet. Not a single dust layer could be found. Only at the lowest site (4950 m a.s.l.) was the drilled glacier ice covered with a dirt layer, which had probably accumulated on the glacier surface over several years. In comparison, density profile and dust layers in a 20 m deep pit in the firn area of the Ötztal Kesselwandferner (3240 m a.s.l.) show clear autumn horizons. In Fig. 5.5.3, the data of the top 11 m is displayed (from Ambach & Eisner, 1966).

However, dirt horizons can be identified in the crevasses in the firn area of the Glaciar Artesonraju (Fig. 5.5.4). Theoretical considerations and knowledge from the drill cores, however, make it imprudent to interpret the dust horizons in crevasse walls as a closed succession of seasons and/or years.

The ice extracted from about a depth of 1.5 m had enclosed air bubbles but no visible grain structure (Fig. 5.5.5). This is a further indication of the fact that the destructive and the regelation metamorphoses play a dominant role throughout the whole year when snow is transformed to firn and then to ice.

Fig. 5.5.2 Extracting a firn core at 5350 m on the Glaciar Artesonraju. Photo: G. Kaser, 6.7.1995.

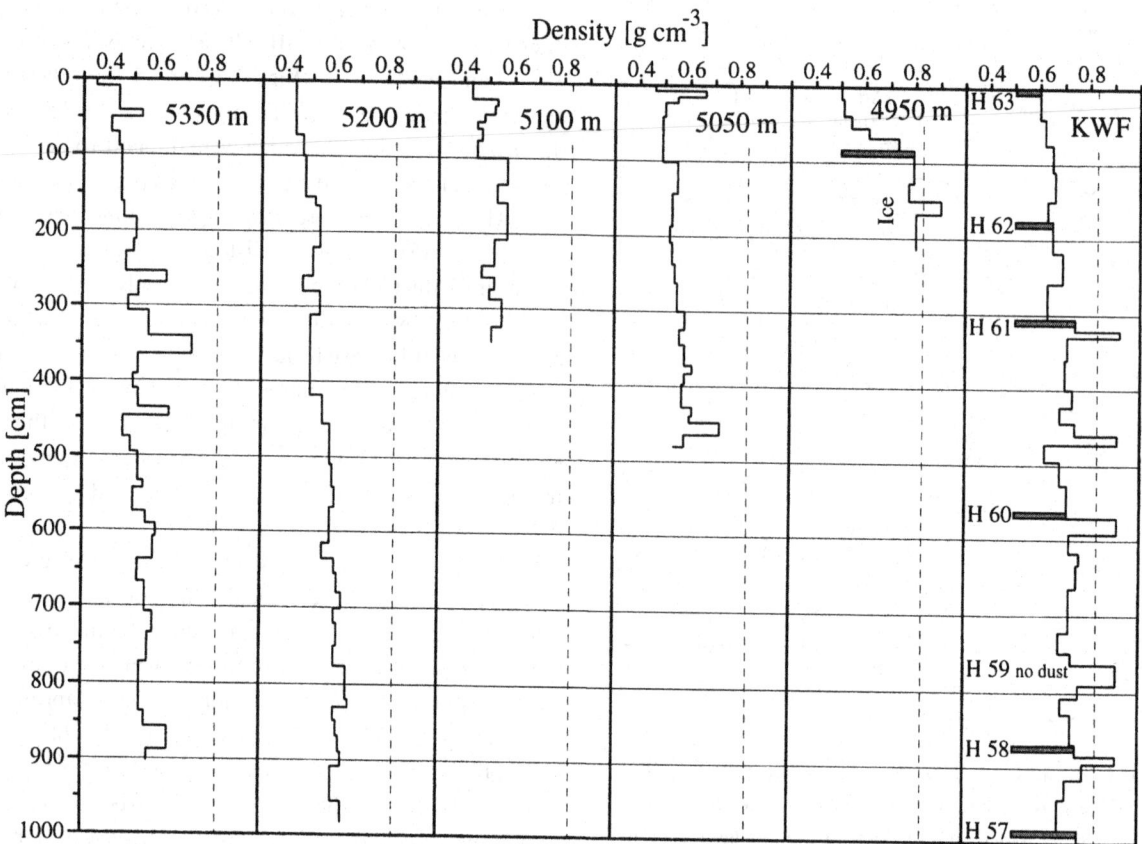

Fig. 5.5.3 The density distribution in the drill cores from the firn
area of the Glaciar Artesonraju. It is compared with the density
profile of a pit in the firn area of the Kesselwandferner (Ambach &
Eisner, 1966). The transition density from firn to ice ($\rho = 0.83$ g
cm^{-3}; Paterson, 1994), the dirt layers (grey) and the annual horizons
are provided.

A week after the measurements were taken on Glaciar
Artesonraju, several drillings were made on Glaciar
Yanamarey in the southern Cordillera Blanca. Pits up to 160 cm
deep were dug next to them in order to examine the top layers
even better. Two dust layers were found at the highest site at
4975 m a.s.l. (Fig. 5.5.6, Table 5.5.1). While the lower layer
(146–148 cm below the actual surface), in conjunction with an
icy lamella and higher density, probably comes from the dry
period in 1994, the upper one (16–33 cm below the actual
surface) could come from a period of fine weather at the beginning of the dry period in 1995, or from wind that brought in
dust. Tables 5.5.1 and 5.5.2 and Figs. 5.5.7 to 5.5.9 show the data
measured and observed from the Yanamarey pit at 4975 m a.s.l.

The metamorphically changed firn grains are rounder than
those in snow pits in the Alps (Fig. 5.5.7). This was also
observed in the drill cores from the Glaciar Artesonraju and
in the other snow pits on the Glaciar Yanamarey.

This can be seen as a clear result of the dominant influence
of destructive metamorphism.

An attempt to measure a temperature profile in the snow
pack in this pit was also made. The temperature measurements
were made with a resistance sensor (PT 100) which was
inserted about 15 cm into the pit wall. Although measurements were taken on the shady side, the radiation from the
opposite, sunny pit wall was very high. Shading the sensor
from this reflected radiation was ineffective. Despite great care,
temperatures slightly above the melting point were measured
when the sun shone in the pit. Only when clouds moved in
front of the sun were realistic measurements possible. This was
the case in the middle part of the pit at a depth of between
30 cm and 70 cm (Fig. 5.5.9).

Upon opening the pit, the entire snow cover was soaked and
the frozen dirt horizon was wet. The dust concentration,
decreasing downwards from the dirt horizon towards the base,
indicates removal and transport by percolating meltwater.

Similar results were observed in lower pits on the Glaciar
Yanamarey. A clear and slightly frozen, but thoroughly soaked
dirt layer was preserved from a previous dry period. This is in
contrast to A. Ames' long-standing attempts to define clear
horizons and, thus, to determine the accumulation (see above).
It seems that a visible layer develops in the dry period in the

Fig. 5.5.4 Dust layers in crevasse walls in the firn area of the
Glaciar Artesonraju. Photo: G. Kaser, 6.7.1995.

Fig. 5.5.5 Glacier ice from the tongue of the Glaciar Artesonraju
(4950 m a.s.l.). Photo: G. Kaser, 7.7.1995.

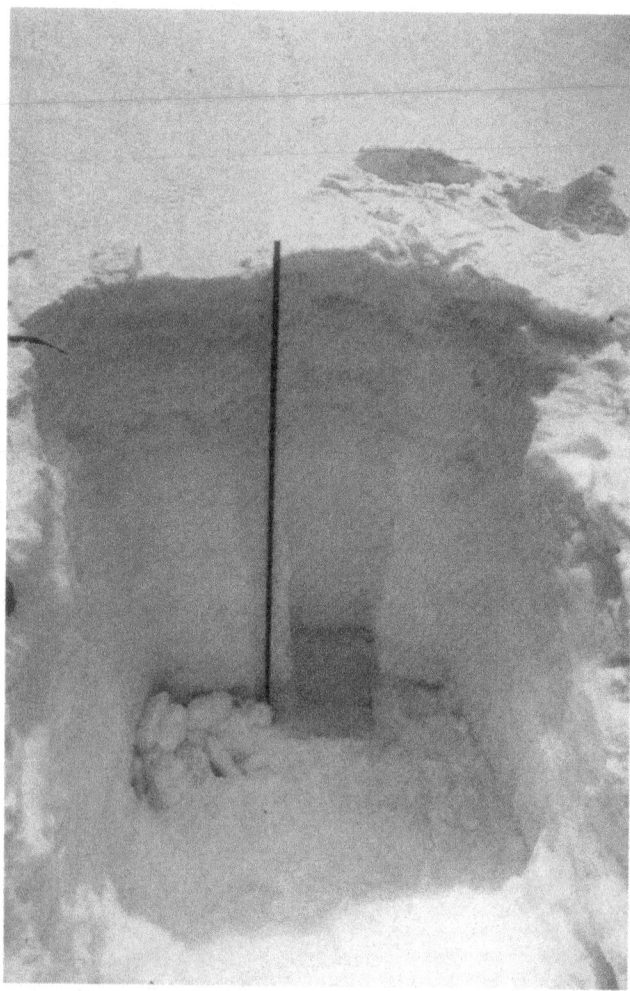

Fig. 5.5.6 Glaciar Yanamarey. Snow pit at 4975 m a.s.l. on July 13, 1995. The dirt layer, recognizable on the pit floor (below 146 cm depth), can probably be assigned to the dry period in 1994. A dust layer situated higher (16–33 cm below the surface) cannot be recognized in the photo. Photo: G. Kaser.

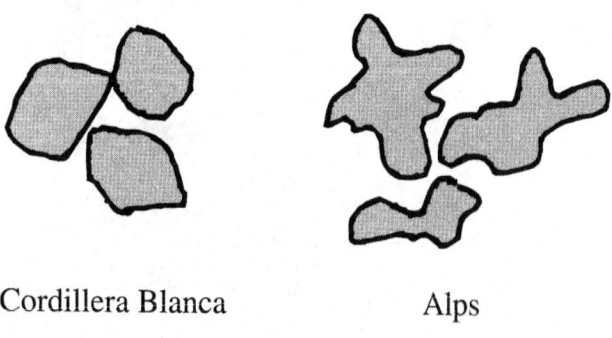

Cordillera Blanca   Alps

Fig. 5.5.7 A schematic depiction of the different grain structures of firn. The tropical form is based on observations on the Artesonraju and Yanamarey glaciers in the Cordillera Blanca, made in July 1995. The grain forms under alpine conditions are depicted and described, among others, by Good (1982).

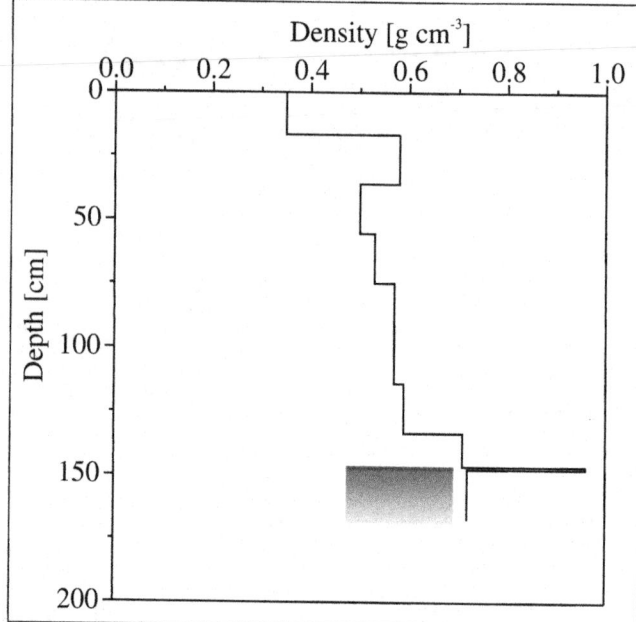

Fig. 5.5.8 Glaciar Yanamarey: snow pit at 4975 m a.s.l. Density profile of the snow cover with a dust layer that decreases towards the base (shaded). (G. Kaser, J. Gómez, July 13,1995.)

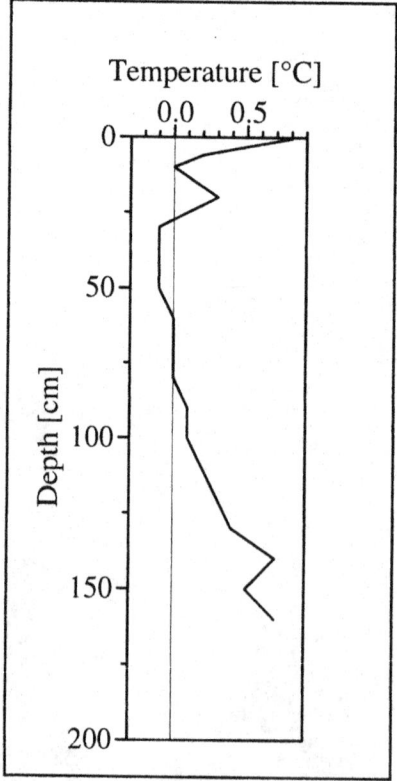

Fig. 5.5.9 Glaciar Yanamarey: snow pit at 4975 m a.s.l. Temperature measurements in the snow cover. High radiation made exact measurements impossible. (July 13, 1995, G. Kaser, J. Gómez.)

Table 5.5.1. *Glaciar Yanamarey: snow pit at 4975 m a.s.l. Stratigraphy of the composition of the snow cover*

Depth (cm)	Stratigraphy
0–1	fresh snow from 12 and 13 July 1995
5	slightly encrusted, clean, grain size = 1 mm
7	fine-grained, white, loosely packed, wet
8	slightly encrusted
15	fine-grained, loosely packed, white, wet
16	loosely packed, clean, wet, grain size = 1 mm
33	encrusted, grain size = 1–2 mm, very fine ice layers, some dust
34	same as above, but loosely packed
44	solid, clean, wet, grain size = 2 mm
45	loosely packed, wet, slightly grey, grain size = 3 mm
146	solid, but not frozen, wet, without layer structure, grain size = 2–3 mm, slightly frozen at the bottom
148	frozen layer, very soiled, many ice lamellas
161	same as above, but dirt becomes less present towards the base
Below	again more loosely packed, coarse-grained, cleaner; similar to above the dirt layer

Notes:
13.7.1995, G. Kaser, J. Gómez.

Table 5.5.2. *Glaciar Yanamarey: snow pit at 4975 m a.s.l. Measurements of density and the water equivalent of the snow cover*

Number from top	Altitude (cm)	Cumul. altitude (cm)	Mass (g)	Volume (cm³)	Density (g cm⁻³)	Water equiv. (mm we)
1	16.5	16.5	157	451	0.35	57
2	19.5	36.0	310	534	0.58	113
3	19.5	55.5	269	534	0.50	98
4	19.5	75.0	282	534	0.53	103
5	19.5	94.5	302	534	0.57	110
6	19.5	114.0	303	534	0.57	111
7	19.5	133.5	313	534	0.59	114
8	13.0	146.5	251	356	0.71	92
Total		**146.5**				**798**
9	3.0	147.5	106	82	0.96	29
10	17.0	167.5	336	465	0.72	123

Notes:
G. Kaser, J. Gómez, July 13, 1995.
Density tube: length = 20.1 cm; diameter = 5.9 cm.

Cordillera Blanca during some years. But, in these dry years, it must also be reckoned that meltwater percolates from the surface into the firn layers of previous years and partly into the drainage system.

It can, therefore, be assumed that large parts of the accumulation zones of tropical glaciers belong to the so-called *percolation zone* (Paterson, 1994 according to Benson, 1961 and Müller, 1962). When penetrating into the snow cover, meltwater from the surface does not encounter cold layers, where it could freeze again. Even if a dust layer develops at the end of the dry period, it is not impermeable, which allows meltwater to penetrate into accumulation layers of previous years. The amount of accumulation which percolates to lower layers cannot be recorded with the common accumulation measurement methods. Paterson (1994) suggests using a tray for such conditions. However, even when using trays, the amount of percolation water that still remains in lower layers cannot be determined. This can only be explained by glacio-hydrological research with tracer experiments and certainly depends on the altitude of the area, its relief, and, above all, the climatic history of the structure of the snow cover of several years.

5.5.3 Ice dynamics on tropical glaciers

The domination of those processes of metamorphism that develop simple, round shapes indicates that the ice structure is different on tropical glaciers than in the mid-latitudes (Fig. 5.5.5). Examinations of the ice structure and of the dynamic behaviour of tropical glaciers, which could possibly be influenced, do not exist. The measured ice movements in the Irian Jaya (Allison, 1976), on the Lewis Glacier (Hastenrath, 1984, 1989; Hastenrath & Kruss, 1982), on the Elena Glacier (Whittow *et al.*, 1963), and on the Yanamarey and Uruashraju glaciers (UGRH–Hidrandina S.A., unpublished data; Hastenrath & Ames, 1995[1,2]) give no indication of a difference of the ice dynamics when compared with the mid-latitudes.

Most glaciers in the Cordillera Blanca appear relatively thin and highly crevassed (Fig. 5.5.10), which could suggest special dynamic behaviour.

The slight thickness, however, can be explained by the relatively steep inclinations. In Fig. 5.5.11, the inclination frequency distribution of the glacier surfaces of the inner Ötz Valley (runoff gauge Vent) and that of the Santa Cruz–Pucahirca group in the Cordillera Blanca are compared. The Ötztal Alps are clearly flatter.

In a simplified version, the basal shear stress on the glacier bed τ_b is calculated using the hydrostatic fundamental

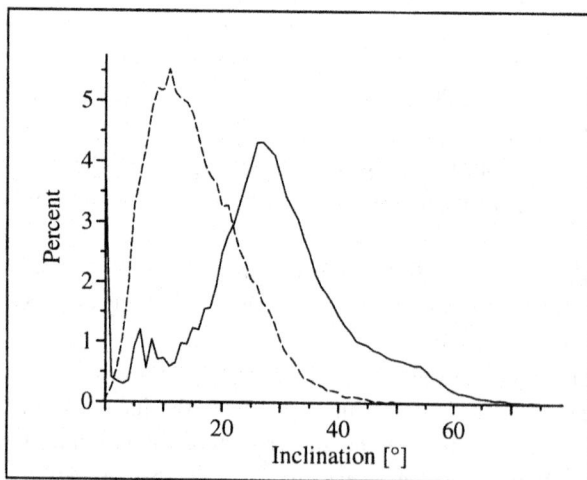

Fig. 5.5.10 Glaciers in the Santa Cruz–Pucahirca group (Cordillera Blanca). Photo: J. Alean, 1980.

Fig. 5.5.11 Frequency distribution of the inclination of the surface of the glaciers in the Rofen Valley/inner Ötz Valley (broken line) and in the Santa Cruz–Pucahirca Massif of the Cordillera Blanca (solid line).

equation from the thickness, inclination, and form of a glacier (Kuhn & Herrmann, 1990):

$$\tau_b = \rho \cdot g \cdot f \cdot h \cdot \sin \alpha \quad [Pa] \qquad (5.5.1)$$

where ρ is the density of ice, g the acceleration of gravity, f a form factor which is close to the relation of the hydraulic radius (area of the cross-section divided by the ice-covered part of the circumference) to the thickness of the glacier; h is the thickness of the glacier and α the angle of inclination of the glacier surface.

The thickness of the ice h can be calculated in defined areas of the glacier:

$$h = \frac{\tau_b}{\rho \cdot g \cdot f \cdot \sin \alpha} \quad [m] \qquad (5.5.2)$$

When all values on the right-hand side of the equation, besides the angle, are constant, then the thickness of the ice is solely a function of the surface inclination. In Table 5.5.3, the corresponding values for the Hintereisferner at 2920 m a.s.l., which is about the mean equilibrium line, are presented.

Fig. 5.5.12, resulting from equation 5.5.2 and the respective values from Table 5.5.3, shows the connection between the angle of inclination of the glacier surface and the ice thickness.

Table 5.5.3. *Values of the key variables in equation 5.5.2 for the Hintereisferner at 2920 m a.s.l.*

τ_b	97 kPa
ρ	910 kg m^{-3}
g	9.8 m s^{-2}
f	0.69

Source: Kuhn & Herrmann (1990).

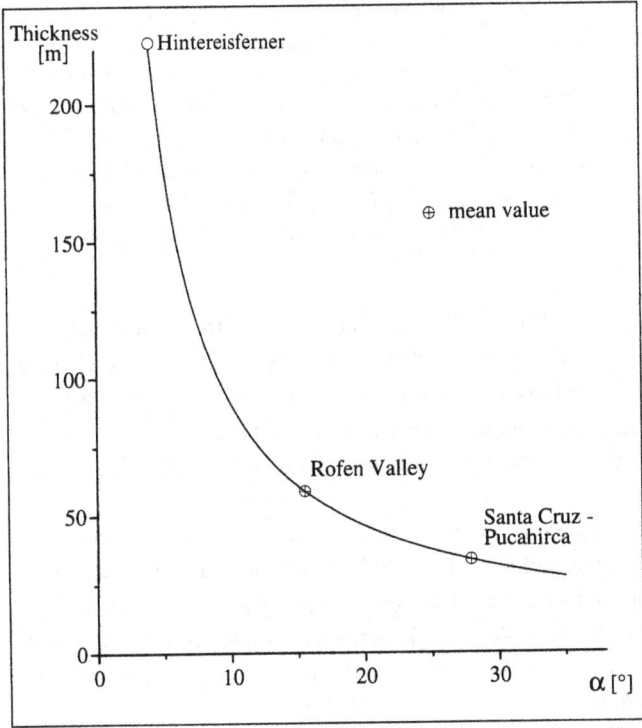

Fig. 5.5.12 The thickness of the ice (h) dependent on the inclination of the glacier surface (α).

The points mark the inclination of the Hintereisferner in the area of its mean equilibrium line as well as the average and most frequent inclinations in the inner Ötz Valley and in the Santa Cruz–Pucahirca Massif in the Cordillera Blanca (from Fig. 5.5.11). It becomes clear that the glaciers of the Cordillera Blanca must be distinctly thinner due to their greater inclination. Measurements of the thickness of the ice are, however, missing, just as for all other tropical mountains.

The thinning out and the heavy fracturing of the glaciers in the Cordillera Blanca are not only due to the faster flow as a consequence of the heavy inclination, but also to the fact that many of the glaciers pour out over vast mountain slopes and are not forced into valley-shaped beds.

In spite of the indications of special structures of the glacier ice, neither the few measured surface speeds nor the observed thickness of the glaciers give any indications of unusual flowing behaviour of tropical glaciers.

5.6 THE INFLUENCE OF THE GLACIERS ON THE WATER BALANCE OF TROPICAL MOUNTAINS

In the mountains of the mid-latitudes, precipitation which falls during the cold season in the form of snow down to the valleys is withheld from the runoff. In spring, first the snow melts at continuously increasing altitudes. In summer, ice is withdrawn from the glacier through ablation and is added to the runoff. The portion of the runoff which is derived from snow and glacier melt is available in the lower valleys, especially during the season of plant growth. The sequence of thermal seasons controls the difference between the annual variation of precipitation and the runoff. The glacier will increase the amplitude of the runoff compared with the annual variation of precipitation.

In the tropics, the influence of the glacier on the variation of the runoff is only possible through moisture seasonality. Accumulation takes place only in the uppermost part of the catchment area. Ablation lasts the whole year round. The glacier will lower the amplitude of runoff compared with the annual variation of precipitation.

Corresponding to the ideal conditions defined in chapter 4 (Fig. 4.5.1) and to the consequent climate and glacier regimes, the water balance regimes can be schematically described for the mid-latitudes, the inner tropics and the outer tropics, respectively.

In the mid-latitudes, glacierization increases the annual amplitude of runoff. Considering the points made above, it is expected that runoff variation is subdued in tropical glacierized regions (Fig. 5.6.1). No typical annual variation of the precipitation exists in the mid-latitudes. Therefore, the precipitation is kept constant in the simplified scheme in Fig. 5.6.1.

Again, only few data are available about tropical mountains which could describe these connections in detail.

In the following figures, the coefficients of runoff (c_Q) and

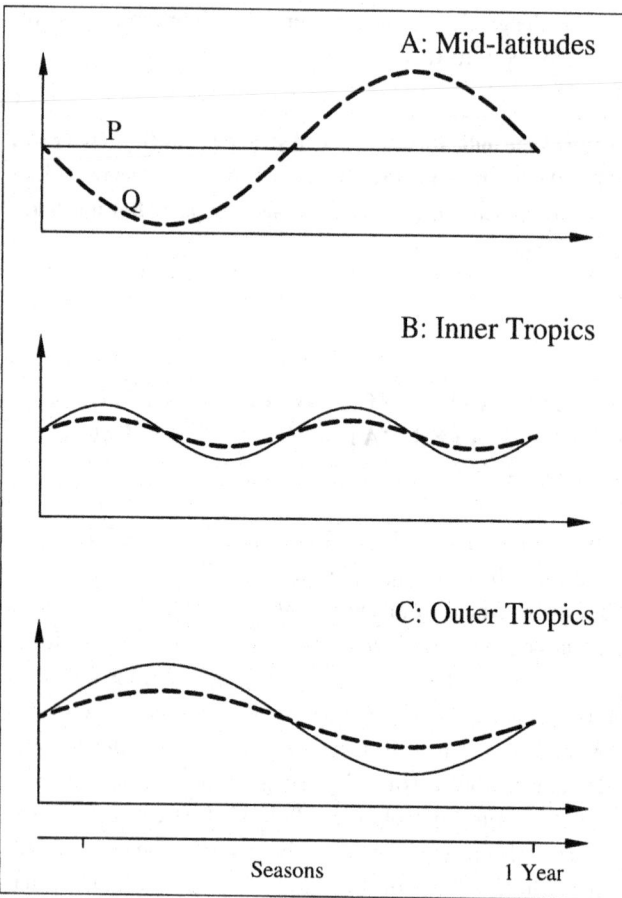

Fig. 5.6.1 A schematic comparison of the seasonal variations of precipitation (thin line) and runoff (broken line) of glacierized catchment areas in the mid-latitudes, the inner and the outer tropics.

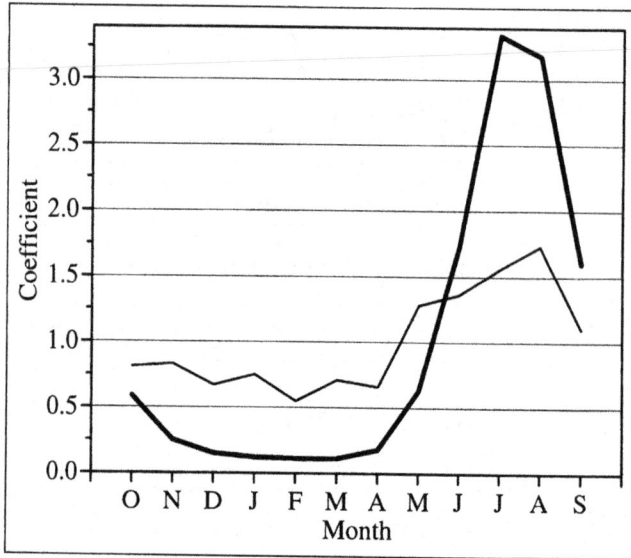

Fig. 5.6.2 Mean runoff coefficients (1964/65–1989/90) (thick line) in the catchment area of the Vent–Rofenache gauging station. The catchment area is 97 km² in size and showed 41% glacierization in 1969. Mean precipitation coefficients (1964/65–1989/90) (thin line) at the Vent weather station (1896 m a.s.l.).

precipitation (c_p) of two alpine catchment areas are compared with those of two catchment areas in the Cordillera Blanca, one on the Rwenzori and one on Mount Kenya. The coefficients describe the respective relation of monthly to mean annual amounts:

$$c_P = \frac{\bar{P}_m}{\bar{P}_a / 12} \qquad (5.6.1)$$

and

$$c_Q = \frac{\bar{Q}_m}{\bar{Q}_a} \qquad (5.6.2)$$

where $\bar{P}_m$ and $\bar{Q}_m$ are the long-term monthly means and $\bar{P}_m$ and $\bar{Q}_a$ the long-term annual means of precipitation and runoff in [mm we or kg m^{-2}] and [m^3 s^{-1}] respectively.

The periods used are not always the same. The data for the catchment areas in the Ötz Valley are from the Hydrographisches Zentralbüro (various years) and were analysed by Raggl (1996). The precipitation data from Mount Kenya

were published by Berger (1989), and the runoff data by Leibundgut *et al.* (1986). The data for the Rwenzori were compiled by Osmaston (1989²). The information from the Cordillera Blanca comes from the Unidad de Glaciologia e Recursos Hidricos and was examined by Niedertscheider (1990).

Figs. 5.6.2 and 5.6.3 show the variation of the coefficients for two catchment areas in the Ötz Valley. The catchment area of the Rofenache at the Vent gauging station covers 97 km², of which about 40% is glacierized. The precipitation was also measured in Vent close to the runoff gauging station (Fig. 5.6.2).

The catchment area of the Ötztaler Ache at the Tumpen gauging station covers 159 km², of which about 2% is glacierized. Precipitation was measured in Längenfeld within the catchment area (Fig. 5.6.3).

In both catchment areas, the increase of the annual amplitude of runoff compared with that of precipitation is distinct because of the seasonal influence of the snow cover and glaciers. They correspond to type A in Fig. 5.6.1.

Only few data series are available for the inner tropics. On the Rwenzori, runoff data from the Mubuku River (1953–63) can be compared with precipitation amounts from a station at Lake Bujuku (1951–54) (Osmaston, 1989²) (Fig. 5.6.4). The catchment area covers about 200 km² and shows a glacierization of about 2 km² (1950; see chapter 6), which is about 1% coverage. Precipitation was recorded close to Lake Bujuku

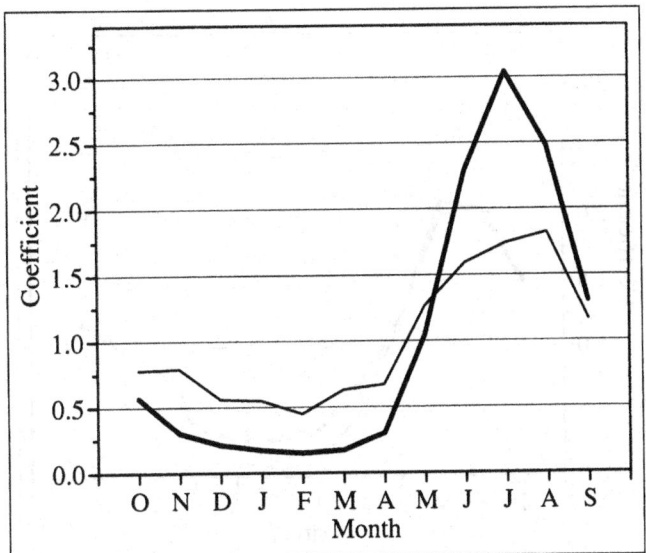

Fig. 5.6.3 Mean runoff coefficient (1964/65–1989/90) (thick line) in the catchment area of the Tumpen–Ötztaler Ache gauging station. The catchment area is 159 km² in size and showed 2 % glacierization in 1969. Mean precipitation coefficients (1964/65–1989/90) (thin line) at the Längenfeld weather station (1179 m a.s.l.).

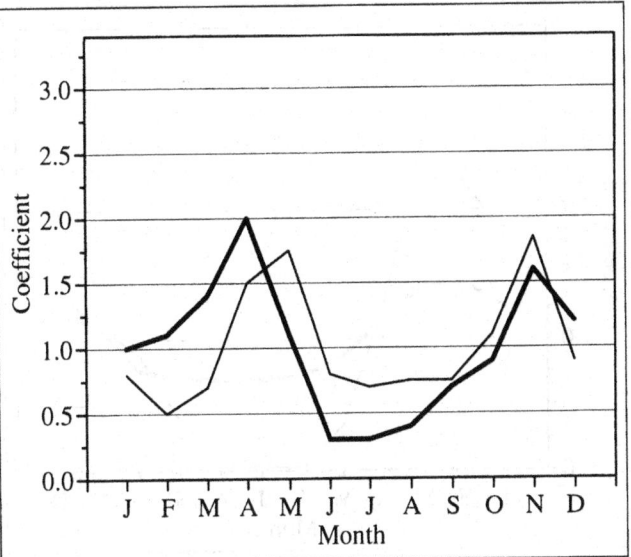

Fig. 5.6.5 Mean runoff coefficients (1961–1982) (thick line) in the catchment area of the Nanyuki–Naro Moru gauging station at the foot of Mount Kenya. The catchment area is 102 km² in size and shows less than 0.6% glacierization. Mean precipitation coefficients (1983–1989) (thin line) at the 'Met-station' (3050 m a.s.l.).

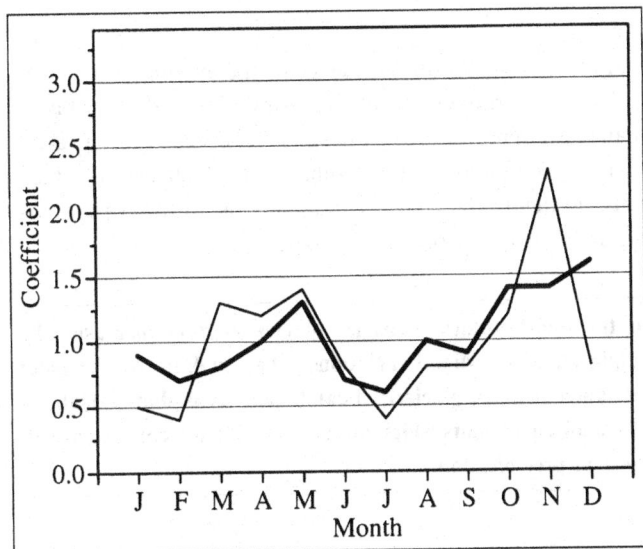

Fig. 5.6.4 Mean runoff coefficients (1953–1963) (thick line) in the catchment area of the Mubuku River. The catchment area is about 200 km² in size and showed a 1% glacierization in 1950 (see chapter 6). Mean precipitation coefficients (1951–1954) (thin line) at Lake Bujuku (3960 m a.s.l.).

at 3960 m in the high basin between Mount Baker, Mount Stanley, and Mount Speke.

The seasonal course of the coefficients corresponds to the expectations for the inner tropics. Two peaks can be seen in the precipitation line which are repeated in the runoff line in a subdued manner (compare Fig. 5.6.1). Besides very low glacierization, extensive bogs in the catchment area may also exert some influence.

On Mount Kenya, corresponding coefficients can be reconstructed (Fig. 5.6.5). However, the catchment area of the Naro Moru in Nanyuki, being 102 km² in size, only shows about 0.5% glacierization, with a glacierized area of 0.56 km² (Hastenrath et al., 1989). The influence of the glaciers on the runoff is therefore very low.

There is no comparable information from the Irian Jaya.

There are several data series available for the outer tropical Cordillera Blanca. The coefficients of two catchment areas are shown in Figs. 5.6.6 and 5.6.7. The extent of glacierization was taken from the glacier inventory of 1970 (Ames et al., 1989). The catchment area of the Rio Llanganuco at the runoff gauging station is 90.0 km² in size, and 36.3 km² or about 40% of it is glacierized. The precipitation was measured close to the runoff gauging station (Fig. 5.6.6).

The catchment area of the Rio Yanayacu at the Querococha gauging station covers 64.4 km², of which 3.5 km² or 5.5% is glacierized. The precipitation was again measured at the Querococha station (Fig. 5.6.7).

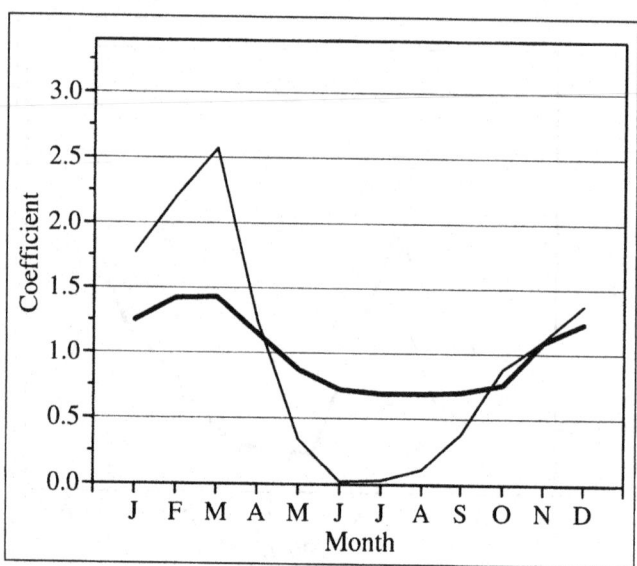

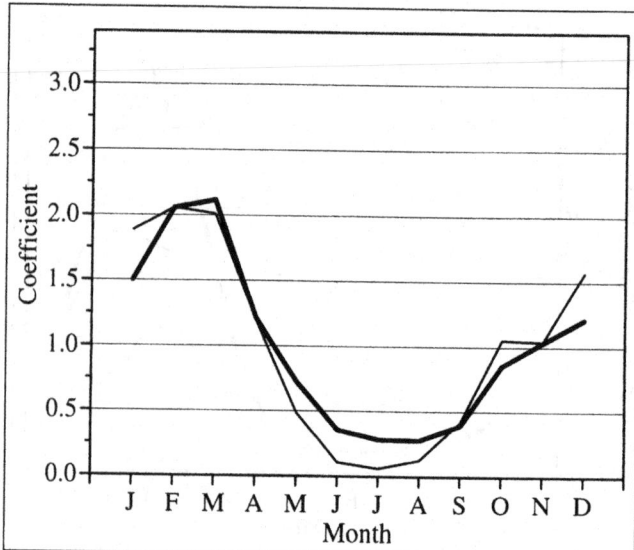

Fig. 5.6.6 Mean runoff coefficients (1963–1968) (thick line) in the catchment area of the Llanganuco–Rio Llanganuco gauging station. The catchment area is 90.0 km² in size and showed a glacierization of 40% in 1970. Mean precipitation coefficients (1953–1986) (thin line) at the Llanganuco weather station (3850 m a.s.l.).

Fig. 5.6.7 Mean runoff coefficients (1963–1968) (thick line) in the catchment area of the Querococha–Rio Yanayacu gauging station. The catchment area is 64.4 km² in size and showed a glacierization of 5.5% in 1970. Mean precipitation coefficients (1953–1986) (thin line) at the Querococha weather station (3980 m a.s.l.).

The example of the heavily glacierized catchment area of the Quebrada Llanganuco shows especially clearly that the annual amplitude of runoff is subdued compared with the amplitude of precipitation. They correspond to type C in Fig. 5.6.1. Part of the precipitation is withheld during the wet period in the accumulation area of the glaciers. However, the damping effect cannot only be ascribed to the glaciers. In both catchment areas, there are lakes above the runoff measurement stations which also contribute to the subduing effect on the runoff variation. Additionally, boggy valley plains (*pampas*) also play a role in the Cordillera Blanca. In the course of the rainy period, they fill with water which evaporates during the dry season, but also partly adds to the runoff. Their influence on the runoff regime has yet to be analysed.

An indication that the influence of the melting of a glacier is clearly subordinate to that of precipitation in the Cordillera Blanca is given by Lliboutry et al. (1977[1]). They show a variation of precipitation at the Laguna Parón (4200 m a.s.l.) which runs parallel to the lake level fluctuations. In the Alps, reservoirs fill during the period of ablation.

In the mid-latitudes, seasonal glacier melting increases the amplitude of the runoff variations. The few data from tropical and outer tropical glacierized catchment areas show variations of runoff coefficients which are clearly subdued compared with those of precipitation.

II

Modern glacier fluctuations in tropical high mountains

Around the start of the twentieth century, exploratory expeditions moved to the tropical high mountains. They brought back reports and photos which showed more or less fresh and distinct moraines not far from the glacier tongues. The tongues had been withdrawing from the maximum extent of the 'Little Ice Age' for some time. In some areas, the maximum extent can be mapped with the help of aerial photographs. The exact point of time of the beginning of the retreat, however, is not known for any tropical high mountain and can only be estimated. Today, tropical glaciers are much smaller than they were 100 years ago (Osmaston, 1961, 1965, 1975; Osmaston & Pasteur, 1972; Hope et al., 1976; Hastenrath, 1981, 1984; Osmaston, 1989[1,2]; Jordan, 1991; Williams & Ferrigno, 1989, 1991 give very detailed but incomplete summaries of the glaciology in single tropical high mountains). On some mountains, an advance had taken place at the beginning of the twentieth century, which again reached, in some cases, the extent of the previous century (Patzelt et al., 1984; Bhatt et al., 1981; Kaser & Georges, 1997) and, in the 1960s and 1970s, small and short advances were measured on a few glaciers (Temple, 1968; Kaser et al., 1990).

In the following chapters, reconstructed 'modern' glacier fluctuations in the twentieth century on the Rwenzori (chapter 6) and in the Cordillera Blanca (chapter 7) will be presented. They are the result of analyses based on terrain models of aerial photos, normal photos, some mainly short-term measurements of tongue changes and reports from earlier expeditions. Almost all expeditions and passing travellers report snow or firn limits. Apart from partly inaccurate altitudes, due to a lack of or poor maps, this information is probably always only relevant for that point of time or season, and cannot be accepted as a climatologically mean snow limit or even as a clearly defined mean equilibrium line. Therefore, these frequently reported snow lines will not be considered in this work.

Possible climatic reasons for the reconstructed modern glacier fluctuations will be discussed together with results from other tropical high mountains (chapter 8).

6 Modern glacier fluctuations on the Rwenzori

6.1 THE RWENZORI MOUNTAINS

The Rwenzori Mountains* are the third highest glacierized mountain massif in Africa after Kibo of Kilimanjaro and Mount Kenya. They are situated slightly north of the equator in the border area of the Democratic Republic of Congo and Uganda, have an extent of about 113 km from north to south and about 48 km at their largest width. Margherita Peak is the highest with an elevation of 5109 m a.s.l. Although both Uhuru Peak on the Kilimanjaro (5895 m a.s.l.) and Batian on Mount Kenya (5199 m a.s.l.) are higher, the Rwenzori is much the largest of the three massifs. While the first two are single, towering volcanoes, the Rwenzori consists of Precambrian rocks which were deformed in the course of the origination of the East African rift system and were raised as a horst (Osmaston, 1989[2]). Six groups of the rugged mountain range had glaciers until recently (Hastenrath, 1984; Osmaston, 1965, 1989[2]; Osmaston & Pasteur, 1972) and moraines show former, much larger glaciations.

The extent of 1955 was reconstructed by Osmaston (1989[2]) from aerial photos. The Lac Gris stage is characterized by a series of not very distinct frontal moraines at an altitude of about 4300 m which the first explorer reported

Table 6.1.1. *The six glacierized mountain groups, their summit altitudes and the extent of their glacierization in 1955*

Mountain group	Highest altitude (m)	Glacier area (km²)
Mount Stanley	5111	2.05
Mount Speke	4891	1.38
Mount Baker	4873	0.73
Mount Emin	4802	0.07
Mount Gessi	4769	0.17
Mount Luigi di Savoia	4665	0.03
Rwenzori total		4.43

Source: Osmaston (1989[2]).

were fresh and directly in front of the tongue (Moore, 1901, 1902; David, 1904, 1909; De Filippi, 1909). It was confirmed by de Heinzelin (1953) and Bergstrøm (1955) through lichenometrical dating and descriptions of soil that the retreat from the last maximum extent started in the second half of the nineteenth century. Thus, this stage corresponds to the world-wide maximum extent of the 'Little Ice Age'. A detailed discussion about extents, dating and possible causes of past glaciation is offered by Osmaston (1989[2]) and by Osmaston in chapter 9 of this book. In the Rwenzori Massif, the recent moraines, if existent at all, are generally not very distinct. Osmaston & Pasteur (1972) suggest that the only minor erosion is partly due to the lack of large temperature variations. The old moraines, however, are very large.

6.2 THE ESSENTIAL FEATURES OF THE CLIMATE OF THE RWENZORI

The Rwenzori are situated within the humid inner tropics (Fig. 6.2.1).

Whittow (1960) analysed synoptic maps from the East Africa Meteorological Department over a twelve month period (September 1958 to August 1959) and obtained a detailed picture of the origin and the character of different air masses:

---

* 'Ruwenzori' was the name given to the range by Henry Stanley in 1890 in his book *In Darkest Africa*, as an approximation to one of the local names, 'Rwenzururu', and this was used for all purposes for a century. In 1990 the Ugandan sector of the range was declared and named the 'Rwenzori Mountains National Park' to be closer to the vernacular name (the actual name Rwenzururu had in the meantime been appropriated by a dissident political movement). Confusingly, the nearby lowland Queen Elizabeth National Park had been renamed 'Rwenzori National Park' in 1973 by Presidential Order to express the anti-British sentiment of Idi Amin but never legally gazetted as such, though there were publications from that time using that name; it reverted to its original name in 1980 by Order of the Board of Trustees.

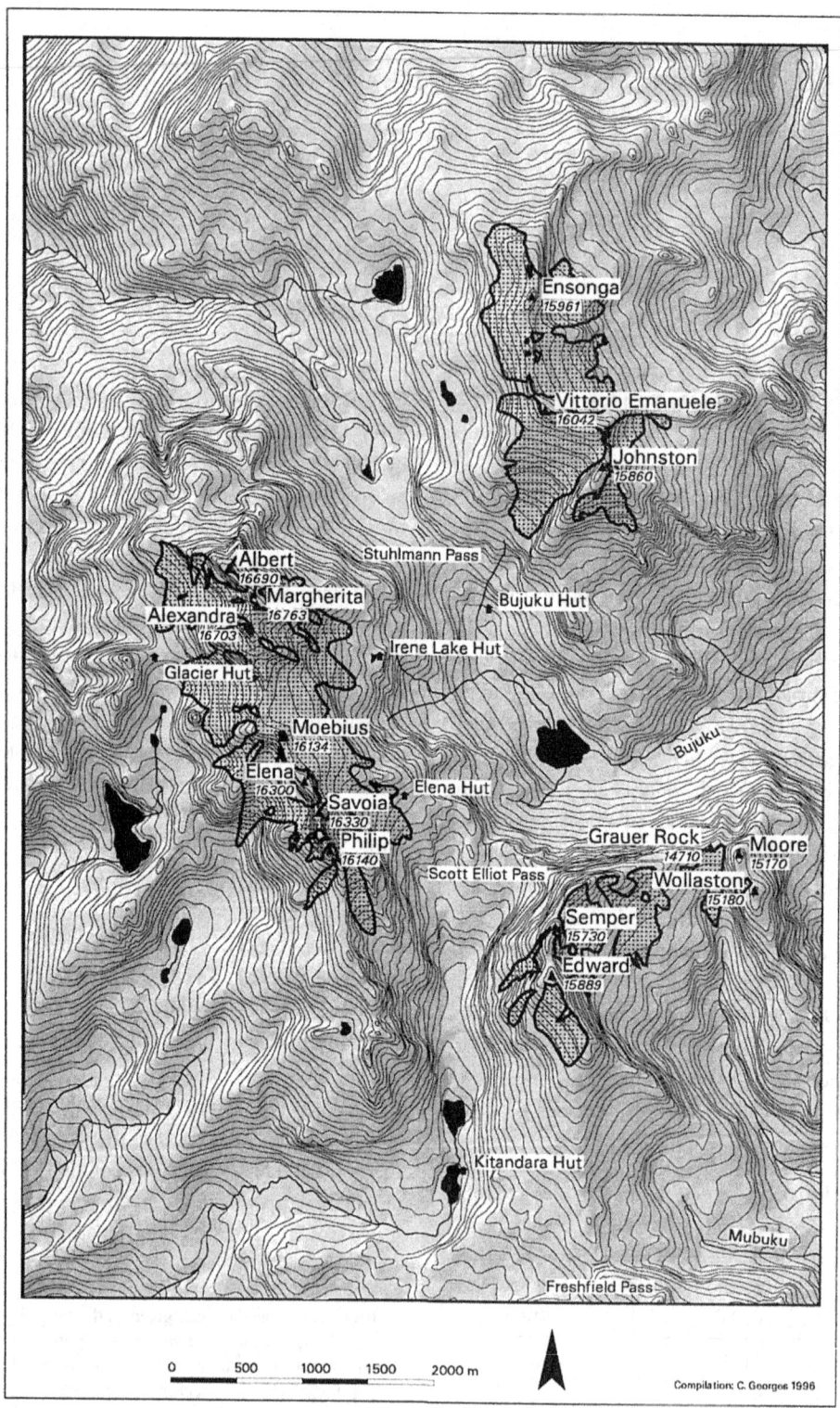

Fig. 6.1.1 The central part of the Rwenzori with its most important peaks, passes and huts. Glacier extent: 1955 (section 6.5); altitudes in ft.; vertical interval of the contour lines: 100 ft., the contour lines are labelled in the detailed maps. Base map: D.L.S.U. (1970).

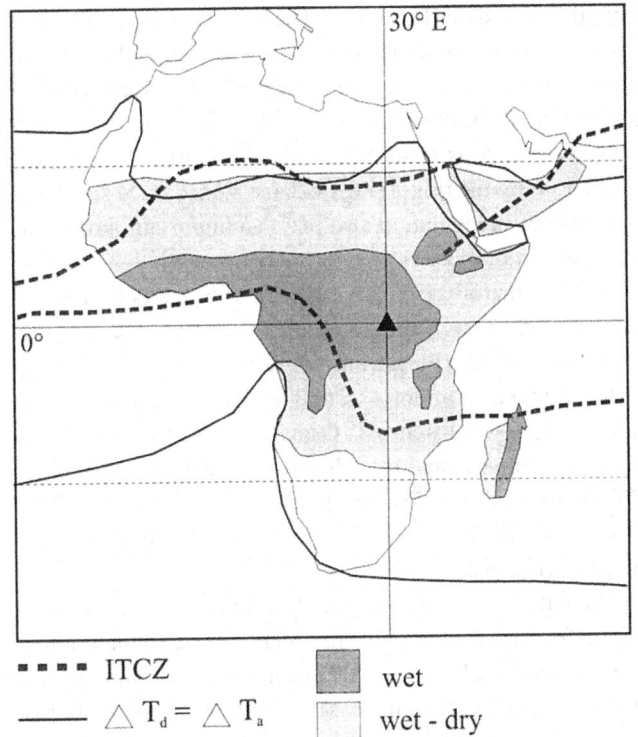

Fig. 6.2.1 The Rwenzori mountain range (▲) in the inner tropics of Africa. (Sources: Lauer, 1975; Liljequist and Cehak, 1984; Paffen, 1967.)

- Between September and December 1958, the Inter Tropical Convergence Zone (ITCZ) was on its way south and the Rwenzori were under the influence of very wet air masses which came over the Congo Basin from the South Atlantic and from the Indian Ocean.

- In January 1959, when the ITCZ was situated south of the mountain range, it was influenced by 'predominantly dry' air masses from North Africa. In February and part of March, the lower altitudes (down to about 4500 m a.s.l.) were still under the influence of quite dry air. Above that, however, there was again wet air.

- In April, a second rainy season started as the ITCZ began moving north. Wet air came over the Congo Basin from the South Atlantic and from the Indian Ocean.

- In June and July 1959, the conditions were opposite to those in January and February. While the lower altitudes were characterized by wet air masses, relatively dry conditions dominated at the peaks.

The sequence of two rainy periods and two periods of reduced precipitation also becomes clear in the annual variation of several precipitation series (Fig. 6.2.2), which had been measured on the Rwenzori in the 1950s (Osmaston, 1989[2]). A dependency of the precipitation on altitude can be discerned

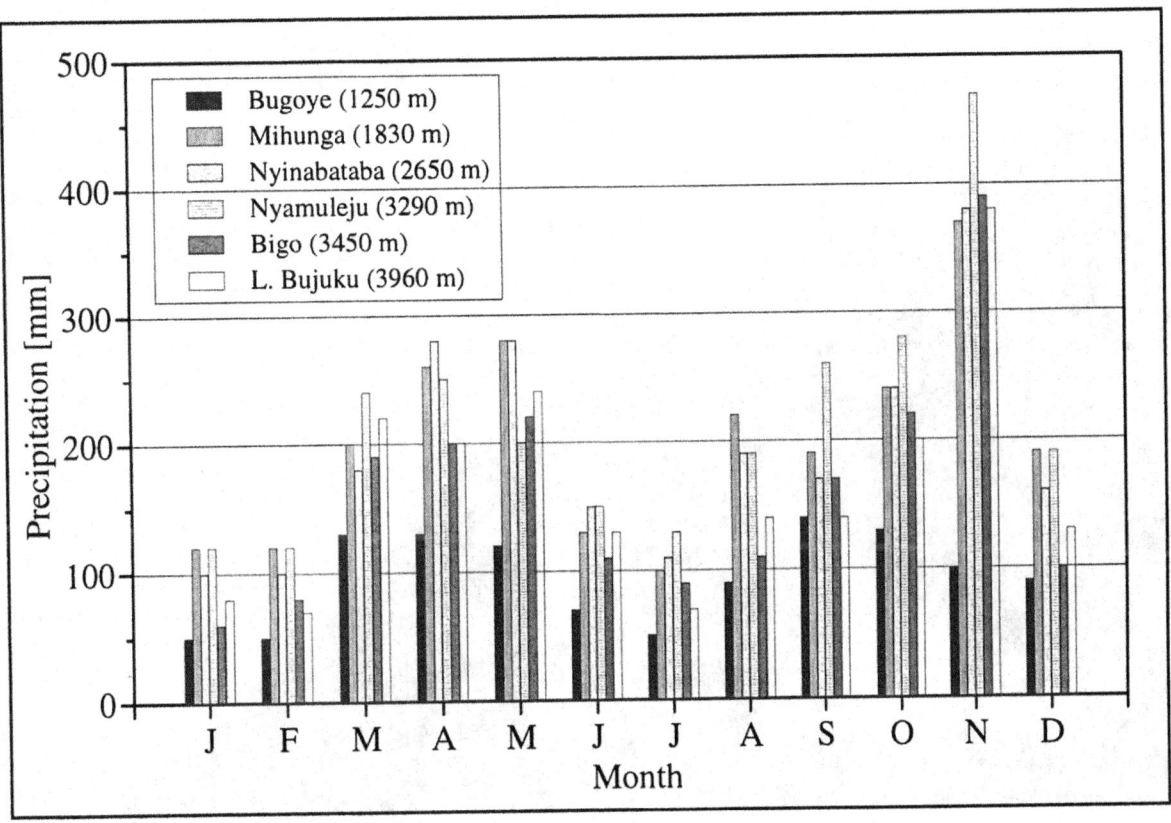

Fig. 6.2.2 Mean annual variation of precipitation from 1951 to 1954 on the Rwenzori (according to Osmaston, 1989[2]).

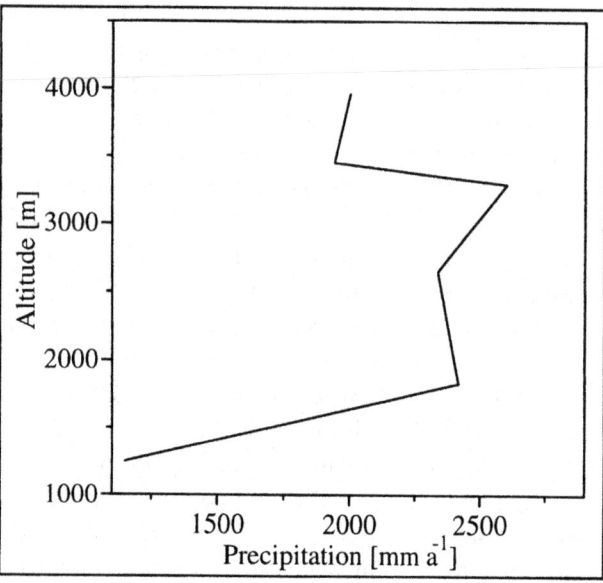

Fig. 6.2.3 The distribution of precipitation in correspondence with altitude on the Rwenzori (according to Osmaston, 1989[2]).

from the same data (Fig. 6.2.3). The highest precipitation falls within altitudes between 1500 m and 3500 m a.s.l. Above that, precipitation decreases again. Selective measurements of accumulation (Bergstrøm, 1955) do not allow for any conclusions as to its long-term dependency on altitude. Thick rime formations on the ridges (Figs. 6.2.4, 6.8.14, 6.8.15) show that this form of precipitation also plays an important part in the mass budget of glaciers.

Other climatological variables, such as temperature and radiation, have been measured on the Rwenzori exclusively over short periods (Bergstrøm, 1955; Whittow et al., 1963). In order to make an attempt to find at least the altitude of the 0°C level at the Rwenzori, Osmaston (1989[2]) included all existing temperature records from different expeditions (Woosnam, 1907; De Filippi, 1909; Hauman, 1933; Bergstrøm, 1955; Osmaston, 1958 and Temple, 1961) in a diagram (Fig. 6.2.5).

The data points are situated close to a line calculated for East Africa by Trapnell & Griffiths (1960). Based on that line, the 0°C level lies between 4600 and 4700 m a.s.l. The decrease in temperature with an increase in height is $\partial T_a/\partial z = -0.0065$ K m^{-1}.

Fig. 6.2.4 Rime on the ridge of Alexandra Peak of Mount Stanley.

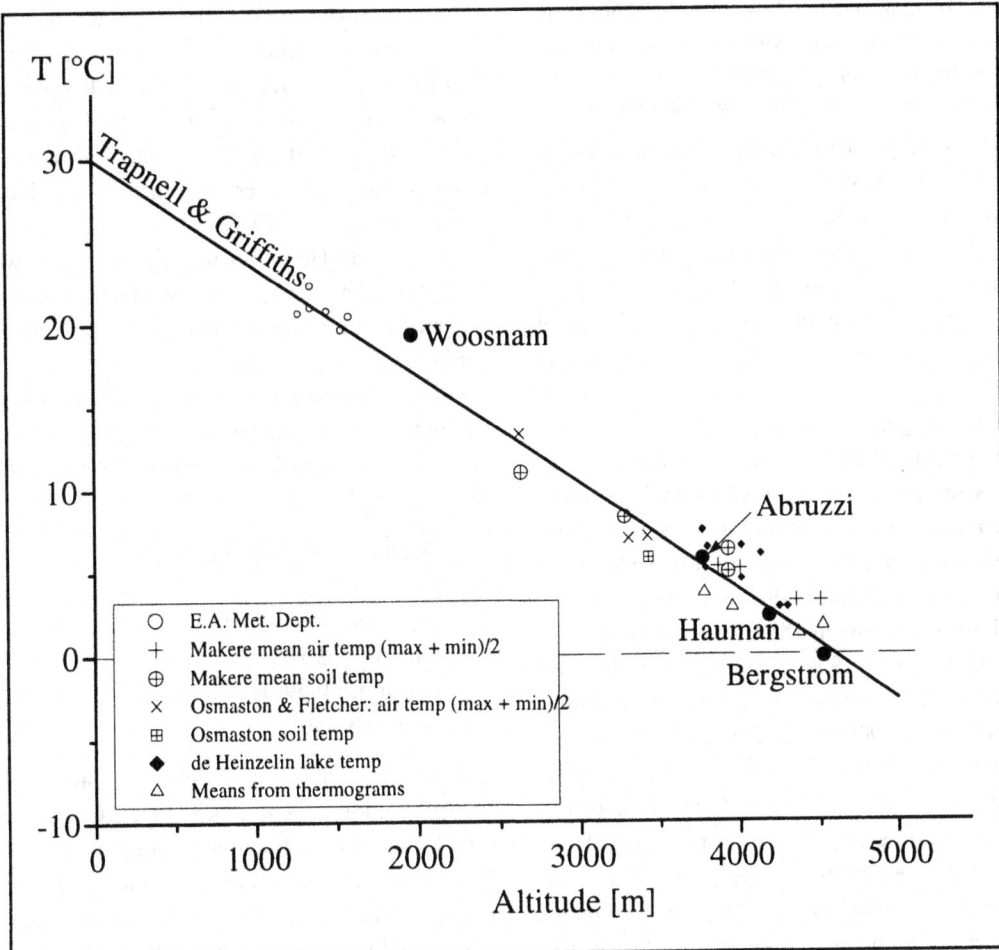

Fig. 6.2.5 Temperatures on the Rwenzori (from Osmaston, 1989[2]).

A detailed description of the mountain range, the climate, the vegetation etc. is given by Osmaston & Pasteur (1972) in their book *Guide to the Ruwenzori.*

6.3 GLACIER RESEARCH ON THE RWENZORI

The 'discovery' of glacierized mountains in East Africa came hand in hand with the European conquest, colonization and missionary work in Africa and happened late compared with when the first reports were made of tropical glaciers in South America and on Irian Jaya. The missionary Rebmann saw the snowy hilltop of the Kibo in 1848 and reported it. His colleague Krapf 'discovered' the glaciers of Mount Kenya in 1849 (Rebmann, 1849[1,2], 1855; Krapf, 1849, 1858, 1860 according to Hastenrath, 1984). In tropical South America, the Spanish conquerors had already seen glacierized mountains in the sixteenth century and, on February 16, 1623, the Dutch captain

Jan Carstensz wrote in his logbook when off the coast of Irian Jaya: '. . . at a distance of about 10 miles, by estimation, into the interior we saw a very high mountain range in many cases white with snow, which we thought a very singular sight, being so near the line equinoctial' (Hope, 1976, according to Wollaston, 1912). It was even more astonishing that the first reports of snow on the Kilimanjaro and Mount Kenya were questioned in Europe.

The Rwenzori had to wait longer for its 'discovery'. After the first European travellers had seen the mountain range from afar in the second half of the nineteenth century, and some had been told of snow and ice (Stanley, Gessi, Mason, Emin Pasha, Casati), Sir Henry M. Stanley's attention was drawn 'to a mountain said to be covered with salt, and I saw a cloud of most beautiful silver colour, which assumed the proportions and appearance of a vast mountain covered with snow' (Stanley, 1890[1]), as described by an African companion on May 24, 1888.

'While I looked towards southeast, thinking about the events of the past month, a boy drew my attention to a cloud strangely formed, which had a very beautiful silver-like colour and had the proportions and the look of a vast mountain covered with snow. . . . I became convinced that I was not only looking at the mere picture of a big mountain but at a solid, real peak whose top was covered with snow.' (Stanley, 1890[2])

In 500 BC, however, Aeschylus had already written that 'Egypt is fed by snow'. In 350 BC, Aristotle asserted that silver mountains were the sources of the Nile, and in 150 BC, Claudius Ptolomeus wrote about 'mountains of the moon' whose snow-fed lakes were the origin of the Nile. In 1154, the Arab geographer Edrisi (Abu' Abdulla Mohamed) reported on the Jebel el Qamar, or the Mountains of the Moon (Osmaston & Pasteur, 1972).

The first pictures of glaciers on the western side of the mountain, later called Mount Stanley, were taken by Stuhlmann on his journey with Mehmed Emin Pascha (E. Schnitzer, the German doctor and Governor of the Equatorial Province) along the western side of the Rwenzori (Stuhlmann, 1894; Schweitzer, 1898; De Filippi, 1909). After a few visits and unsuccessful ascents (Stairs in 1889, Stuhlmann in 1891, Scott-Elliot in 1895, Moore in 1900 [he succeeded in climbing Mount Baker], Johnston in 1900, David in 1904, Freshfield in 1905, and Wollaston in 1906 [according to Osmaston & Pasteur, 1972]), the Duke of Abruzzi led his successful expedition to the mountain range (De Filippi, 1909). Besides the ascent of all the important peaks, the first 'topographical and geological map of the Rwenzori Range' was drawn and Vittorio Sella, the photographer of the expedition, brought a large number of excellent photographic records to Europe. The expedition not only reported a general retreat of the glaciers from an extent that had left behind thin moraines not long ago, but also of indications that referred to an earlier, larger glacier extension.

After two visits in 1926 and 1927, Humphreys succeeded in the first flight over the mountain range in 1931. His pictures showed for the first time that, besides the known peaks, no others were glacierized (Humphreys, 1927, 1933).

A Belgian expedition visited the western side (then the Belgian Congo) of the Rwenzori in 1932 (de Grunne, 1933; de Grunne et al., 1937; Michot, 1933). Synge (1937) published pictures of the rarely visited Thomson Glacier on the southern flanks of Mount Luigi di Savoia. In 1937/38, a Deutscher Alpenverein expedition succeeded in completing the first survey of a contour map of the Central Rwenzori. Mount Speke, Mount Baker, and the eastern side of Mount Stanley are depicted at a scale of 1:25000 (Eisenmann, 1939; Stumpp, 1952).

A further Belgian expedition concentrated on the retreat of the glaciers on the western side of Mount Stanley: Alexandra, West Stanley, Moebius, and West Elena (de Heinzelin 1951, 1952, 1953).

The most extensive examination after the research done by the Abruzzi expedition was made by the 'British Ruwenzori Expedition 1952'. Over short periods, Bergstrøm, who was on the Rwenzori at the same time as de Heinzelin and Menzies, made the most detailed glaciological measurements and observations of that era on the Elena Glacier of Mount Stanley (Bergstrøm, 1955).

In 1956, de Heinzelin was again on the western side of Mount Stanley. His work on the western side is the most comprehensive and detailed up to the present (de Heinzelin, 1962).

On the Ugandan side of the mountain range (then a British colony), the first Makerere College Ruwenzori Expedition (M.C.R.E.) started in December 1957. Five further expeditions followed:

December 1957–January 1959
June 1958
December 1958–January 1959
June–July 1959
December 1959–January 1960
June–July 1961

The first expeditions took place within the framework of the International Geophysical Year (IGY) in 1957/58. As the establishment of a long-term programme was intended, the first journey was mainly an exploratory expedition in order to find suitable glaciers which could be reached quite easily, and which could be worked on without too much danger. Elena Hut (4511 m a.s.l.) (Fig. 6.1.1) was made the base for all further work. It was mainly concentrated on the examination of the glaciers Savoia and Elena (Mount Stanley), Moore (Mount Baker), and Speke (Mount Speke). Changes of the tongues concerning length and thickness, as well as accumulation, ablation, and climate parameters, were measured and old tongue extensions were reconstructed (Whittow & Shepherd, 1958; Whittow, 1959, 1960; Whittow et al., 1963).

Temple, a member of the Makerere Expeditions, continued part of the glaciological work up to 1967 (Temple, 1968). In the following decades, the politically restless conditions made only sporadic and short visits to the Rwenzori possible. In 1983, a British expedition collected material and data from Mount Gessi and Mount Emin, but these were taken from the group during a police search. They were never recovered (A. Hughes-d'Aeth, unpublished report, 1985).

6.3.1 Aerial photographs, maps and summarizing publications

After the first aerial photographs by Humphreys, a series of vertical **aerial photographs** were taken in the 1950s:

Table 6.3.1. *Different area details of the glaciers on the Rwenzori (km^2)*

		1906 Filippi[a]	c. 1955 Hastenrath[b]	c. 1955 Osmaston[c]	'Present' WGMS[d]	c. 1990 Noggler[e]
1	Mt Stanley	2.34	1.04	2.05		1.39
2	Mt Speke	2.14	0.78	1.38		0.75
3	Mt Baker	1.61	0.39	0.73		0.32
4	Mt Emin		0.08	0.07		} 0.16
5	Mt Gessi		0.18	0.17		
6	Mt L. Savoia		0.04	0.03		
1–6	Total		2.51	4.43	4.96	2.62
1–3	'Central Rwenzori'	6.09	2.21	4.16		2.46

Notes:

[a] From the map of the 1906 Abruzzi expedition (De Filippi, 1909) planimeterized.

[b] Hastenrath (1984, appendix 3) on the basis of maps and historical pictures. His 'recent glaciation' covers a period of time from 1891 (Stuhlmann, 1894) to 1967 (Temple, 1968).

[c] Osmaston (1989^2) on the basis of the D.O.S. 1:25000 map, as well as aerial photographs by MacLachlan (1951 and 1952) and Hunting Aerosurveys (1955).

[d] WGMS (1989): World Glacier Inventory, Status 1988.

[e] Noggler (1992): Present glacier extents from photographs by different authors.

- December 11, 1951: Haward MacLachlan: Only the higher, non-glacierized parts of the Nyamugasani and Nyamwamba valleys were photographed.
- April 4 and 20, 1952: R.A.F. Mosquito: two strips of aerial photographs 82D/555 and 587, about 1:15000 (at 13000 ft.). The pictures have a weak contrast and cover only a part of the mountain range.
- September 17 and 27, 1952: Haward MacLachlan: central peaks, about 1:16000 (at 13000 ft.), suitable for details, strongly distorted outwardly.
- June and October 1955: Hunting Aerosurveys: whole mountain range. No. 15UG 13, 14, 29, 31, 33. These pictures were the main basis for the maps of the Directorate of Overseas Surveys (D.O.S.), about 1:30000 (at 13000 ft.).

Additionally, there are oblique photographs by MacLachlan from 1952.

In addition to smaller sketches in reports of expeditions, the following **maps** of the glacierized mountains of the Rwenzori were made:

- 'Topographical and geological map of the Ruwenzori Range 1:40000'. This map was made by Roccati within the framework of the expedition of the Duke of Abruzzi in 1906 (De Filippi, 1909).
- 'Central group of the Ruwenzori-Mountain 1:25000', expedition in 1937/38 of the Deutscher Alpenverein, Sektion Stuttgart (Stumpp, 1952).
- Between 1957 and 1961, the D.O.S. (Directorate of Overseas Surveys) published four maps of the Rwenzori

at a scale of 1:50000 on the basis of aerial photographs from 1955.
- In 1962, the 'Central Ruwenzori 1:25000' of the D.O.S. was issued. A further print was published by the Uganda Government Department of Lands and Surveys (1970).

Besides the work by Whittow *et al.* (1963), two further publications were published with extensive **summaries** of the state of the knowledge on this subject:

- In 1972, a Rwenzori guidebook was issued by Osmaston & Pasteur. In addition to an alpine-tourist description, this guidebook contains an extensive summary of scientific results of the research.
- An extensive review and a new interpretation of the glaciological studies on the Rwenzori were published by Osmaston (1989^2).

Hastenrath (1984) published a glacier inventory of the Rwenzori on the basis of an analysis of aerial photographs from 1955, and based on his own mappings of the terrain from 1974.

6.3.2 Modern glacier extents

Noggler (1992) combined all the photographs of the glaciers that could be found and tried to reconstruct the glacier fluctuations for the whole area since the expedition of the Duke of Abruzzi in 1906 up until 1990. Information on the area given by different authors is compiled in Table 6.3.1.

On the map of the Abruzzi expedition, from which the glacier extensions from 1906 were measured, only single peaks are measured trigonometrically. Therefore, the positions and the extents of the glaciers are inaccurate. The analyses by Osmaston (1989[2]) are the most accurate and those by Noggler (1992) definitely show the correct magnitude of order of the glaciation around 1990. In comparison, values given by Hastenrath seem too low. Osmaston (1989[2]) has already criticized these values, giving reasons for his objections. The values in the World Glacier Inventory 1988, on the other hand, seem to be clearly too high.

In the following chapter, an attempt will be made to comprehend the changes of the glaciation of the Rwenzori as accurately as possible with the help of all available material and based on a Geo Information System (GIS). Because of the availability of the material, three glacier extensions are appropriate for the reconstruction of the respective surface areas: those in 1906, around 1955, and around 1990. Four single glaciers, the Elena and the Savoia on Mount Stanley, the Speke on Mount Speke, and the Moore on Mount Baker are examined more closely. As they were objects of earlier research (Bergstrøm, 1955; Whittow & Shepherd, 1958; Whittow, 1959, 1960; Whittow et al., 1963; Temple, 1968), more and better information is available and makes detailed analyses possible. Initial results and interpretation attempts have already been published (Kaser & Noggler, 1991, 1996; Kaser, 1993).

6.4 THE 1906 GLACIER EXTENT

As a basis for all reconstruction and the analyses of glacier extents, contour lines were digitized from the map 'Central Ruwenzori 1 : 25 000' (D.L.S.U., 1970). They correspond to the situation in 1955, the year the aerial photographs were taken (Hunting Aerosurveys, 1955), which are the main basis of the map. The sector was chosen with the intention of covering the three main groups of the Rwenzori, Mount Stanley, Mount Speke, and Mount Baker. The three other groups (Emin, Gessi, and Luigi di Savoia) were not included because of the extremely small amount of information available about the present glacier surfaces. Their glacierization was already minimal earlier (Table 6.3.1) and, according to information by Guy H. Yeoman (1996, personal communication), only a few relict ice patches were left on Mount Gessi in 1984. The selected sector will henceforth be called 'Central Rwenzori Massif'. The scale in feet (1 ft = 0.3048 m), which was used on the base map, will be maintained due to reasons of accuracy and will only be changed in the following figures and tables if it seems necessary in order to compare values with results from other areas or for an overall view.

In a first attempt, the extents of the glacier surfaces on Mount Stanley, Mount Speke, and Mount Baker were digitized from the 'Topographical and geological map of the Ruwenzori Range 1:40000' (De Filippi, 1909). The points measured in 1909, the peaks Margherita (Mount Stanley), Vittorio Emanuele (Mount Speke), as well as Wollaston and Edward (Mount Baker) corresponded exactly with the points in the D.O.S. map. However, In Fig. 6.4.1, the two surveys are superimposed, permitting a comparison of the glaciers with the contours, and it becomes obvious that vast parts of the glacier surfaces are situated in an unrealistic shape and position in the terrain.

The high-quality photographs by Vittorio Sella, the photographer of the Abruzzi expedition, show a large part of the Rwenzori glaciers (De Filippi, 1909; Fiory-Ceccopieri, 1981; Fantin, 1968; Mantovani, 1996). Using these photographs, which even show some glaciers from different angles, the 1906 glacier extent for each of the three massifs was drawn into the contour line model (Figs. 6.4.2–6.4.4). Fig. 6.4.5 offers an overall view, which also outlines the differences from Fig. 6.4.1. Those parts that are not shown in the photographs were best fitted into the topography according to the 1906 map. The western flank of Mount Stanley was not visited by the Abruzzi expedition, so the only help in this area was the photograph Stuhlmann had taken from the upper Butagu Valley in 1891, and which was reproduced by De Filippi (1909). Slight adjustments in the glacier limits were made while piecing together the three reconstructions of 1906, 1955 and 1990.

The photograph on page 59 in the publication by Mantovani (1996) shows the tongue of the Moore Glacier on Mount Baker in high quality and at a favourable angle for reconstruction. This photograph proved an important correction to the extent published by Kaser (1993) and Kaser & Noggler (1996). As stated above, the extent in 1955 also had to be slightly corrected in order to match the improved extent in 1906. Although the evaluation may be imprecise in one or the other detail, the inaccuracy of the whole glacier area is estimated to be <5%. The values for the individual massifs are shown in Table 6.4.1 and are compared with those from the 1906 Abruzzi map.

The new results are clearly higher on Mount Stanley. On Mount Speke and Mount Baker the differences in the extent of the glacier areas are insignificant and within the limit of the accuracy of the evaluation. However, the position and the shape of the individual glacier tongues are clearly different from those on the Abruzzi map (compare Fig. 6.4.1 with Fig. 6.4.5). The new evaluation offers a more realistic picture of the glacierization around the turn of the century and is, moreover, good background information for the reconstruction of younger glacier extents.

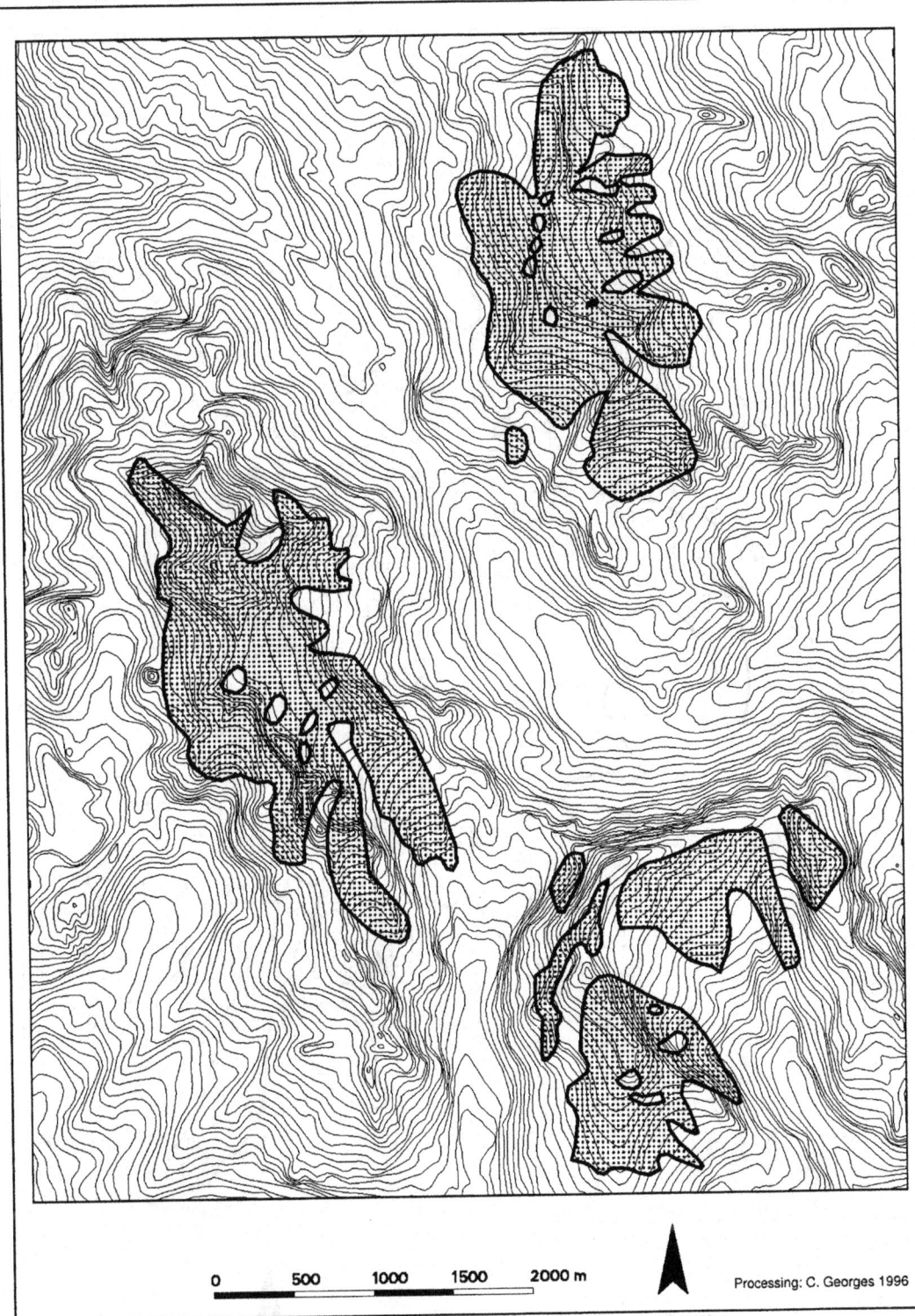

Fig. 6.4.1 The glacier area from the 'Topographical and geological
map of the Ruwenzori Range 1:40 000', extent in 1906 (De Filippi,
1909) over the contour lines of the map 'Central Ruwenzori
1:25 000' (D.L.S.U., 1970); vertical interval of the contour lines:
100 ft. For contour heights see Figs. 6.4.2 to 6.4.4.

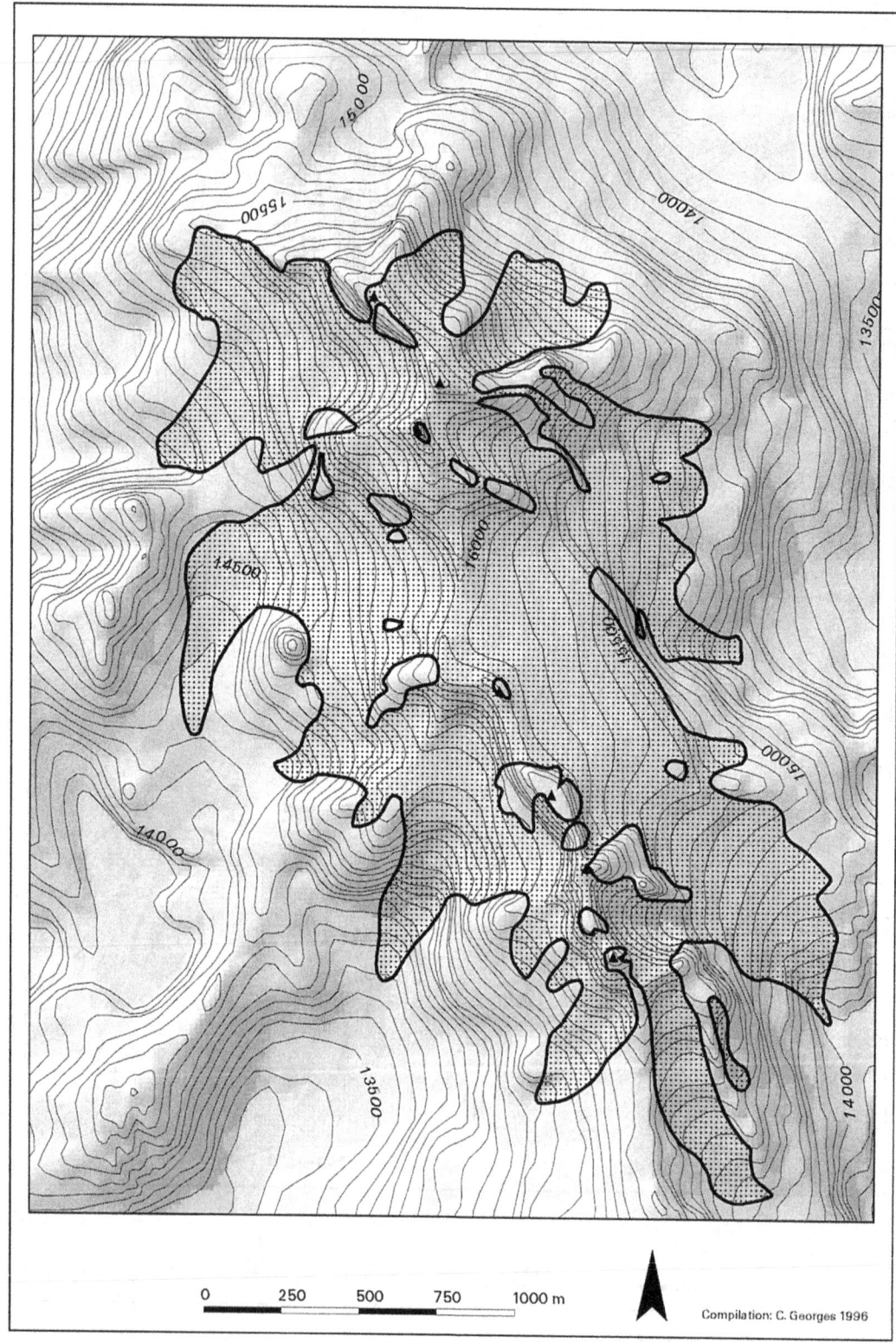

Fig. 6.4.2 The reconstructed glacier area on Mount Stanley in 1906.
The contour lines represent the glacier surface in 1955. Base map:
D.L.S.U. (1970); vertical interval of the contour lines: 100 ft.

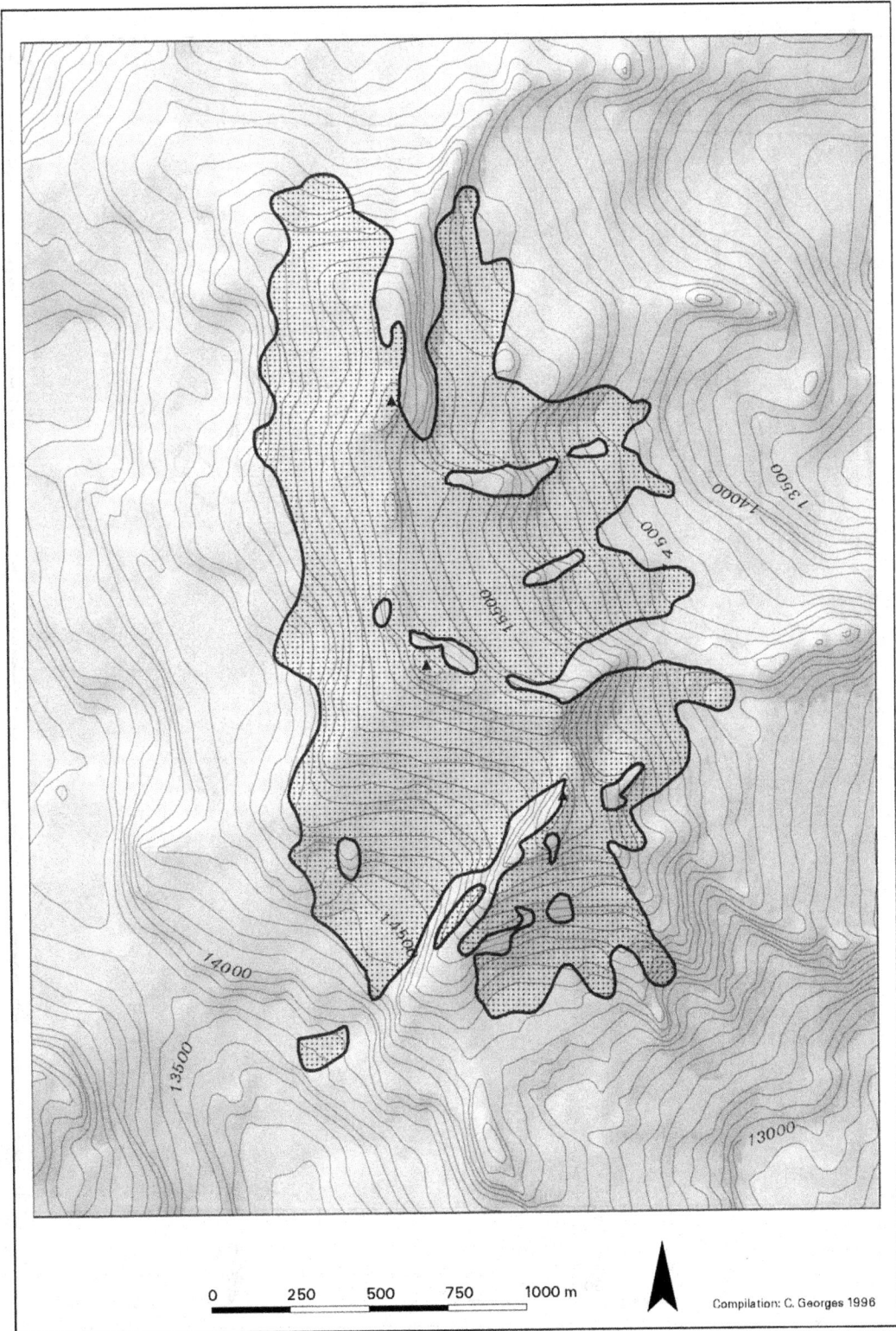

0   250   500   750   1000 m

Compilation: C. Georges 1996

Fig. 6.4.3 The reconstructed glacier area on Mount Speke in 1906.
The contour lines represent the glacier surface in 1955. Base map:
D.L.S.U. (1970); vertical interval of the contour lines: 100 ft.

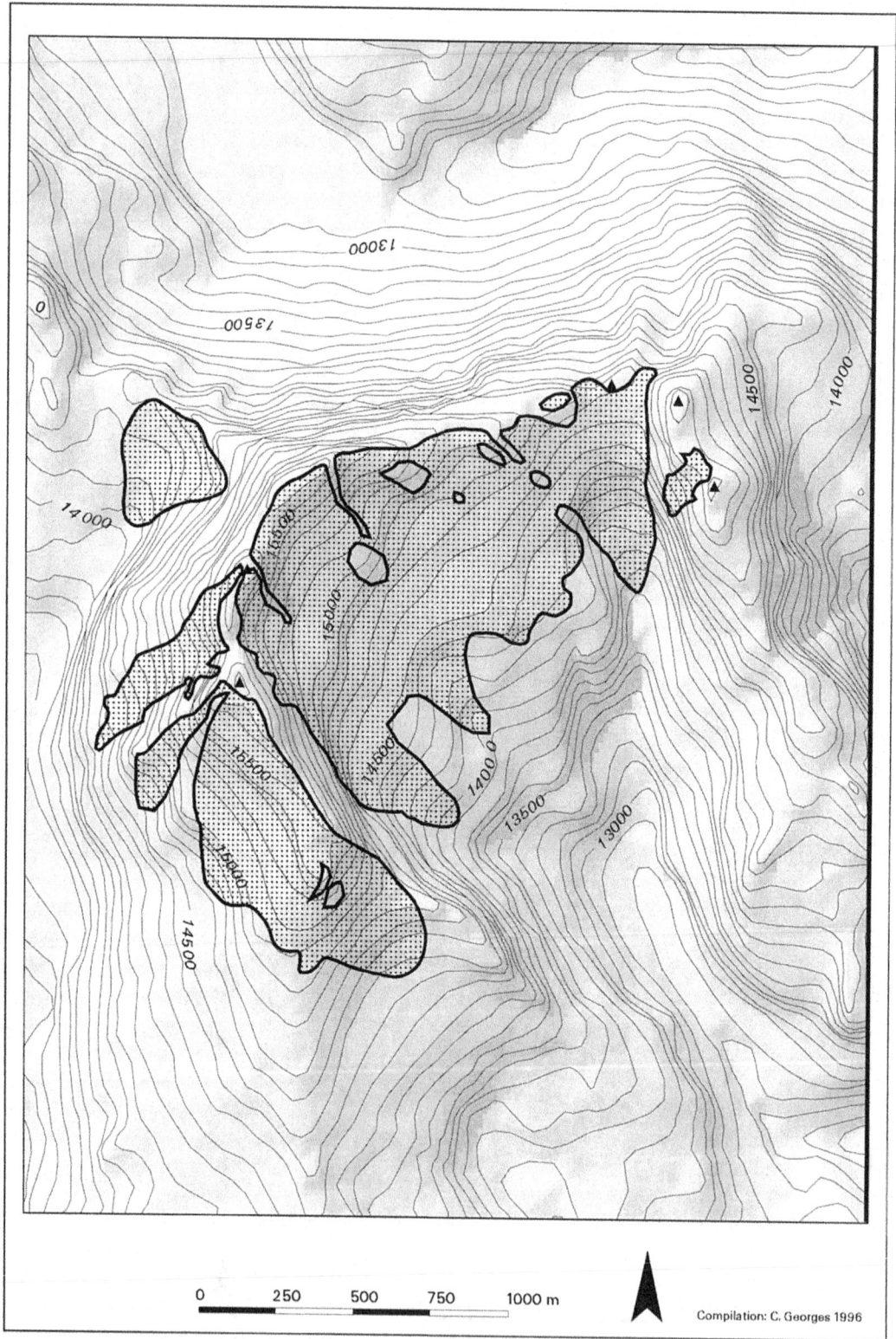

0   250   500   750   1000 m

Compilation: C. Georges 1996

Fig. 6.4.4 The reconstructed glacier area on Mount Baker in 1906.
The contour lines represent the glacier surface in 1955. Base map:
D.L.S.U. (1970); vertical interval of the contour lines: 100 ft.

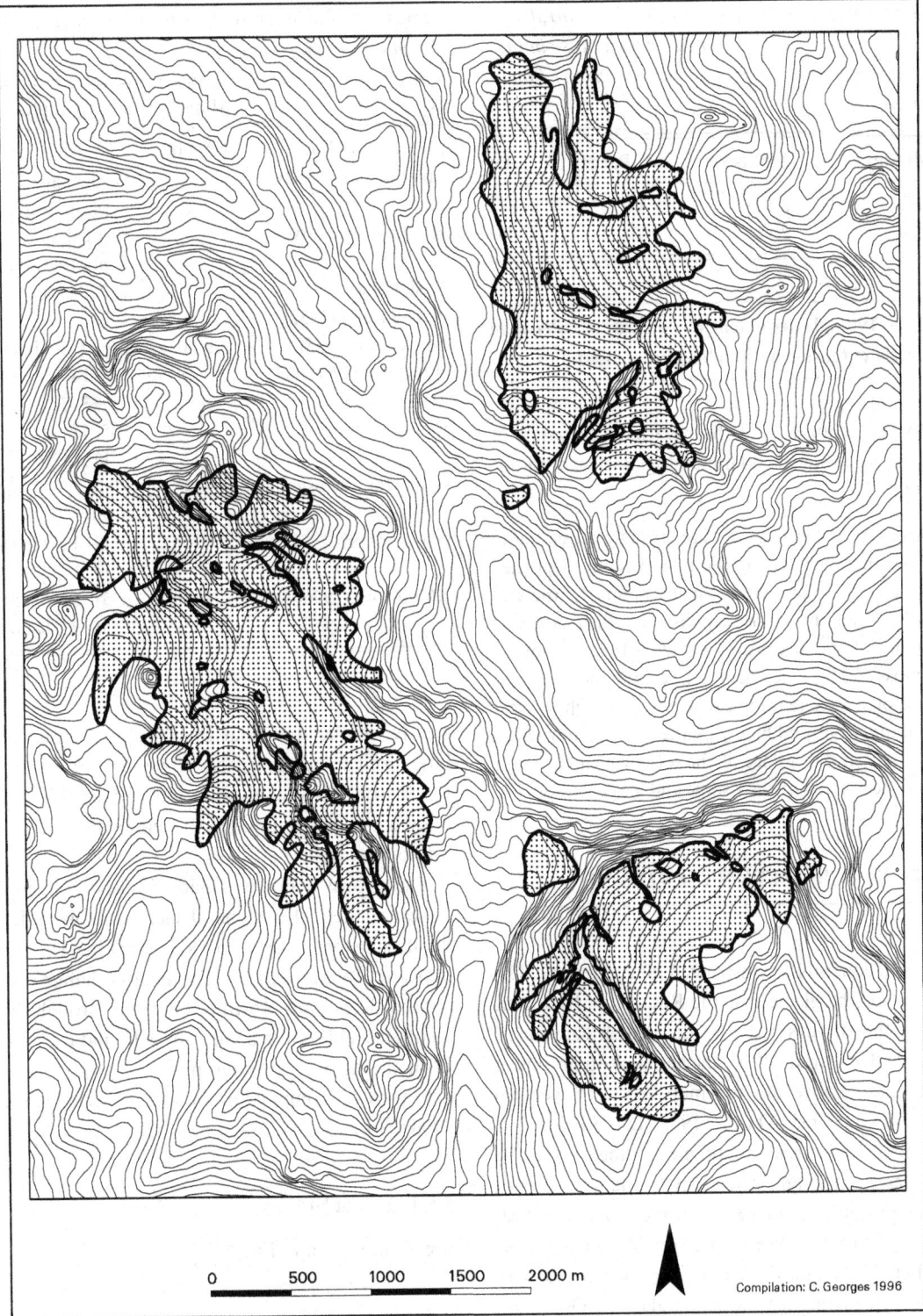

0   500   1000   1500   2000 m

Compilation: C. Georges 1996

Fig. 6.4.5 The reconstructed glacier area of the Central Rwenzori
Massif in 1906. The contour lines represent the glacier surface in
1955. Base map: D.L.S.U. (1970); vertical interval of the contour
lines: 100 ft. For contour heights see Figs. 6.4.2 to 6.4.4.

Table 6.4.1. *The reconstructed glacier areas in the Central Rwenzori Massif (in km²) in 1906 compared with those in the 'Topographical and geographical map of the Ruwenzori Range 1:40 000' (1906 A)*

	Mt Stanley	Mt Speke	Mt Baker	Total
1906	2.85	2.19	1.47	6.51
1906 A	2.34	2.14	1.61	6.09

It is clear that the 1906 glacier extent was already smaller than that at the greatest extension not too long before that date.

Alessandro Roccati reports in his 'Summarizing overview of the geological, petrographical and mineralogical observations' (De Filippi, 1909):

> . . . all glaciers on the Ruwenzori are strongly retreating at present. This can be proved by the only recently deposited moraines which can be observed at many points, by the vast stretches along which the rocks are polished at their sides and at the front end of the glaciers, by stretches that are not yet covered by moss and lichens, whereas the overabundant occurrence of these plant genera is typical even for the highest areas of the mountain range, and finally by the white colour which shows very often on the surface of the rocks which have only shortly before been freed of their snow and ice cover.

6.5 THE GLACIER EXTENT AROUND 1955

This is the most frequently reconstructed glacier extent. The best survey data for the glaciers of the Rwenzori Mountains are provided by the surveys made by the British Ruwenzori Expedition 1952 (Bergstrøm, 1955), the aerial photographs of various flights, and, above all, the map based on vertical photographs taken in 1955 (D.O.S., 1962; D.L.S.U., 1970). The map provides contour lines suitable for measuring the extent of the glaciers, though their exact boundaries are not clear on normal prints. The outer areas of the original vertical aerial pictures (Fig. 6.5.5), however, are very blurred and portions of the light rock areas, already described by Roccati (1909), are hard to distinguish from the glaciers. Osmaston (1989²) pointed out these difficulties and tried to explain the major differences from Hastenrath's evaluations (1984). In order to be able to assess his evaluations (1989²), Osmaston's glacier extents for Mount Baker were digitized and superimposed upon the contour lines and the coverage of the 1906 glacier extent (Fig. 6.5.1). Although the 1955 glacierization is

Table 6.5.1. *The data used for the reconstruction of the glacier extent on Mount Stanley around 1955*

Sources	Maps	Photographs	Sketches
D.L.S.U. (1970)	1:25 000		
MacLachlan (1952)		vertical air photographs	
Hunting Aerosurveys (1955)		vertical air photographs	
Busk (1954)		1, 2, 3, 5, 6	
de Heinzelin (1952)			p. 139
Bergstrøm (1955)	Fig. 8	Fig. 2, 4, 5b, 6	
Whittow *et al.* (1963)		Fig. 14	Fig. 18
Temple (1968)		Fig. 3, 7	
Fantin (1968)		105, 111, 114, 115, 119, 120, 125	
Osmaston & Pasteur (1972)	Fig. 5, 7	Plate 5, 6, 7, 8, 9, 10	
Noggler (1992)		8, 11, 12, 15, 16, 17, 18	
Mantovani (1996)		p. 51, 52, 53, 54, 55, 56, 58, 60	

clearly smaller than that in 1906, some areas appear to be larger and some parts seem to lie in improbable terrain. For this reason, a new analysis was attempted. Photographs taken between the beginning and the end of the 1950s were used to improve results.

The reconstruction was made by using different sources and in several steps. First, the glacier limits were determined by hand using the D.O.S. 1:25 000 map (D.L.S.U., 1970). They are indicated on the map only through the change in colour of the contour lines from brown to blue and through the not clearly distinguishable white areas. These boundaries were digitized. With the help of photographs, corrections were made. The data used are compiled in the following tables and results are shown for each group.

6.5.1 Mount Stanley

See Table 6.5.1 and Fig. 6.5.2.

6.5.2 Mount Speke

See Table 6.5.2 and Fig. 6.5.3.

6.5.3 Mount Baker

See Table 6.5.3 and Fig. 6.5.4.

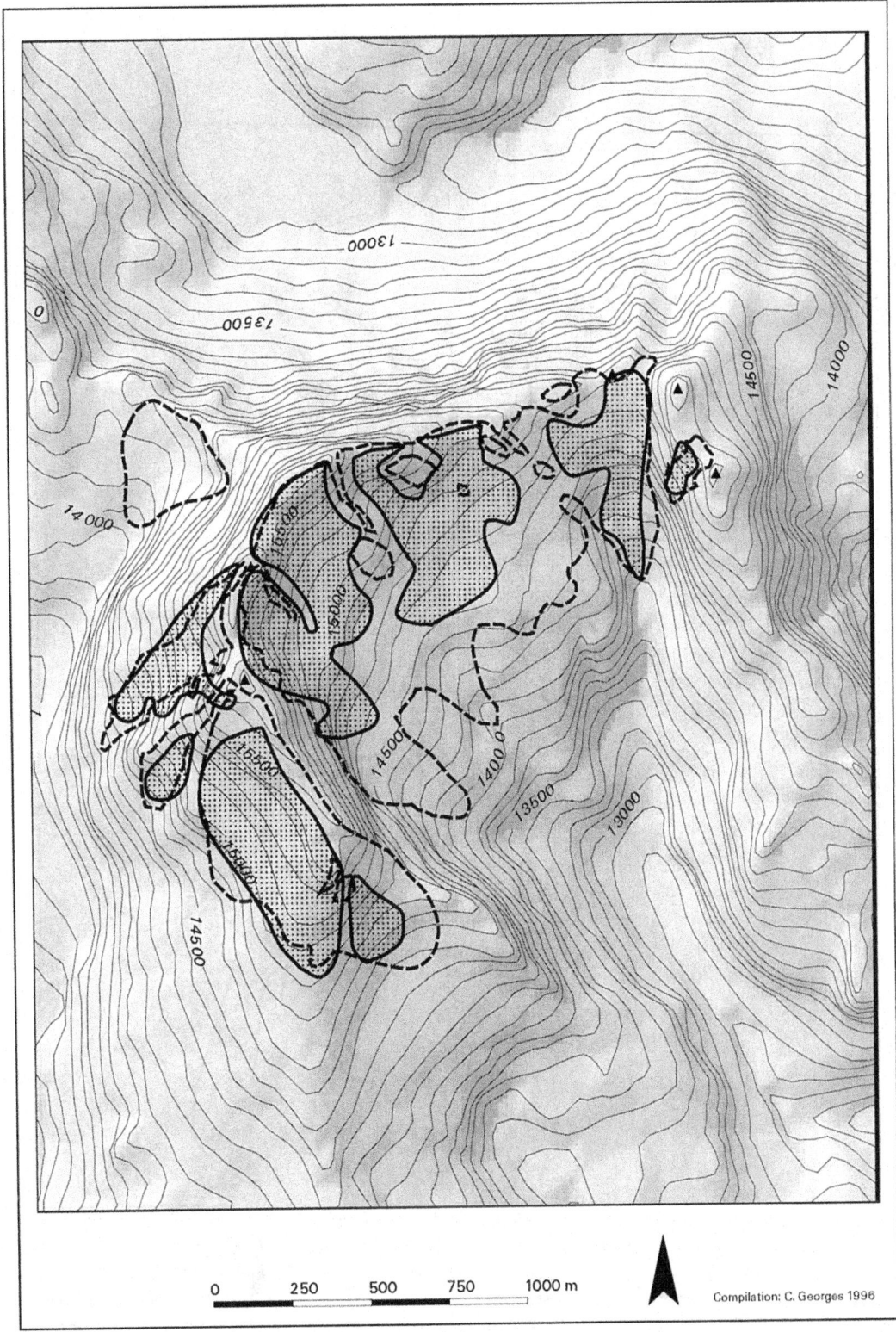

Fig. 6.5.1 Mount Baker. The analysis by Osmaston (1989[2]) for 1955
is superimposed upon the the 1955 contour lines and the 1906
glacier limits. Base map: D.L.S.U. (1970); vertical interval of the
contour lines: 100 ft.

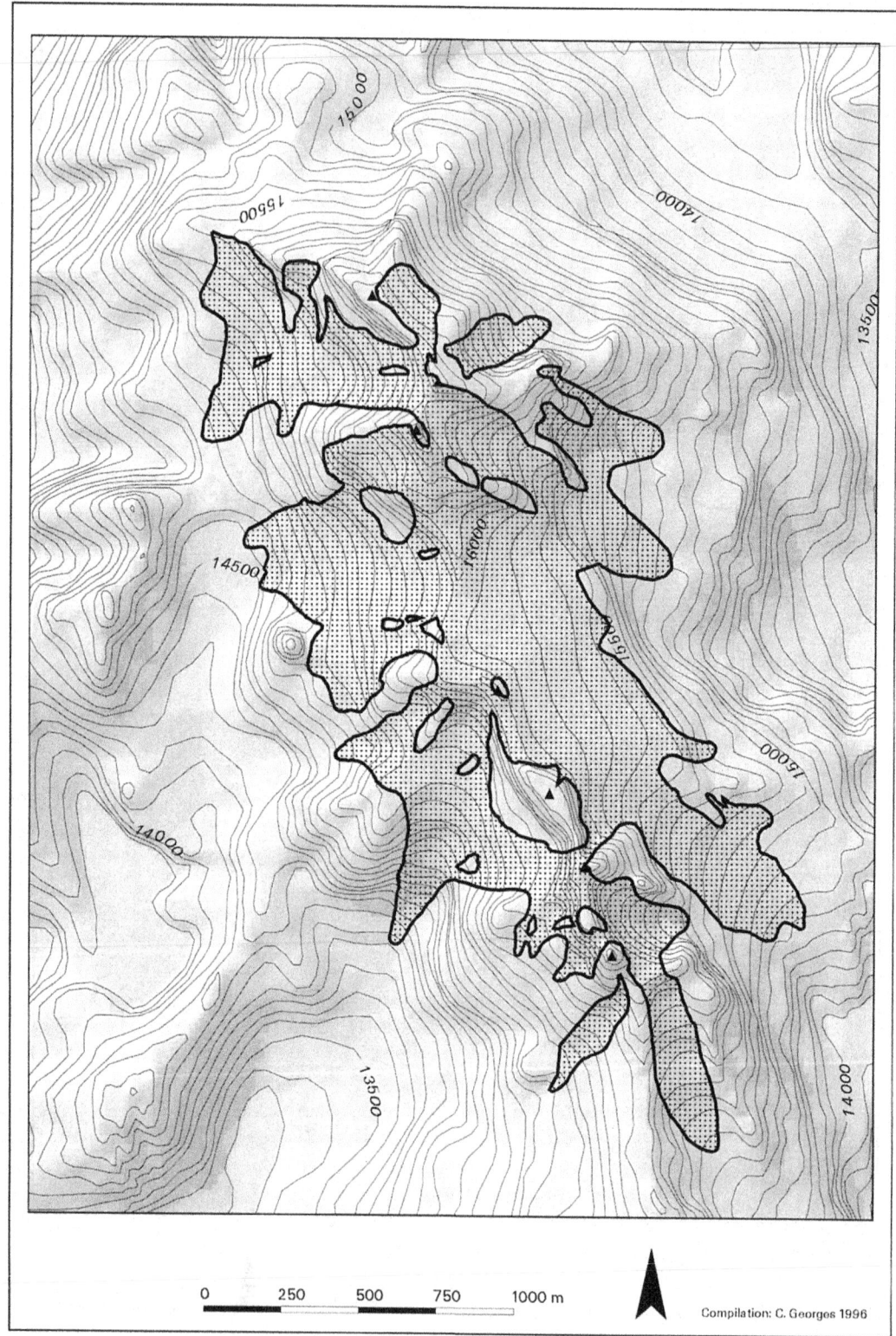

Fig. 6.5.2 Mount Stanley. The reconstructed glacierization around
1955. The contour lines represent the contemporary glacier surface.
Base map: D.L.S.U. (1970); vertical interval of the contour lines:
100 ft.

Fig. 6.5.3 Mount Speke. The reconstructed glacierization around
1955. The contour lines represent the contemporary glacier surface.
Base map: D.L.S.U. (1970); vertical interval of the contour lines:
100 ft.

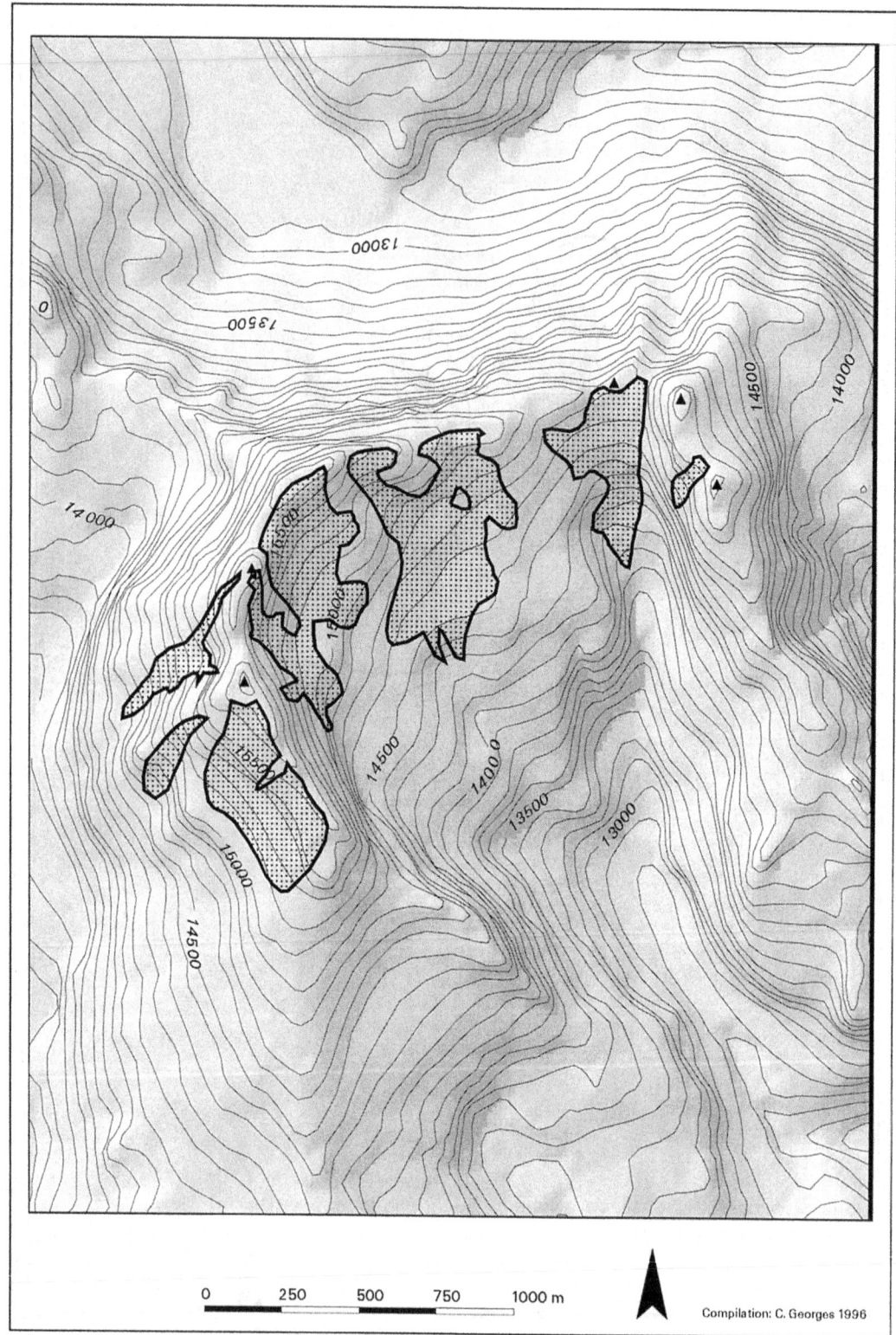

Fig. 6.5.4 Mount Baker. The reconstructed glacierization around
1955. The contour lines represent the contemporary glacier surface.
Base map: D.L.S.U. (1970); vertical interval of the contour lines:
100 ft.

Table 6.5.2. *The data used for the reconstruction of the glacier extent on Mount Speke around 1955*

Sources	Maps	Photographs	Sketches
D.L.S.U. (1970)	1:25 000		
MacLachlan (1952)		vertical air photographs	
Hunting Aerosurveys (1955)		vertical air photographs	
Busk (1954)		2	
Whittow (1959)		1, 3, 4	Fig. 1, 2, 3
Whittow *et al.* (1963)		Fig. 3	
Temple (1968)		Fig. 2, 4, 6	
Osmaston & Pasteur (1972)		Plate 4, 11, 13	
Noggler (1992)		38	

Table 6.5.3. *The data used for the reconstruction of the glacier extent on Mount Baker around 1955*

Sources	Maps	Photographs	Sketches
D.L.S.U. (1970)	1:25 000		
MacLachlan (1952)		vertical air photographs	
Hunting Aerosurveys (1955)		vertical air photographs	
Busk (1954)		2	
Whittow (1959)		1, 3, 4	Fig. 1, 2, 3
Whittow *et al.* (1963)		Fig. 3	
Temple (1968)		Fig. 8, 9, 10	
Osmaston & Pasteur (1972)		Plate 4, 11, 13	
Noggler (1992)		38	
Mantovani (1996)		p. 49, 51, 57, 59	

6.5.4 The Central Rwenzori Massif

The extent of the glaciers in the Central Rwenzori massif around 1955 is depicted in Fig. 6.5.6. As had already occurred for 1906, the largest uncertainties during the reconstruction appeared in the southwest of Mount Stanley and in the northeast of Mount Speke, as it is only possible to get an imprecise view of these areas. However, it was attempted to position the tongue realistically into the topography with the help of the contour line model. The inaccuracy is estimated to be <5% for the total glacierization. In Table 6.5.4, the area values are compiled for the three individual groups and compared with those given by Osmaston (1989[2]), Hastenrath (1984) and the extent in 1906.

Table 6.5.4. *The reconstructed glacier areas (in km^2) in 1955 in comparison with those given by Osmaston (1989[2]) (O) and Hastenrath (1984) (H), as well as those in 1906*

	Mt Stanley	Mt Speke	Mt Baker	Total
1955	1.88	1.31	0.62	3.81
1955 O	2.05	1.38	0.73	4.16
1955 H	1.04	0.78	0.39	2.21
1906	2.85	2.18	1.47	6.51

The new evaluation shows a glacier extent which is 8.5% lower than the one stated by Osmaston (1989[2]).

6.6 THE GLACIER EXTENT AROUND 1990

Cartographic work during two journeys to the Rwenzori (G. Kaser and B. Noggler: January–February 1990; G. and H. Kaser: July 1991) prompted the reconstruction of the glacier extent in the Central Rwenzori Massif around 1990. The first attempt was made by Noggler (1992). A new attempt was made with the help of the digital contour line model, the newly drawn glacier extents of 1906 and around 1955, and additional photographs. In comparison with the reconstructions of the two previous glacier extents, the conditions for 1990 were unfavourable:

- There are no contour lines for the glacier bed that became exposed between 1955 and 1990.
- There are no aerial pictures.
- The photographs were mainly taken by mountain climbers that were using the same few routes.
- Cartographic work in the area concentrated only on the Speke Glacier and the Elena Glacier.

As in the previous reconstructions, the highest uncertainties are on the eastern side of Mount Speke and on the western side of Mount Stanley. After the political situation in Uganda stabilized at the end of the 1980s, the trekking route from Ibanda through the Bujuku Valley, over the Scott Elliot Pass to the Kitandara Lakes, over the Freshfield Pass into the Mubuku Valley and back to Ibanda (Fig. 6.1.1) is again often in use. Access to Alexandra and Margherita (Mount Stanley), Vittorio Emanuele (Mount Speke), and Edward (Mount Baker) peaks was reopened. The glaciers that can be seen from these routes and view points are often photographed and can be quite easily reconstructed.

The glaciers that were not seen were drawn according to the relief and in accordance with the documented glaciers and

Fig. 6.5.5 Air photo of central Rwenzori showing Mount Stanley
(lower left), Mount Baker (lower right) and Mount Speke (upper
centre). East and sun are to the top to facilitate relief perception.
Savoia Glacier is extreme bottom centre; Elena Glacier is next above
it; Speke Glacier is to the left of long narrow shaded rock-face;
Moore Glacier is in shadow on extreme right; Lake Bujuku at
centre. DOS/Huntings 15.UG 13 No.046, f = 6 in, alt. 28 000 ft,
June 1955. This series of air photos was the source of the D.L.S.U.
(1970) map used for Fig. 6.5.6.

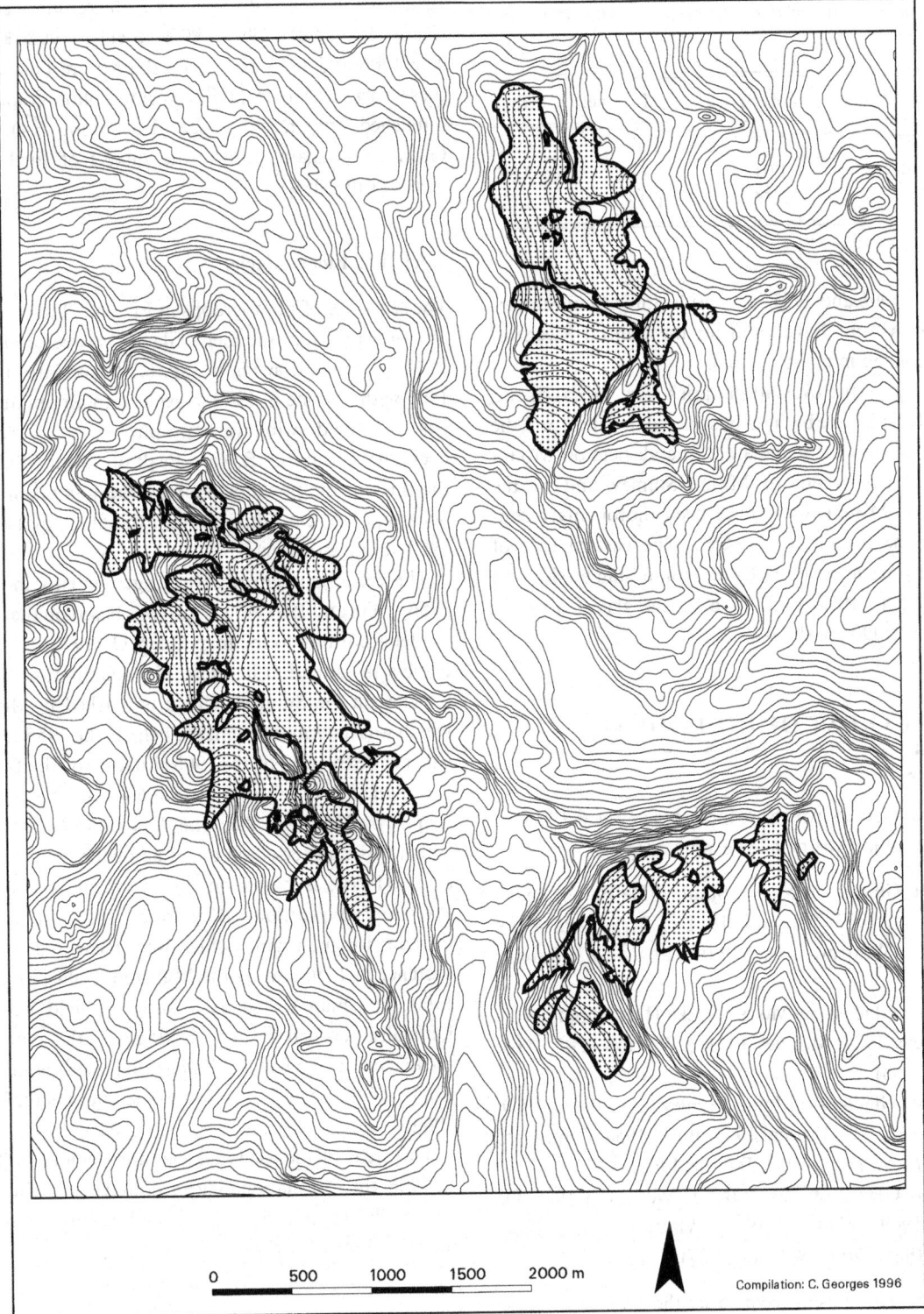

0   500   1000   1500   2000 m

Compilation: C. Georges 1996

Fig. 6.5.6 Central Rwenzori Massif. Reconstructed glacierization
around 1955. The contour lines represent the contemporary glacier
surface. Base map: D.L.S.U. (1970); vertical interval of the contour
lines: 100 ft. For contour heights see Figs. 6.5.3 to 6.5.5.

Table 6.6.1. *The reconstructed glacier areas of 1906, 1955, and 1990 (km²)*

	Mt Stanley	Mt Speke	Mt Baker	Total
1906	2.85	2.18	1.47	6.51
1955	1.88	1.31	0.62	3.81
1990	1.00	0.56	0.12	1.67

their extents in 1906 and 1955. The results for Mount Stanley, Mount Speke, Mount Baker, and for the Central Rwenzori Massif are depicted in figs. 6.6.1–6.6.4.

The accuracy of the evaluation varies. The Elena, Speke, and Moore glaciers have been reconstructed in detail, and the inaccuracy is below 3% of the area. The glaciers on Mount Baker, on the eastern side of Mount Stanley, and on the western side of Mount Speke have been thoroughly reconstructed and the inaccuracy is <5%. The eastern side of Mount Speke and the western side of Mount Stanley show an inaccuracy of 5 to 10%. In Table 6.6.1 the glacier areas of 1906, 1955, and 1990 are shown.

6.7 THE RETREAT OF INDIVIDUAL GLACIERS

Four individual glaciers, which are indicated in Fig. 6.7.1, were repeatedly the subjects of detailed examination:

- **The Elena Glacier** on Mount Stanley (De Filippi, 1909; Humphreys, 1927; Menzies, 1951[1,2]; Bergstrøm, 1955; Busk, 1957; Whittow, 1960; Whittow *et al.*, 1963; Temple, 1968).
- **The Speke Glacier** on Mount Speke (De Filippi, 1909; Humphreys, 1927; Menzies, 1951[1,2]; Whittow *et al.*, 1963; Temple, 1968).
- **The Savoia Glacier** on Mount Stanley (De Filippi, 1909; Humphreys, 1927; Firmin, 1945; Busk, 1954; Osmaston, 1961; Whittow *et al.*, 1963; Temple, 1968).
- **The Moore Glacier** on Mount Baker (Wollaston, 1908; De Filippi, 1909; Whittow, 1959; Whittow *et al.*, 1963; Temple, 1968; Drake & Jones, 1987).

This offers opportunities to reconstruct the changes of these four glaciers in detail. In 1990, the Speke Glacier was the goal of cartographic work and that for Elena Glacier followed in 1991. Photographs taken by mountain climbers from Edward Peak offer a good view of the Moore Glacier. The Savoia Glacier can be seen from the front from the Freshfield Pass and is often photographed. The results of field work and recon-

structions, as well as discussions of the results, were published (Kaser & Noggler, 1991, 1996; Noggler, 1992; Kaser, 1993). The 1906 and 1955 extents of the Moore Glacier shown in these works had to be modified using a photograph from 1906, published only recently (see also section 6.4). The Speke Glacier was again recently the subject of further measurements (Talks, 1993).

6.7.1 The Elena Glacier

The sequence of photographs in Figs. 6.7.2 to 6.7.4 shows the dramatic retreat of the Elena Glacier between 1906, 1960, and 1994, while the drawing in Fig. 6.7.5 compares the extent in 1906 with the one in 1990 as seen from an oblique view. The bases for the drawing were two photographs which had both been taken from the Edward Peak on Mount Baker just like the ones in the previous figures. One was taken by Vittorio Sella in 1906 (De Filippi, 1909), the other by H. Wagner in 1990 (archives Kaser).

The reconstruction of five tongue extents in 1906, 1952, 1956, 1960, and 1991 (Fig. 6.7.6) was based on the contour line model according to the map D.L.S.U. (1970) 1:25000, which depicts the situation in 1955. The delimitation of the Elena Glacier on the Stanley plateau is marked by the ice divide which was determined with the help of the 1955 contour lines. The reconstruction of the 1906 situation of the Elena Glacier was made as described in section 6.4 with the help of photographs by Vittorio Sella. The depiction of the 1952 extent, which was transferred on to a map of the tongue by Whittow *et al.* (1963), is based on measurements from the British Rwenzori Expedition (Bergstrøm, 1955). This map also includes tongue extents which were measured by Smith and Fletcher in 1956 and by the Makerere College Expedition in 1960. The map served as a basis for field work in 1991. Difficulties occurred when working with this map because of the inaccurate scale which Osmaston (1989[2]) pointed out. The length of the scale bar is not 800 ft but is actually less than 400 ft.

The areas of the Elena Glacier are included in Table 6.7.1. In 1906, the Coronation Glacier flowed into the tongue of the Elena Glacier from the west. Therefore, two values are given for the extent in 1906.

6.7.2 The Speke Glacier

The following sequence of photographs (Figs. 6.7.7 to 6.7.11) shows the retreat of the Speke Glacier from 1952 to 1960, 1968, 1974, and 1991 (see also Osmaston, 1961).

Four different extents of the Speke Glacier were determined in a manner similar to that used on the Elena Glacier. The basis

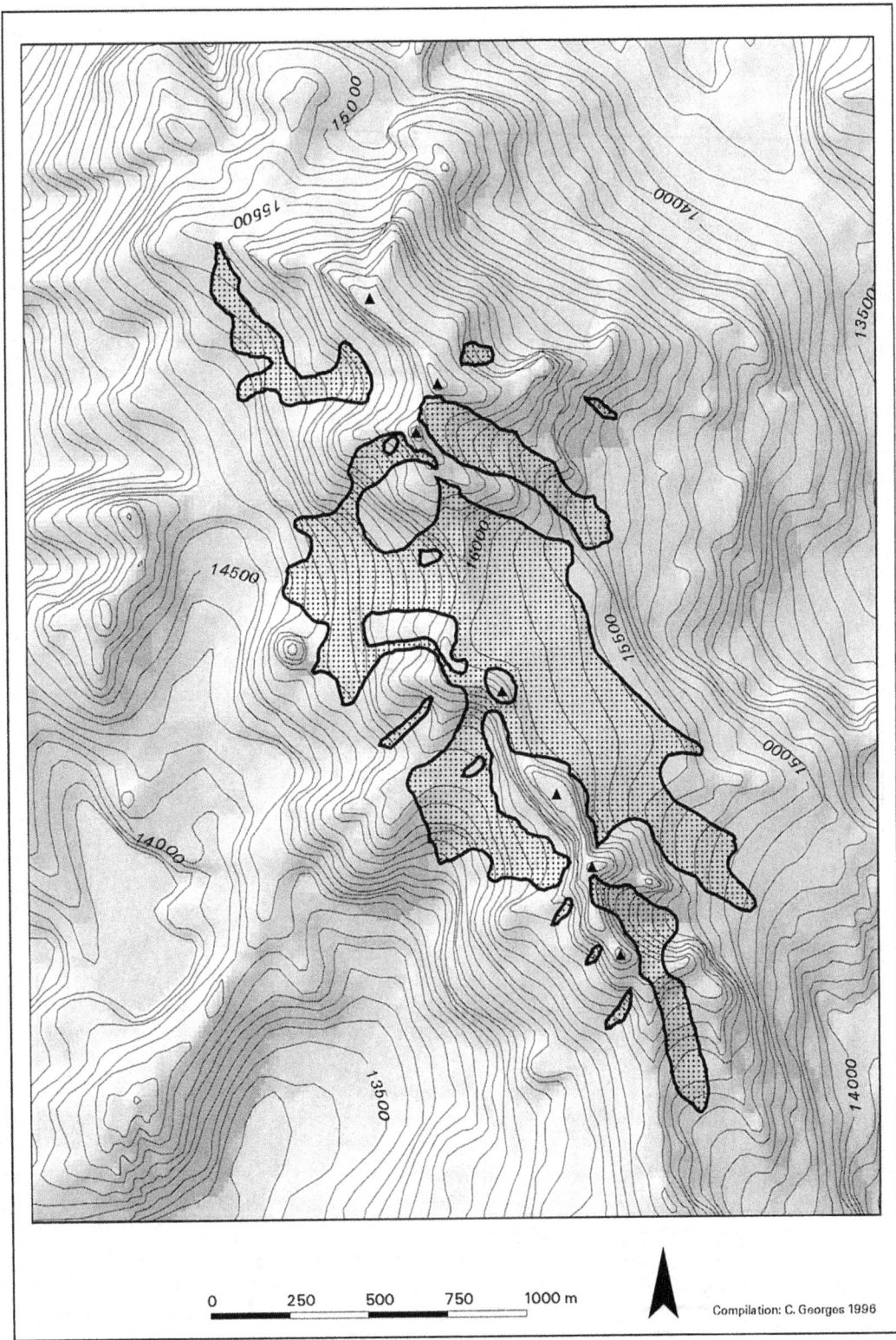

Fig. 6.6.1 The reconstructed glacier area of Mount Stanley in 1990.
The contour lines represent the glacier surface in 1955. Base map:
D.L.S.U. (1970); vertical interval of the contour lines: 100 ft.

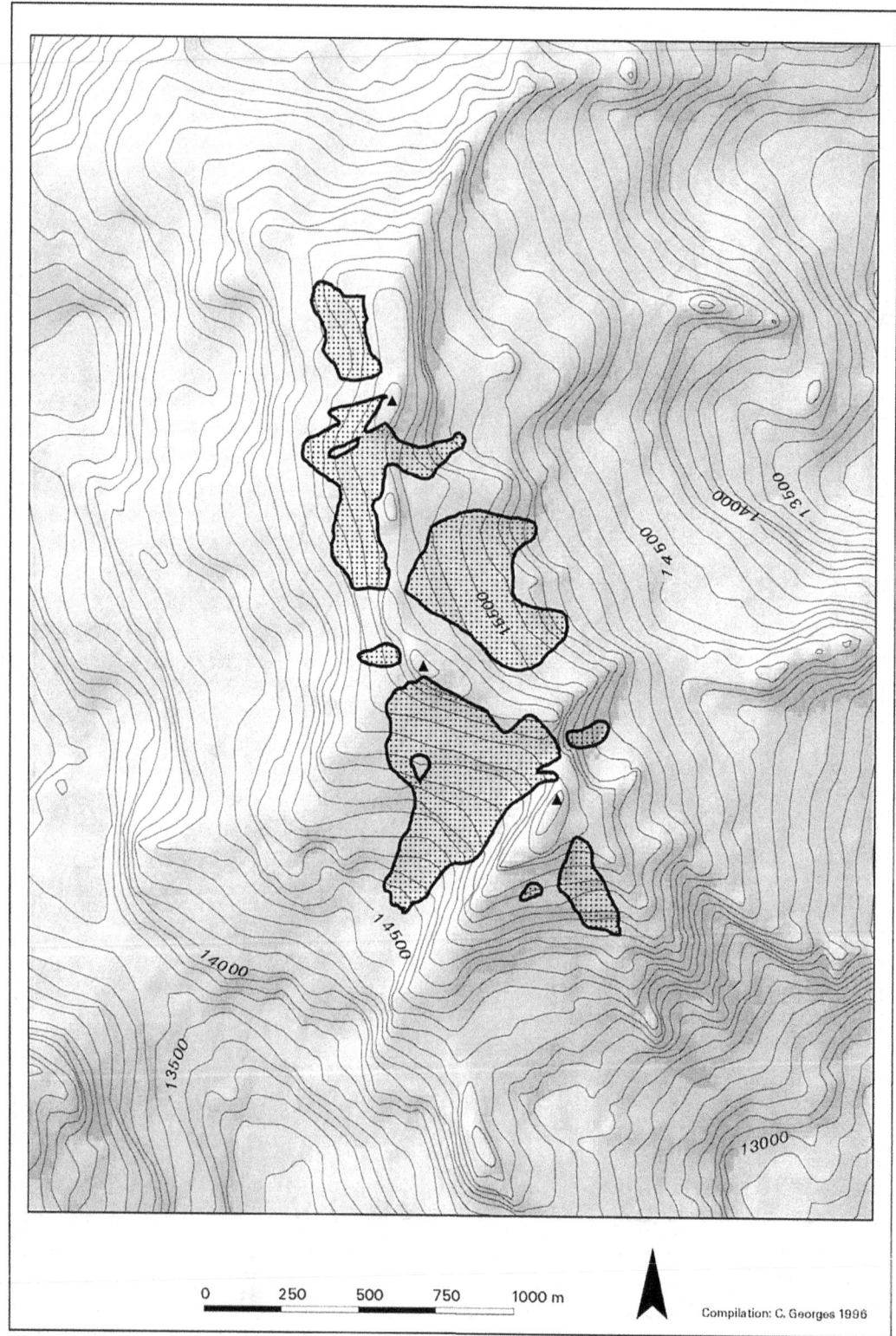

0   250   500   750   1000 m

Compilation: C. Georges 1996

Fig. 6.6.2 The reconstructed glacier area on Mount Speke in 1990.
The contour lines represent the glacier surface in 1955. Base map:
D.L.S.U. (1970); vertical interval of the contour lines: 100 ft.

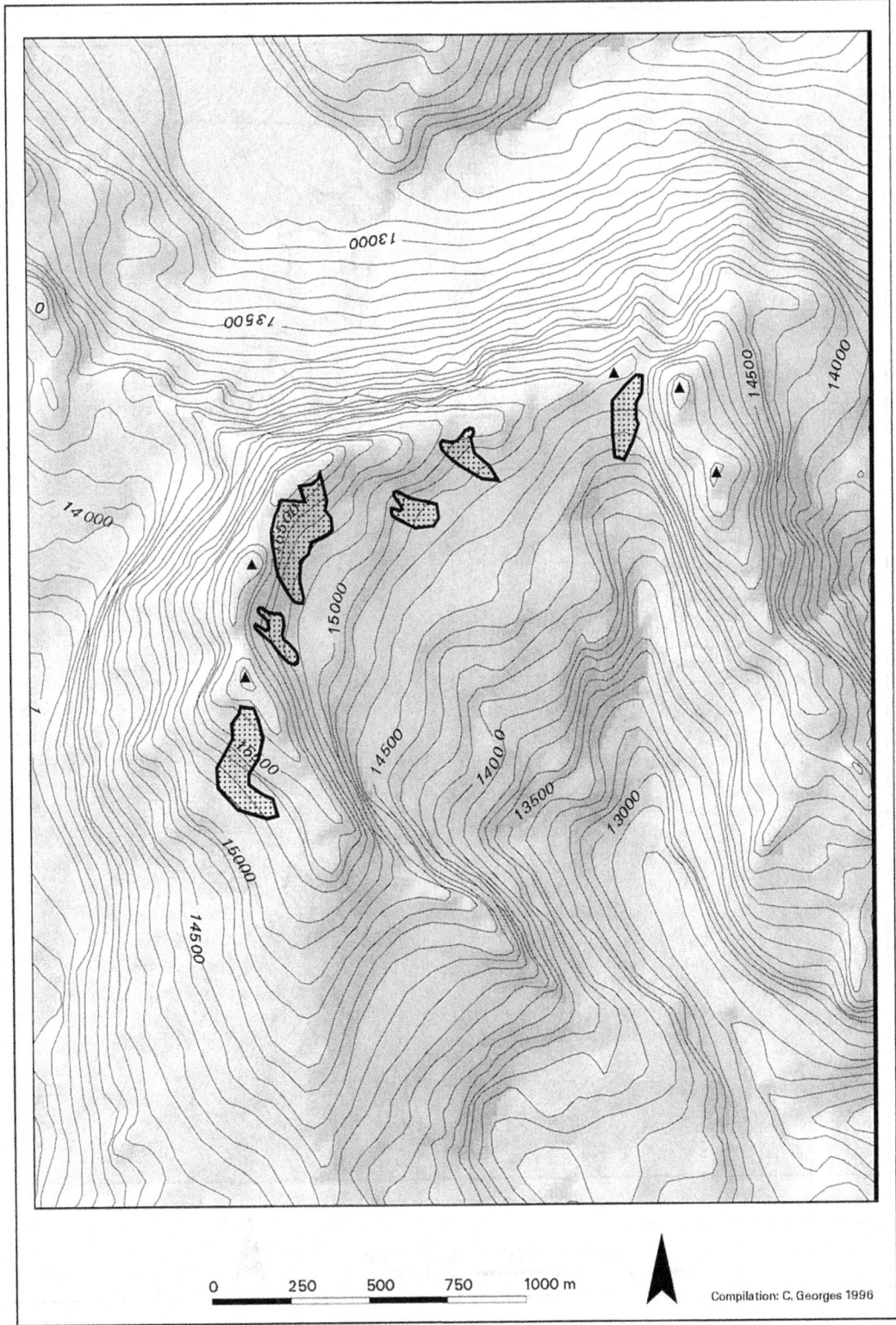

Fig. 6.6.3 The reconstructed glacier area on Mount Baker in 1990.
The contour lines represent the glacier surface in 1955. Base map:
D.L.S.U. (1970); vertical interval of the contour lines: 100 ft.

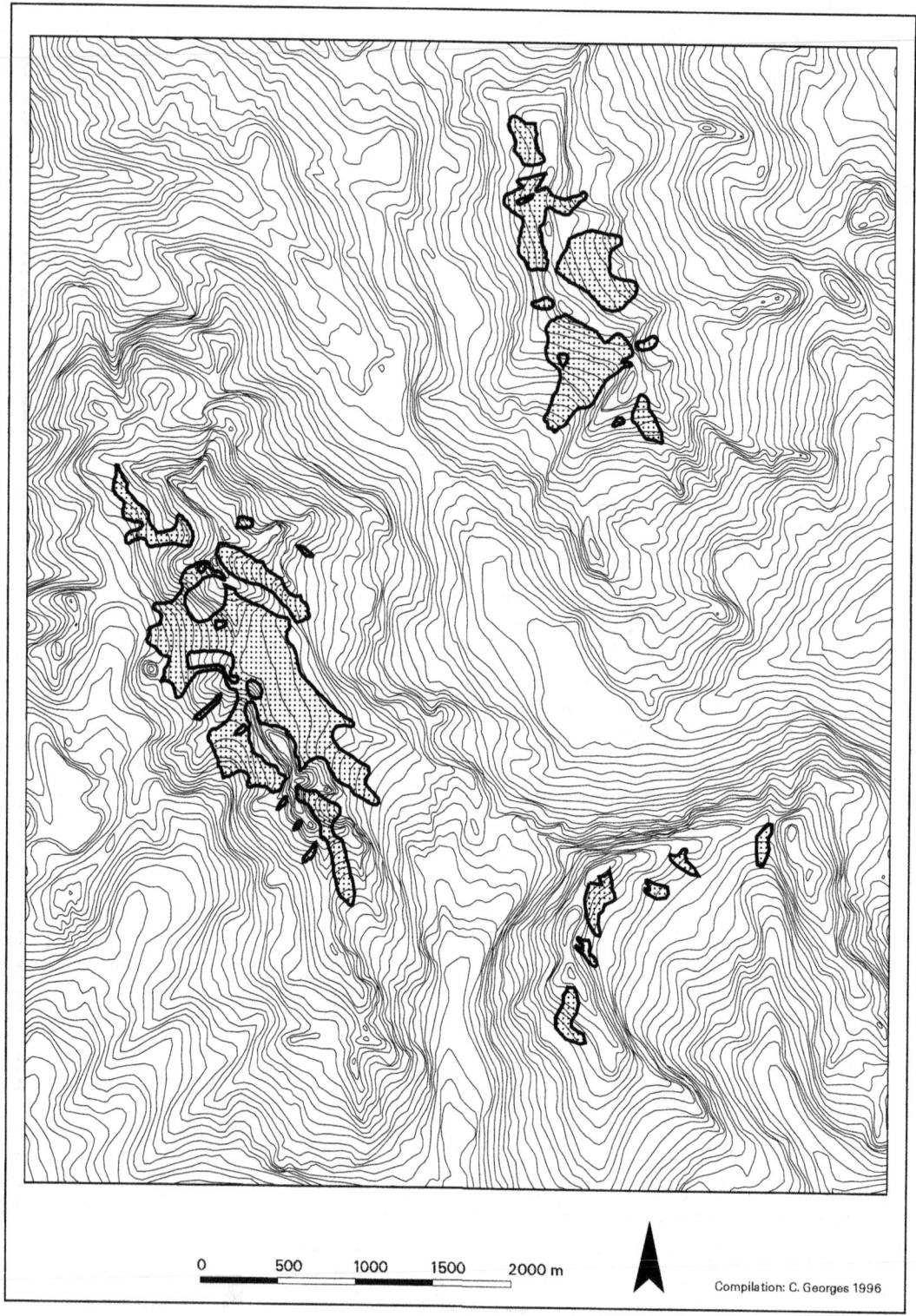

0   500   1000   1500   2000 m

Compilation: C. Georges 1996

Fig. 6.6.4 The reconstructed glacier area of the Central Rwenzori
Massif in 1990. The contour lines represent the glacier surface in
1955. Base map: D.L.S.U. (1970); vertical interval of the contour
lines: 100 ft. For contour altitudes see Figs. 6.6.1 to 6.6.3.

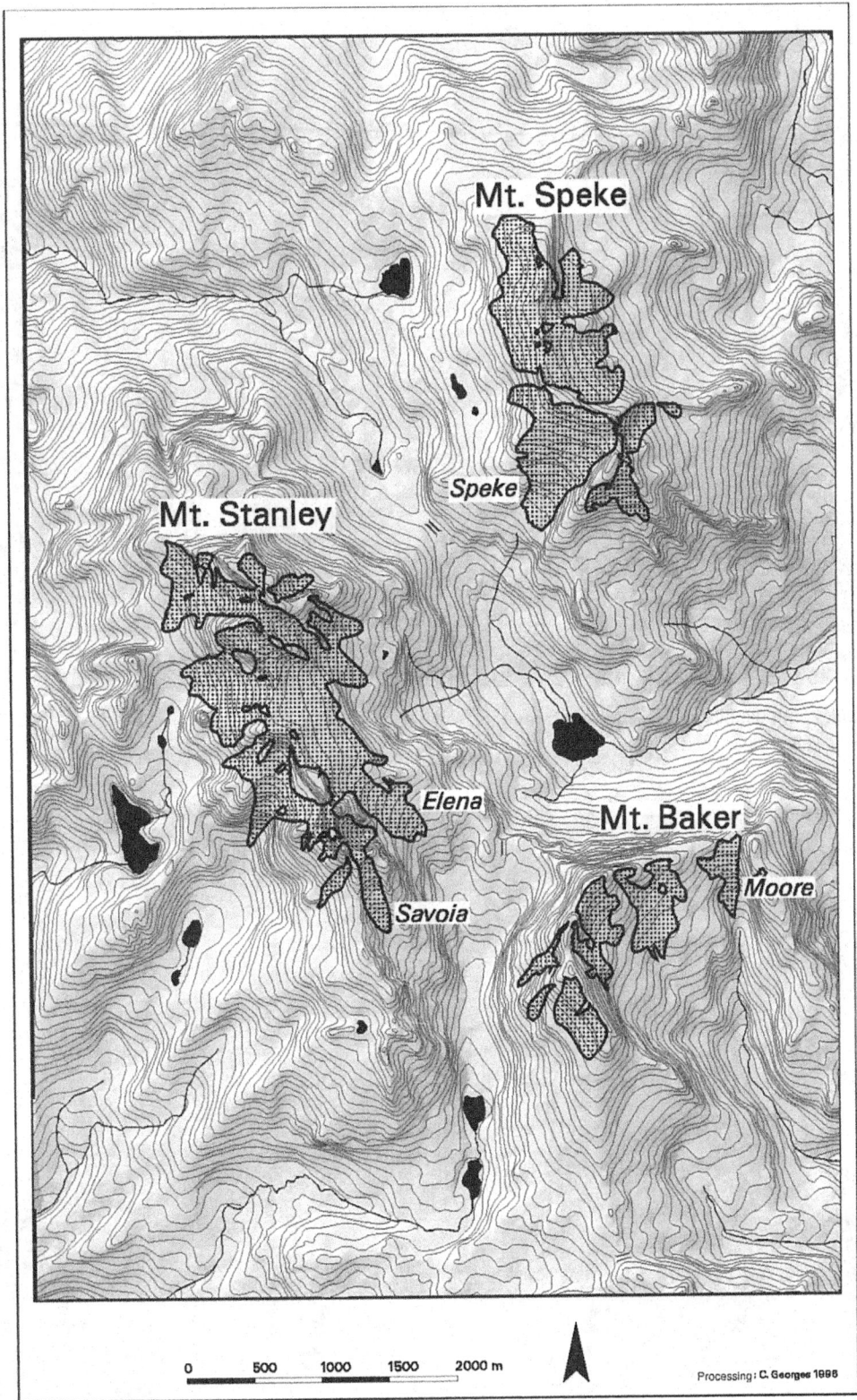

Fig. 6.7.1 The central part of the Rwenzori. The four glaciers discussed in detail are indicated. Glacier extent: 1955 (section 6.5); vertical interval of the contour lines: 100 ft. Base map: D.L.S.U. (1970). For contour altitudes see Figs. 6.6.1 to 6.6.3.

Fig. 6.7.2 Mount Stanley with the Elena Glacier as seen from Mount Baker. Photo: V. Sella, 1906.

Fig. 6.7.3 Mount Stanley with the Elena Glacier as seen from Mount Baker. Photo: J.B. Whittow (archives) 1960.

Fig. 6.7.4 Mount Stanley with the Elena Glacier as seen from
Mount Baker. Photo: P. Glogg, 1994.

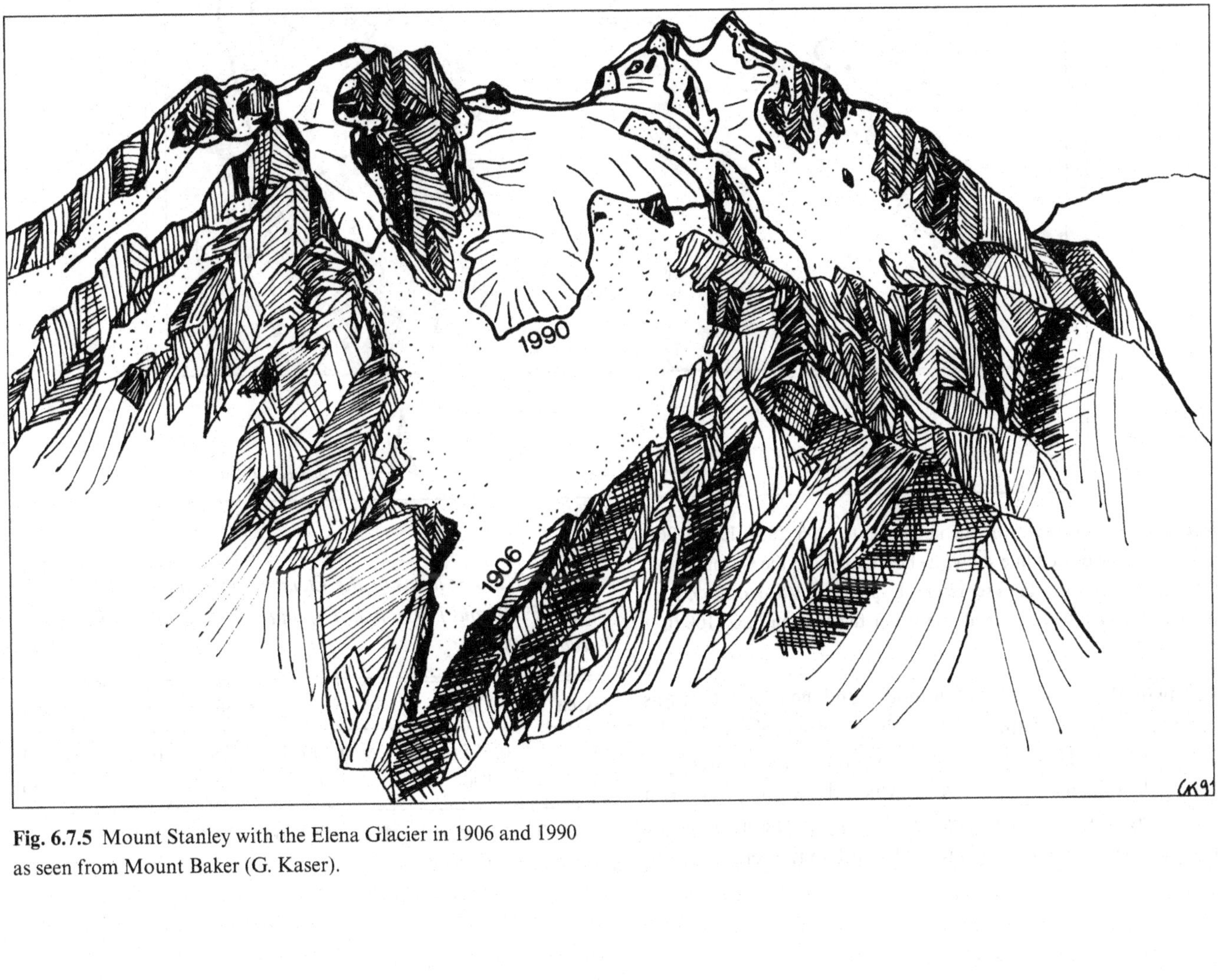

Fig. 6.7.5 Mount Stanley with the Elena Glacier in 1906 and 1990
as seen from Mount Baker (G. Kaser).

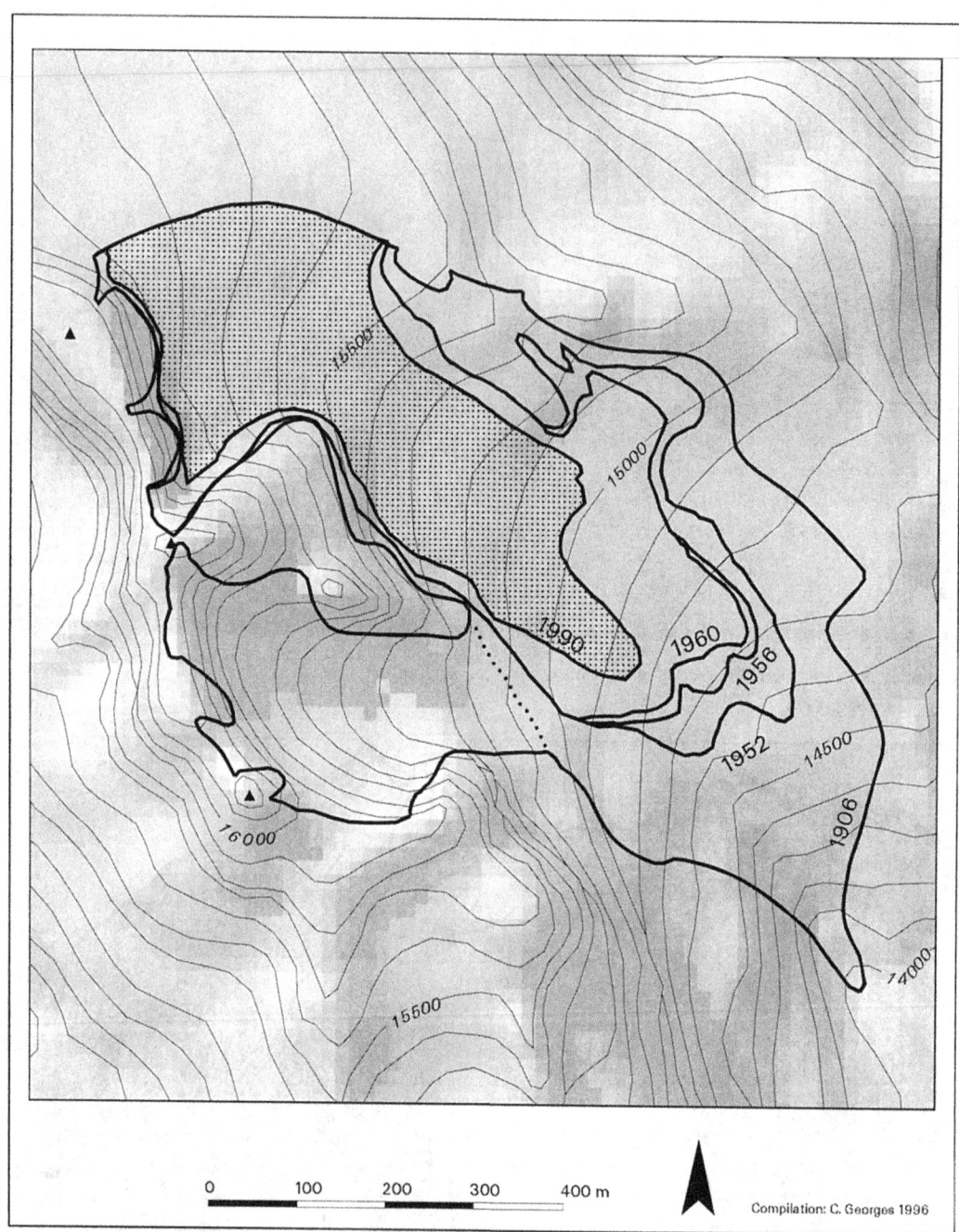

Fig. 6.7.6 The extent of the Elena Glacier in 1906, 1952, 1956, 1960, and 1991. The dotted line shows its boundary with the Coronation Glacier. The contour lines represent the glacier surface in 1955. Base map: D.L.S.U. (1970); vertical interval of the contour lines: 100 ft.

was again the 1955 contour line model, and the 1906 extent was reconstructed with the help of the photographs taken by Vittorio Sella. However, there is no photograph available of the Speke Glacier for this year in which the whole glacier can be clearly seen. The bedrock topography in the area of the tongue does not allow much room for different interpretations.

Table 6.7.1. *The areas of the Elena Glacier in 1906, 1952, 1956, 1960, and 1991 (in km²,), and as a percentage of the area in 1906*

Area	1906	1952	1956	1960	1991
km²	0.315 (0.383)	0.200	0.178	0.173	0.113
%	100	63	56	55	36

Note: The value in brackets refers to the Elena and the Coronation Glacier.

Fig. 6.7.7 Speke Glacier. Photo: 1952 (Bergstrøm, 1955).

Fig. 6.7.8 Speke Glacier. Photo: 1960 (J.B. Whittow).

Fig. 6.7.9 Speke Glacier. Photo: 1967? (Temple, 1968).

The tongue definitely did not extend over the steep rock face and had no contact with the avalanche debris cone situated below.

'. . . Three melt streams emerge from the present glacier terminus to plunge over a steep rock wall before cascading over 300 m down to the Bujuku hut on the valley floor. The amount of debris and excavation at the foot of the main gully may indicate that at an earlier stage the avalanched ice coming down was sufficient to produce a small regenerated glacier at the foot of the rock wall. The photographs which the Duke of Abruzzi and Humphreys took of this gully both show a prominent ice or snow cone at its foot, but this had disappeared by the time Menzies visited it in 1949.' (Whittow *et al.*, 1963)

By contrast, the basin above the mentioned rock wall was completely filled with glacier ice (Figs. 6.7.7 to 6.7.10). Figs. 6.7.12 to 6.7.16 show the retreat of the tongue of the Speke Glacier between 1959 and 1990. The lake, filling the foreground in 1990, was not yet visible on a photograph taken in 1977 by Lichtenegger & Lichtenegger (1978).

Whittow *et al.* (1963) plot the 1950 (cited from Menzies, 1951[1,2]), 1958, 1959 and 1960 tongue extents in a map. This map was the basis for field work in January/February 1990

Fig. 6.7.10 Speke Glacier. Photo: 1974 (Hastenrath, 1984).

Fig. 6.7.11 Speke Glacier. Photo: 1991 (G. Markl).

Fig. 6.7.12 The tongue of the Speke Glacier. Photo: 1961 (Whittow *et al.*, 1963). Marks D, C and B, painted by the 'Makerere College Ruwenzori Expedition', are indicated with arrows.

Table 6.7.2. *The areas of the Speke Glacier in 1906, 1950, 1960, 1990, and 1993 (in km²), and as a percentage of the area in 1906*

Area	1906	1950	1960	1990	1993
km²	0.453	0.372	0.369	0.212	0.198
%	100	82	81	47	44

(Kaser & Noggler, 1991) (Fig. 6.7.17). The members of a Sir Roger Manwood's School expedition again mapped the tongue of the Speke Glacier in August 1993 (Talks, 1993) (Fig. 6.7.18). However, information is partly missing for the 1950 and 1993 extents around the upper areas of the glacier and, consequently, unlike other extents, the area changes only include the change of the tongue.

In Fig. 6.7.19, the reconstructed extents of the Speke Glacier are shown. In Table 6.7.2, the corresponding absolute and relative area values are shown.

Whereas the Elena Glacier in 1960 had only about half the area it had in 1906, the Speke Glacier retained about 80%. The reason is mainly found in the different shape and position of

Fig. 6.7.13 The tongue of the Speke Glacier. Photo: January 1990 (Kaser & Noggler, 1991). Marks D (front), C and B are indicated with arrows. Old glacier extents from previous photographs are drawn in: solid line: June 1958; broken line: July 1961 (both after Whittow *et al.*, 1963); broken and dotted line: January 1964 (Hastenrath, 1984).

Fig. 6.7.14 The tongue of the Speke Glacier. Photo: 1959 (Whittow *et al.*, 1963). Marks B and C are indicated with arrows.

Fig. 6.7.15 The tongue of the Speke Glacier. Photo: January 1990 (Kaser). Marks D, C and B, as well as G (left) are indicated with circles. Old glacier extents from previous photographs are drawn in: solid line: June 1958; broken line: July 1961 (both after Whittow *et al.*, 1963); broken and dotted line: January 1974 (Hastenrath, 1984).

Fig. 6.7.16 The tongue of the Speke Glacier. Photo: January 1990 (Kaser & Noggler, 1991). Mark G is indicated. Old glacier extensions from previous photographs are drawn in: solid line: June 1958; broken line: July 1961 (both after Whittow *et al.*, 1963).

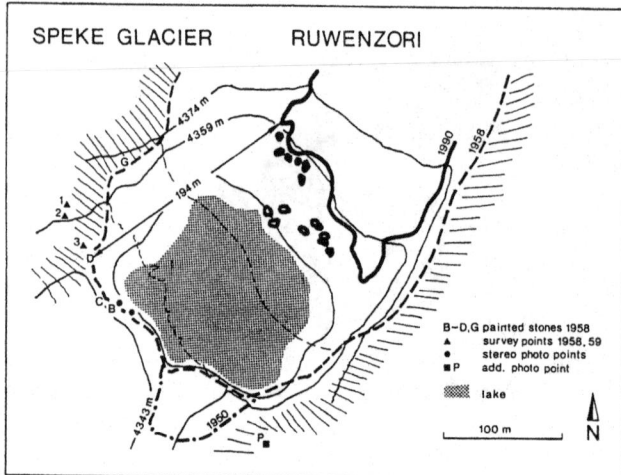

Fig. 6.7.17 The tongue of the Speke Glacier (Kaser & Noggler, 1991).

the tongue bed. Fig. 6.7.20 shows a comparison of the profiles of the Elena and Speke Glaciers, sketched with the help of the contour line model and the reconstructed areas.

The bed of the Elena Glacier allows changes of its tongue that correspond to the fluctuations of its volume. The bed of the Speke Glacier provides a basin for its tongue in which only distinct volume losses cause corresponding changes in the areas and the lengths. Consequently, the retreat of the front of the tongue was minimal up to 1960 despite obvious volume losses which are clearly indicated from a series of photographs published by Osmaston (1961). Around 1960, a small, partial advance on the bedrock edge in front of the Speke Glacier accumulated a small moraine embankment. Up to 1966, the tongue had retreated about 10 m behind that moraine (Temple, 1968). Hastenrath (1984) estimates a retreat of the tongue of 30 to 40 m between 1958 and 1974. Information is missing from that date until our visit in 1990. The tongue of the Speke Glacier had greatly retreated and a lake had formed in the bedrock basin above the rock wall.

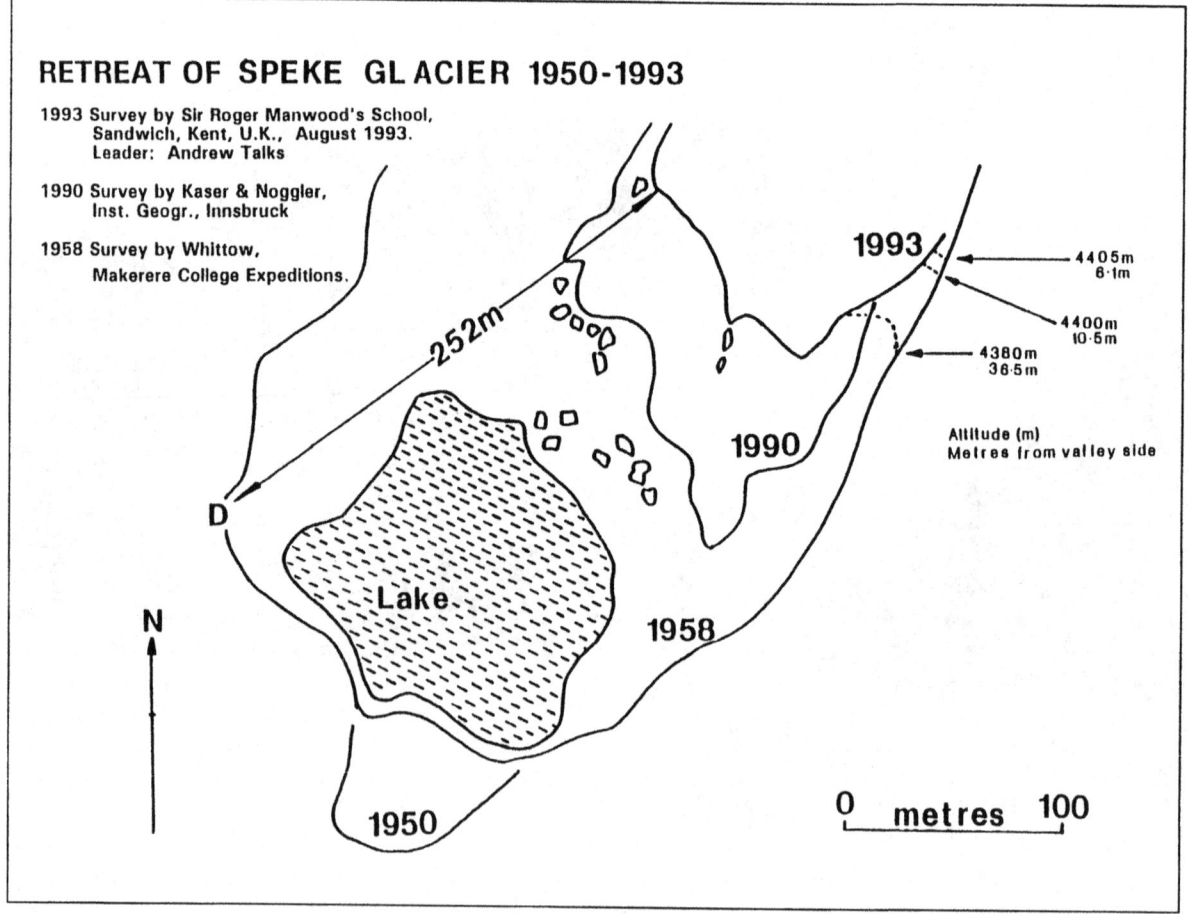

Fig. 6.7.18 The tongue of the Speke Glacier (Talks, 1993).

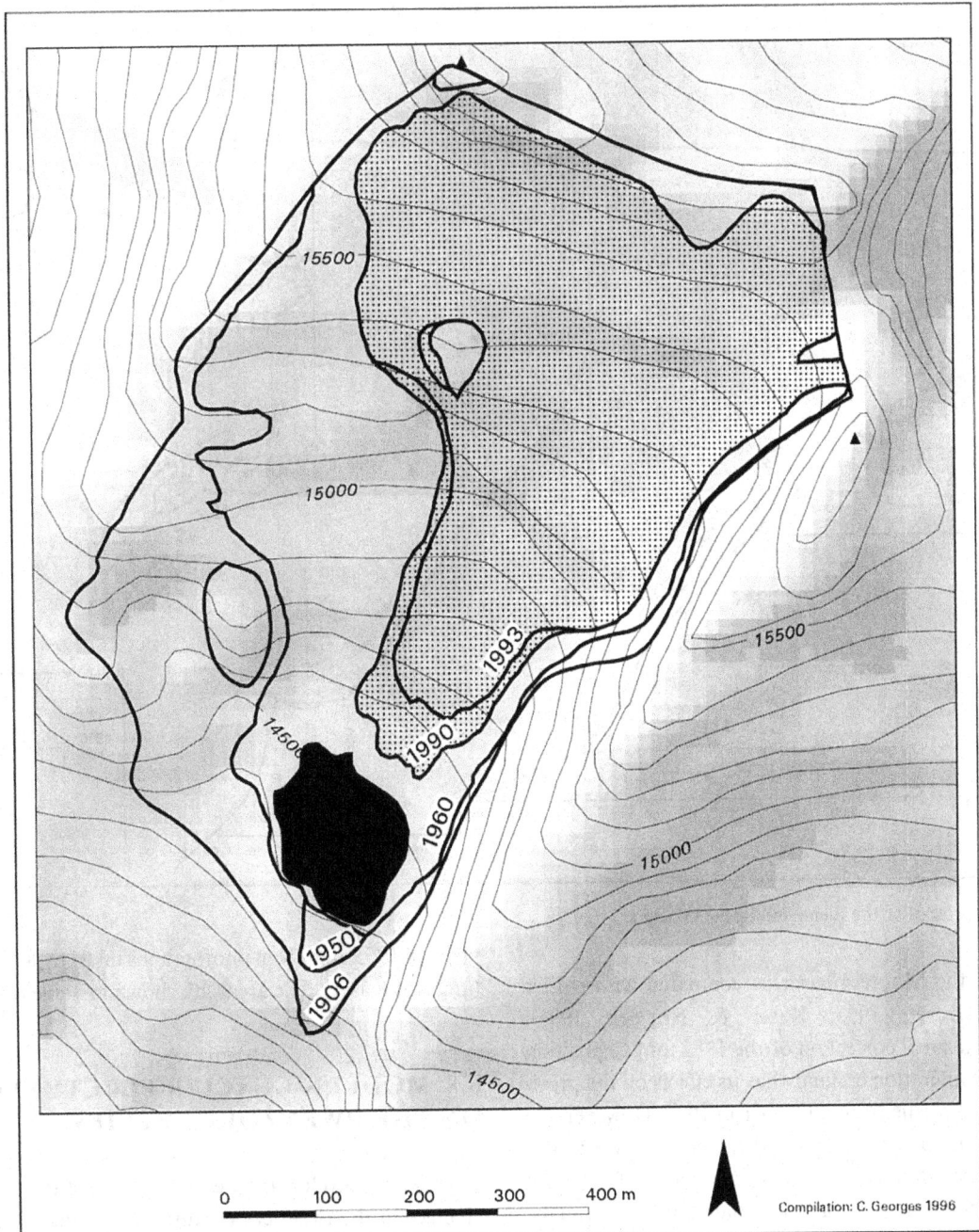

Fig. 6.7.19 The extent of the Speke Glacier in 1906, 1950, 1960, 1990, and 1993. Base map: D.L.S.U. (1970); vertical interval of the contour lines: 100 ft. The contour lines represent the glacier surface in 1955.

6.7.3 The Moore Glacier

For the Moore Glacier on Mount Baker (Fig. 5.3.2) the records are not as detailed in time as for the two other glaciers. However, a clearly defined glacier basin with a gorge into which the tongue squeezes, as well as photographs in which the glacier can be well delimited and additional information about the tongue extents make it possible to draw three extents of this glacier almost exactly. The three glacier extents in Fig. 6.7.21 were reconstructed using photographs taken by V. Sella in 1906 (De Filippi, 1909), by H. MacLachlan taken in 1951 (Osmaston & Pasteur, 1972, plate 11), and by K. Gabl taken in 1991 (Kaser & Noggler, 1996). A photograph taken by Vittorio Sella and only recently published by Mantovani (1996, p. 59) shows the tongue of the Moore Glacier in 1906 in a clear reproduction and at an angle useful for the reconstruction. This photograph created the necessity for a correction of

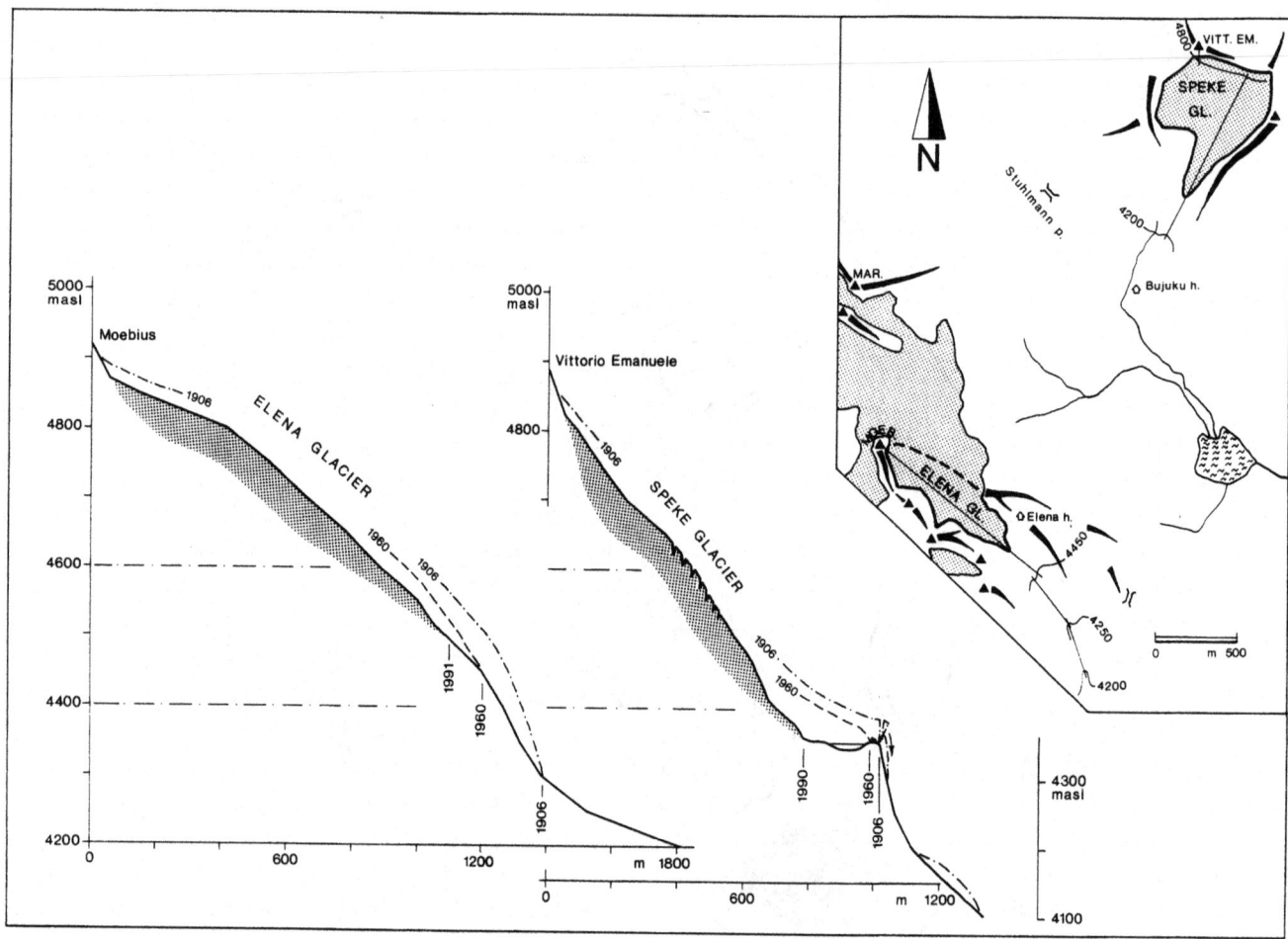

Fig. 6.7.20 The profiles of the Elena and Speke Glaciers.

the tongue of the Moore Glacier as compared with earlier publications (Kaser, 1993; Kaser & Noggler, 1996). Consequently, a small correction of the 1955 tongue position was made. The alteration certainly has its effects on the previously given area of the Moore Glacier in 1906. However, the effects on Mount Baker in 1906 and the entire Rwenzori in 1906, as well as the difference due to the correction of 1955 (see also sections 6.4 and 6.5), are so small that they can be omitted. The areas for the Moore Glacier are shown in Table 6.7.3.

6.7.4 The Savoia Glacier

The photographs in Figs. 6.7.22 to 6.7.24 show the retreat of the Savoia Glacier on Mount Stanley between 1906, 1960, and 1991. The process of the reconstruction of area extents is the same as that for the other glaciers. The three glacier extents in Fig. 6.7.25 were reconstructed using photographs taken by V. Sella in 1906 (De Filippi, 1909), from aerial pictures from the R.A.F. Mosquito taken in 1952, from Hunting Aerosurveys

taken in 1955, and from photographs taken by K. Gabl in 1991 (archives Kaser). The areas are shown in Table 6.7.4.

6.8 MODERN GLACIER FLUCTUATIONS ON THE RWENZORI – REVIEW

In order to develop the best possible picture of the glacier changes in the twentieth century, the results of the previous chapters are summarized in tables and graphs. The maps in Figs. 6.8.1 to 6.8.4 show the extents of the glaciers, at the times they were researched, on the individual mountains as well as on the Central Rwenzori Massif in direct comparison. Tables 6.8.1 and 6.8.2, as well as Fig. 6.8.5, show the changes of the glacier areas of the individual glaciers between 1906 and 1993. In Table 6.8.3, the changes on Mount Stanley, Mount Speke, Mount Baker, and in the Central Rwenzori Massif from 1906 to 1955 to 1990 are shown and graphically depicted in Fig. 6.8.6. In order to give a complete picture for the entire Rwenzori Range, values for Mount Emin, Mount Gessi, and Mount Savoia, reported by Osmaston (1989[2]), are included in

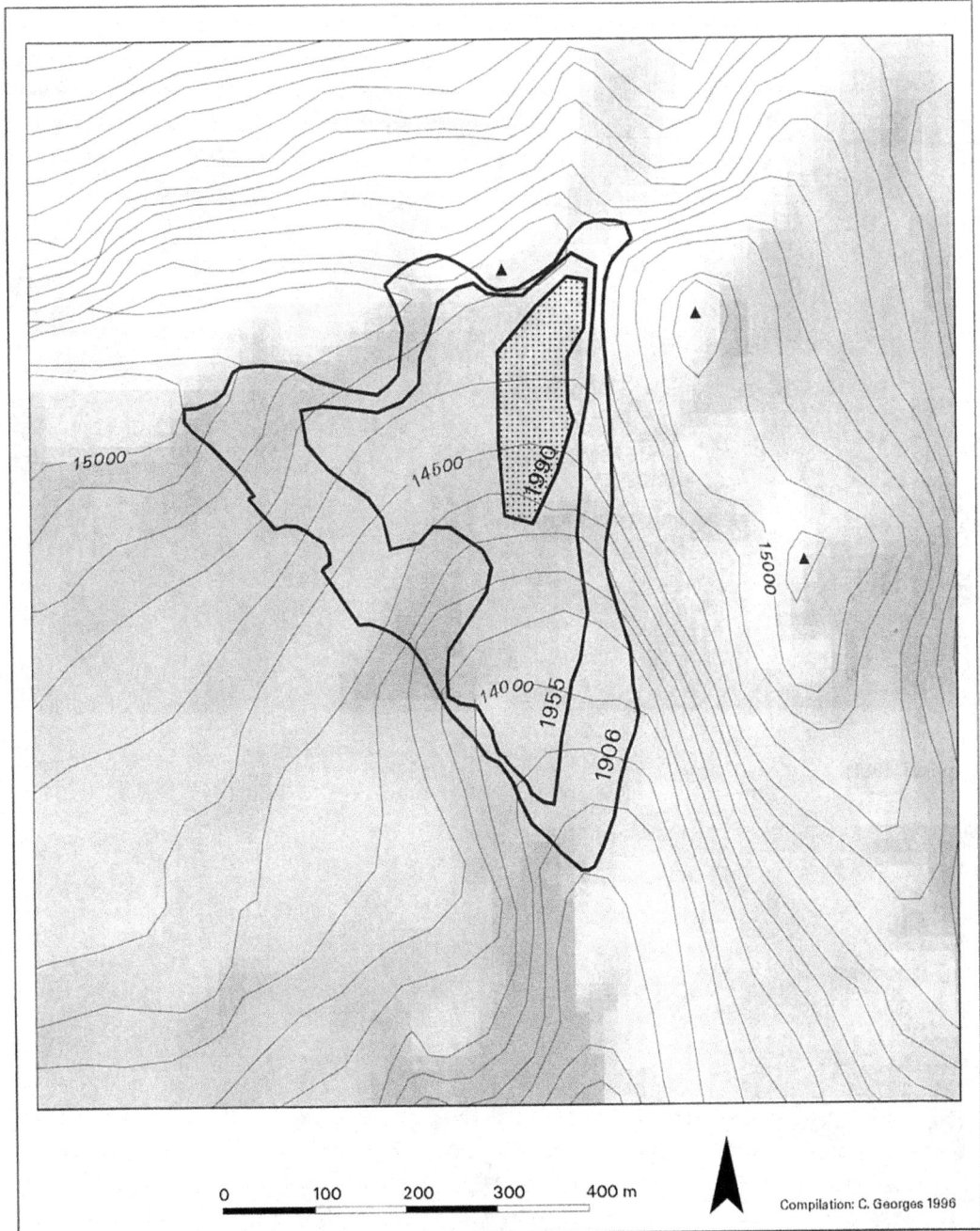

Fig. 6.7.21 The extent of the Moore Glacier in 1906, 1955, and 1991. Base map: D.L.S.U. (1970); vertical interval of the contour lines: 100 ft. The contour lines represent the glacier surface in 1955.

Table 6.7.3. *The areas of the Moore Glacier in 1906, 1951, and 1991 (in km^2), and as a percentage of the area in 1906*

Area	1906	1951	1991
km^2	0.166	0.088	0.016
%	100	53	10

Table 6.8.3. A sum of about 1.0 km^2 for all three in 1906 is a rough estimate from De Filippi (1909); the assumption that by 1993 the remnants of former glaciers had entirely disappeared is based on various reports.

The retreat in the twentieth century was drastic on all glaciers of the Rwenzori. The slightly delayed retreat of the Speke Glacier has already been discussed (section 6.7.2). The Elena and Savoia Glaciers behaved very similarly and the Moore Glacier, which is situated lower, experienced the highest losses. Of the three mountain ranges of the Central Rwenzori, Mount

Fig. 6.7.22 Savoia Glacier (left).
Photo: V. Sella 1906 (Fiory-Ceccopieri, 1981).

Fig. 6.7.23 Savoia Glacier. Photo: 1960 (J.B. Whittow).

Fig. 6.7.24 Savoia Glacier. Photo: 1991 (K. Gabl).

Baker has lost the most glacier area. In 1990, only 8% of the glacier cover at the beginning of the century remained, whereas 25% remained on Mount Speke and 35% on Mount Stanley. The different relative retreats of the glaciers are mainly due to their area–altitude distributions. Fig. 6.8.7 shows the hypsographic curves over the glacier areas of Mount Stanley, Mount Speke, and Mount Baker in 1955, corresponding to the photogrammetric evaluation of the contour lines.

The following considerations are made under the assumption of similar mass balance profiles on the three glacierized areas.

The curve on Mount Stanley is clearly above the one on Mount Speke and remains constantly around 500 ft ≈ 150 m above the one on Mount Baker. If the glaciers of Mount Stanley in 1955 had been in balance with the climate, then large parts of the glaciers of Mount Baker would have been below the equilibrium line if this is assumed to be at the same elevation as on Mount Stanley. An altitude for an equilibrium line, ELA_{ST}, calculated for the hypothetically steady-state case on Mount Stanley, with an accumulation area ratio of $AAR = 0.8$, is drawn into Fig. 6.8.7 and illustrates how little glacier area on Mount Baker would have been above that line. An

accumulation area ratio possibly chosen smaller would even aggravate the situation. However, because the glaciers of Mount Stanley were not in balance and the equilibrium line must have been clearly higher, the following recession ratios were appropriate. In 1990, Mount Stanley had retained about half of the 1955 glaciation. Mount Speke only had two-fifths, and Mount Baker only one-fifth of its 1955 glaciation remaining. But even at the beginning of the 1990s, the glaciers on the Rwenzori were not in balance with the prevailing climate and were retreating further. If the 1990 conditions continue, the rapid disappearance of the glaciers on Mount Baker is likely, and on Mount Speke the remains of the glaciers will also retreat into high, shady niches. Even an ELA surface sloping upwards from east to west, as estimated by Osmaston (1965, 1989[2]), does not seem to correct this tendency.

In order to describe modern glacier fluctuations in the Rwenzori mountains, in addition to the reconstructed area changes, measured and reconstructed changes of the lengths of individual tongues, as well as notes and comments from individual expeditions are available. The numbers in {curly brackets} refer to those given in Fig. 6.8.9.

{1} In 1906, the glaciers of the Rwenzori were already retreating from an advance, of which the end date is not exactly

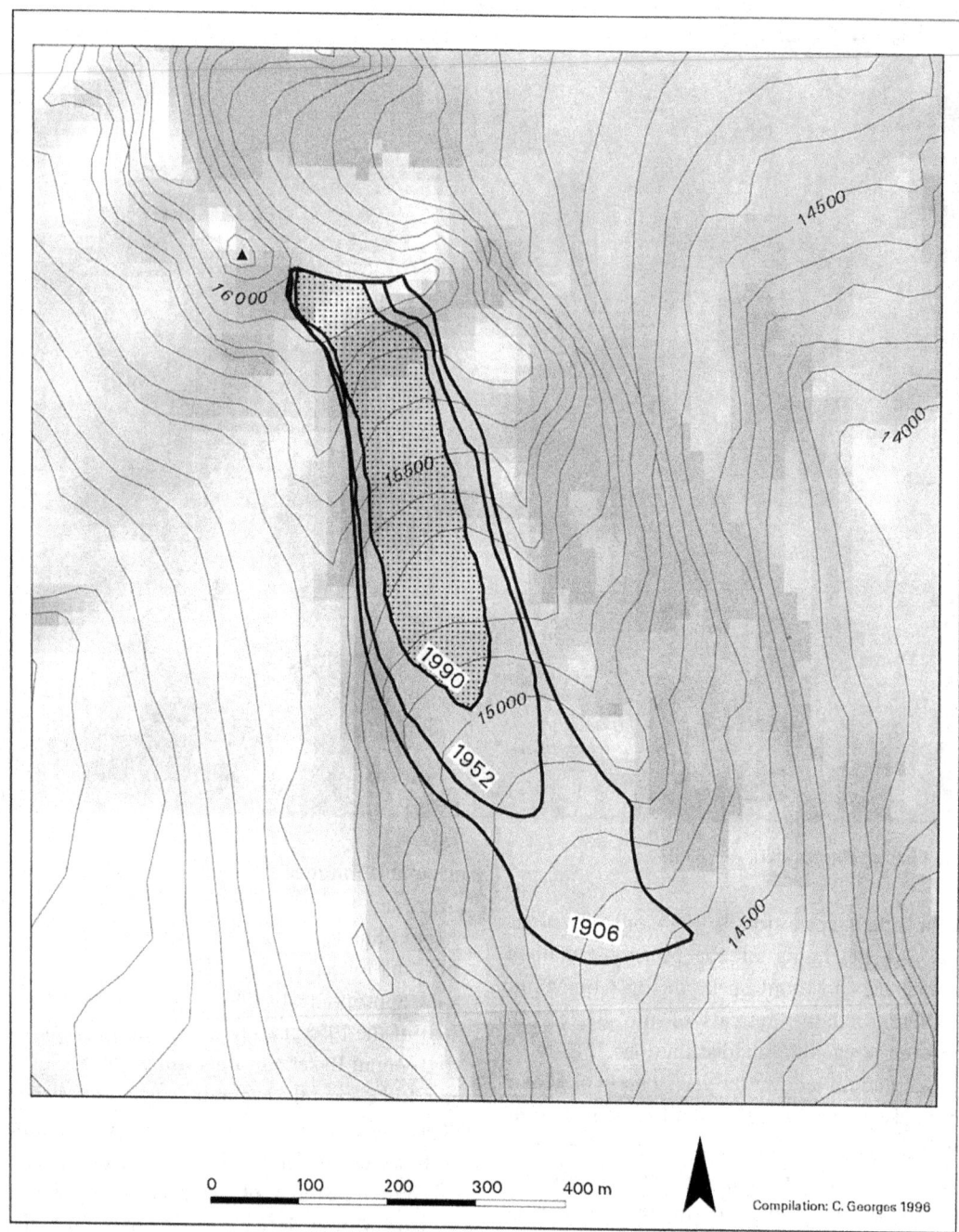

0   100   200   300   400 m

Compilation: C. Georges 1996

Fig. 6.7.25 The extent of the Savoia Glacier in 1906, 1952, and 1990. Base map: D.L.S.U. (1970); vertical interval of the contour lines: 100 ft. The contour lines represent the glacier surface in 1955.

Table 6.7.4. *The areas of the Savoia Glacier in 1906, 1955, and 1991 (in km²), and as a percentage of the area in 1906*

Area	1906	1955	1991
km²	0.121	0.077	0.043
%	100	64	36

known. However, the rock in front of the terminus was still free of lichen (Roccati, 1909). An indication of how quickly lichen colonizes snow-free rocks is given by an observation by Humphreys in 1927. When visiting the Elena Glacier, the cairns of the Abruzzi expedition were already covered with lichen. The advance must have ended not long ago. A description and evaluation of the frontage of the Elena Glacier is given by Bergstrøm (1955).

{2} Denotes the decrease of the glacier areas from 1906 to 1955 to 1990 as a percentage of the 1906 surface area in the Central Rwenzori Massif (from Fig. 6.8.6).

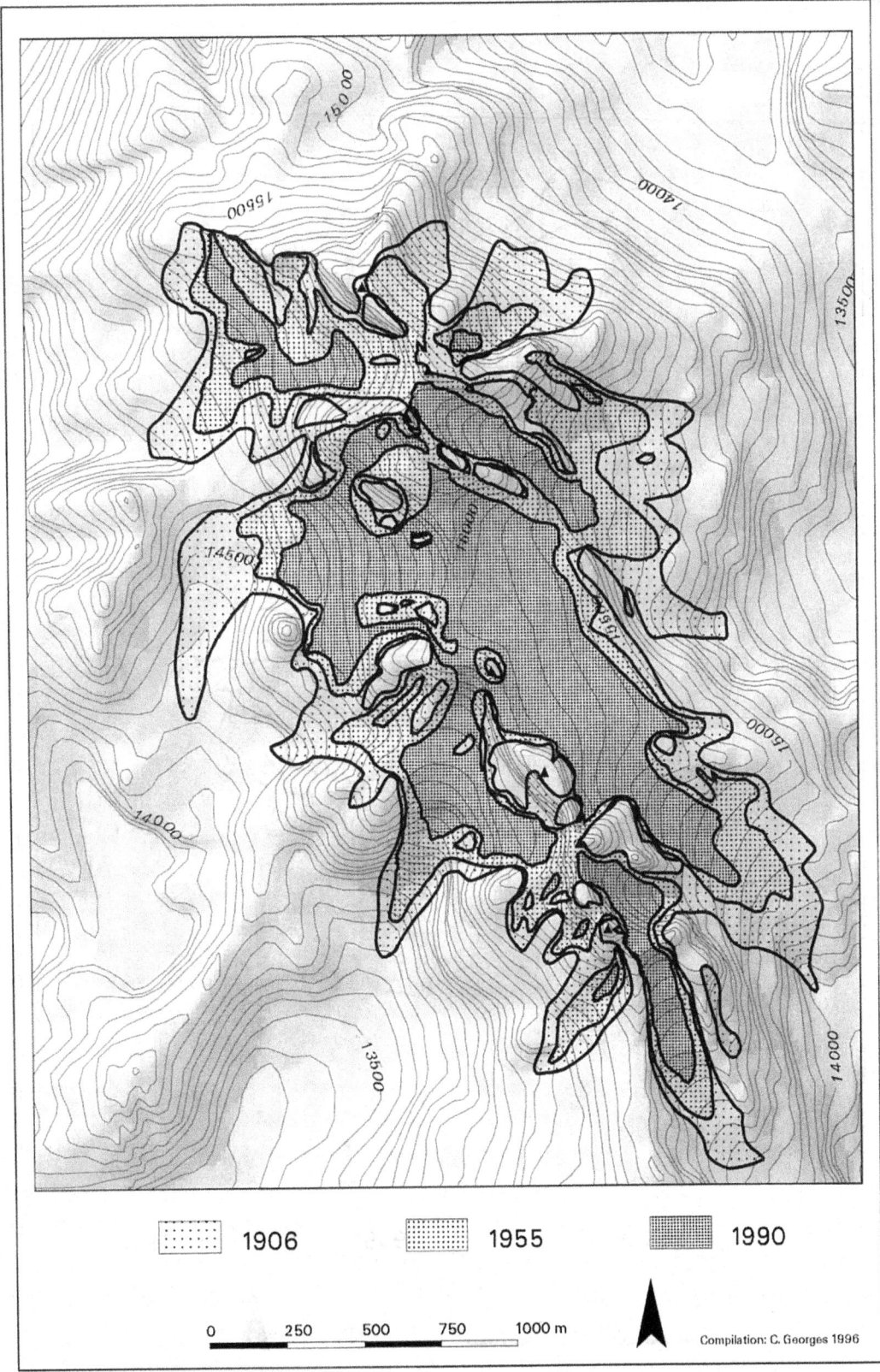

Fig. 6.8.1 The reconstructed glacier areas on Mount Stanley. Base map: D.L.S.U. (1970); vertical interval of the contour lines: 100 ft. The contour lines represent the glacier surface in 1955.

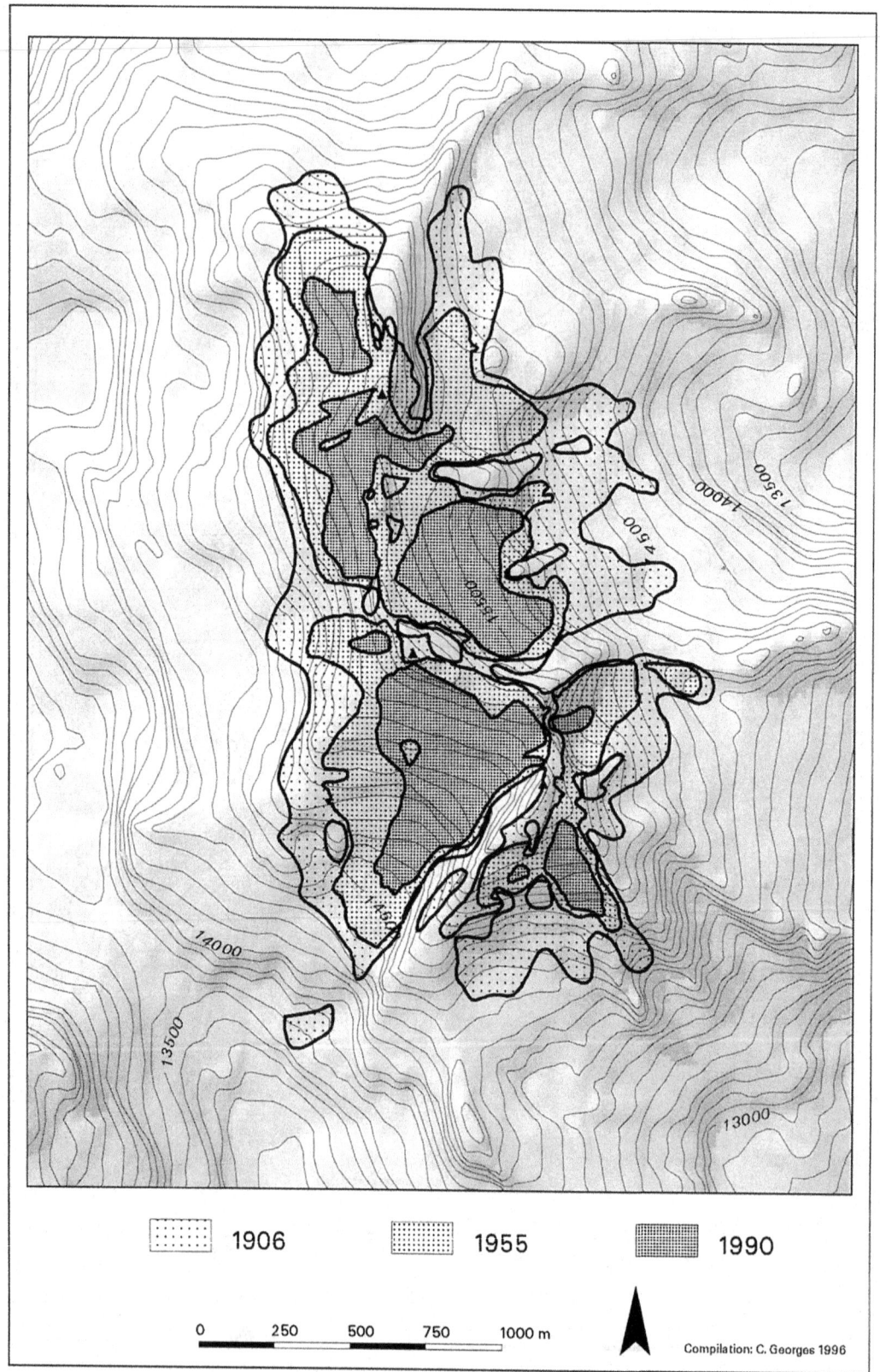

Fig. 6.8.2 The reconstructed glacier areas on Mount Speke. Base
map: D.L.S.U. (1970); vertical interval of the contour lines: 100 ft.
The contour lines represent the glacier surface in 1955.

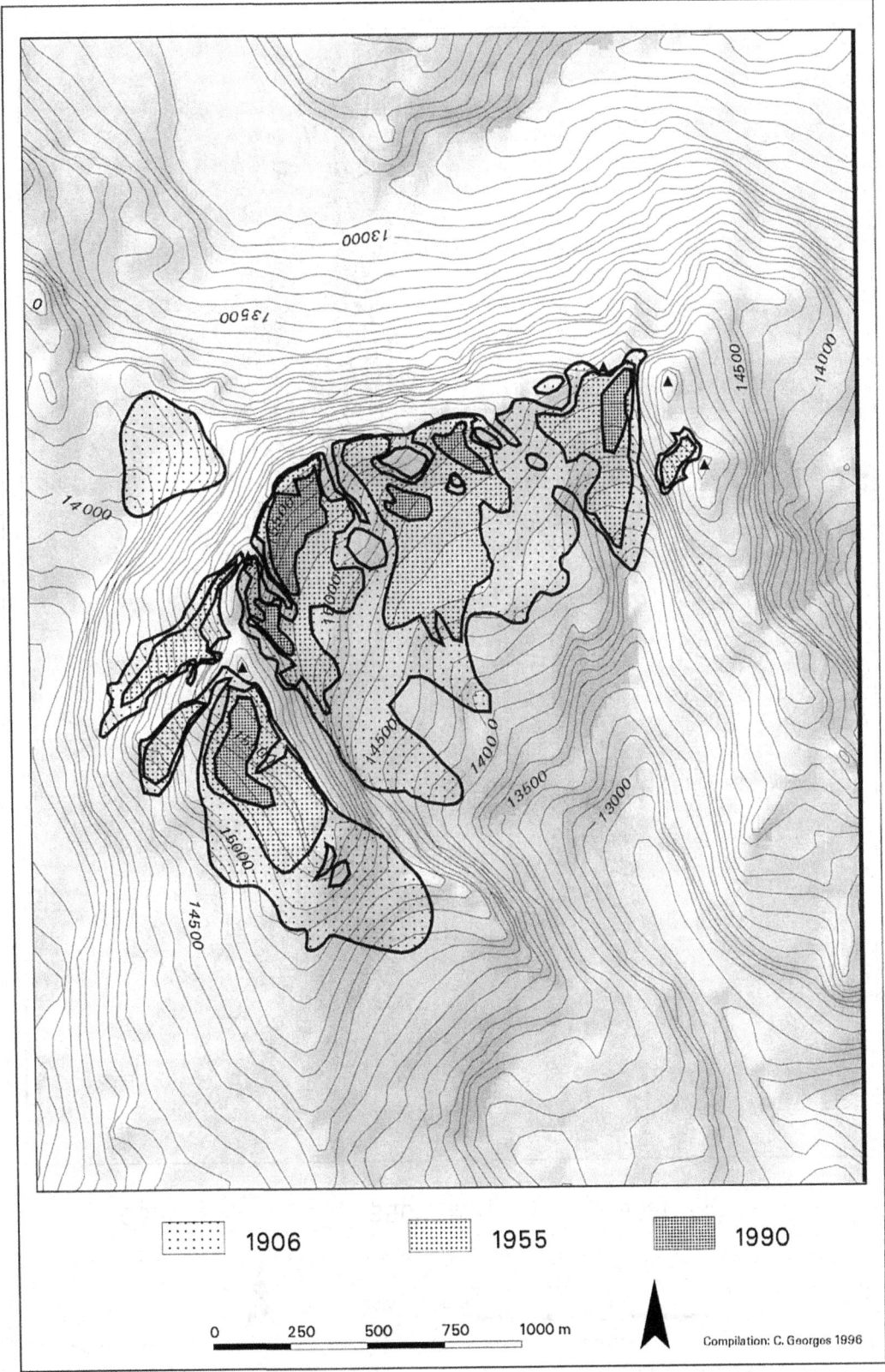

Fig. 6.8.3 The reconstructed glacier areas on Mount Baker. Base
map: D.L.S.U. (1970); vertical interval of the contour lines: 100 ft.
The contour lines represent the glacier surface in 1955.

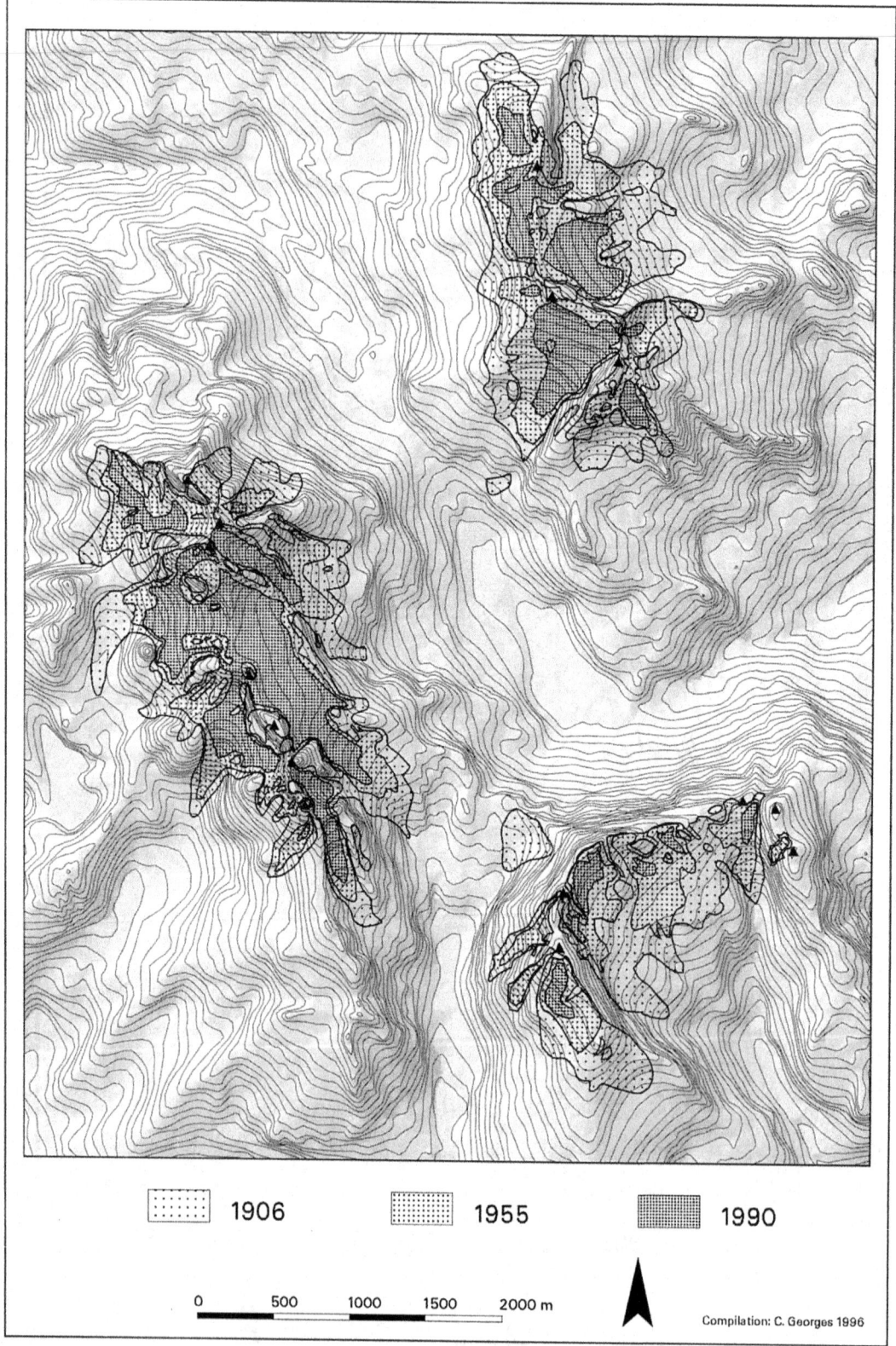

Fig. 6.8.4 The reconstructed glacier areas in the Central Rwenzori
Massif. Base map: D.L.S.U. (1970); vertical interval of the contour
lines: 100 ft. The contour lines represent the glacier surface in 1955.
For contour heights see Figs. 6.8.1 to 6.8.3.

Table 6.8.1. *The areas of the Elena, Speke, Moore, and Savoia glaciers (km²)*

	1906	1950	1951	1952	1955	1956	1960	1990	1991	1993
Elena	0.315			0.200		0.178	0.173		0.113	
Speke	0.453	0.372					0.369	0.212		0.198
Moore	0.166		0.088						0.016	
Savoia	0.121				0.077				0.043	

Table 6.8.2. *The areas of the Elena, Speke, Moore, and Savoia glaciers as a percentage of the area in 1906*

	1906	1950	1951	1952	1955	1956	1960	1990	1991	1993
Elena	100			63		56	55		36	
Speke	100	82					81	47		44
Moore	100		53						10	
Savoia	100				64				36	

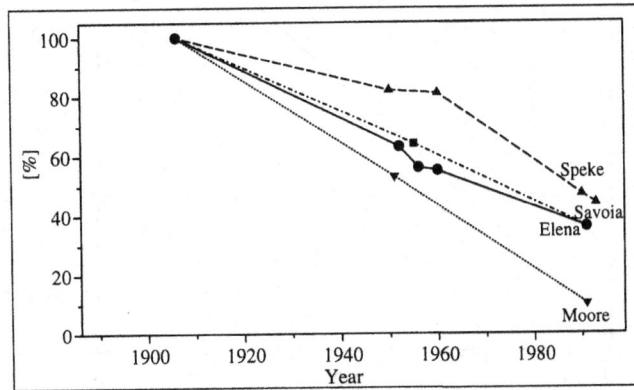

Fig. 6.8.5 The areas of the Elena, Speke, Moore, and Savoia glaciers as a percentage of the 1906 area.

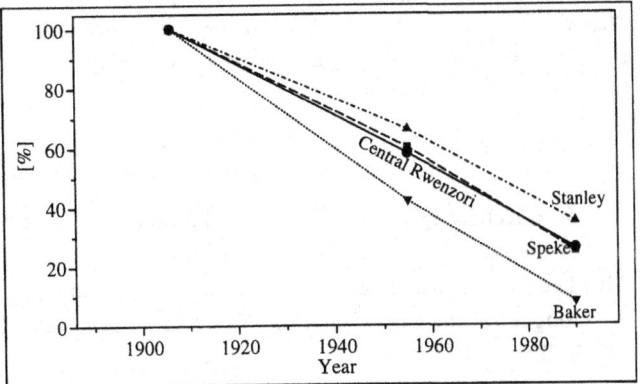

Fig. 6.8.6 The glacier areas of Mount Stanley, Mount Speke, Mount Baker, and of the Central Rwenzori Massif in 1906, around 1955, and around 1990 as a percentage of the 1906 area.

Table 6.8.3. *The glacier areas of Mount Stanley, Mount Speke, Mount Baker, and of the Central Rwenzori Massif in 1906, around 1955 and around 1990 (in km²) and as a percentage of the 1906 area*

	1906		c.1955		c.1990	
	(km²)	(%)	(km²)	(%)	(km²)	(%)
Mt Stanley	2.85	100	1.88	66	1.00	35
Mt Speke	2.19	100	1.31	60	0.56	25
Mt Baker	1.47	100	0.62	42	0.12	8
Central Rwenzori	6.51	100	3.81	58	1.68	26
Mt Emin			0.07		0	
Mt Gessi	1.00	100	0.17	27	0	0
Mt Savoia			0.03		0	
Rwenzori total	7.51	100	4.08	54	1.68	22

Notes: The values for Mt Emin, Mt Gessi and Mt Savoia are quoted from Osmaston (1989[2]). Their entire disappearance by the 1990s is assumed.

Temple (1968) published changes of the lengths on the Elena, Speke, and Savoia Glaciers, which had been measured between June 1958 and January 1967. They show a small advance which had taken place on the **Elena Glacier** between June 1959 and March 1962 {3} and on the **Speke Glacier** between July 1961 and March 1962 {4}. On the Speke Glacier, this phase of slight tongue changes left a thin line of large rocks as a frontal moraine.

{5} Kaser & Noggler (1991) have summarized the changes of the length of the tongue of the **Speke Glacier** between January 1958 and January 1990 (Fig. 6.8.8).

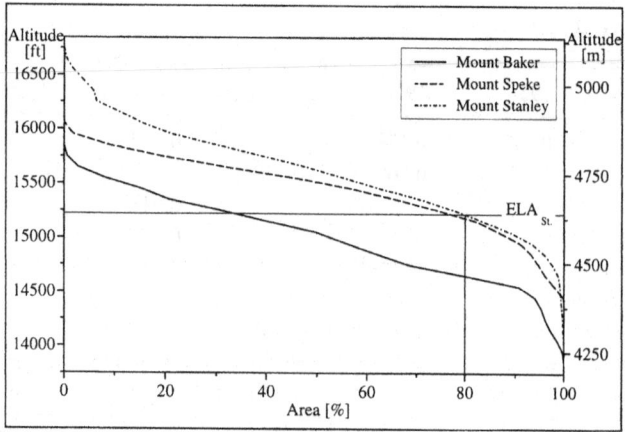

Fig. 6.8.7 The hypsographic curves of the glacierization in 1955 on Mount Stanley, Mount Speke, and Mount Baker. An equilibrium line on hypothetically steady-state Mount Stanley (ELA_{ST}) was determined with an accumulation area ratio of AAR = 0.8 (see section 5.4).

For the **Speke Glacier**, Whittow *et al.* (1963) have compiled the following information:

{6} A photograph taken by Humphreys in 1926 shows a tongue lobe, which was still there in 1949, reaching far down (photograph by Menzies).

{7} In the 'past 30 years' (about 1930 to 1958), Whittow *et al.* (1963) infer an increasingly strong retreat of the Speke Glacier.

{8} During the time of observation by the Makerere College Expeditions, the tongue retreated, except for small advances in the E-sector, and reduced in thickness (Menzies 1950: 36.4 m, Whittow *et al.* 1960: 18.2 m) as well as in width (1950: 292 m, 1960: 237 m). The advances reported by Temple (1968) might, therefore, only refer to parts of the tongues of the Speke and also the Elena Glaciers.

{9} Five photographs, which show the Speke Glacier from a similar angle – one photograph from 1952 (Bergstrøm, 1955; Fig. 6.7.7), one photograph from 1960 taken from the archives of J. B. Whittow (Noggler, 1991; Fig. 6.7.8), one around 1967 (without date in Temple, 1968; Fig. 6.7.9), one from 1974 (Hastenrath, 1984; Fig. 6.7.10) and one photograph from 1977 (Lichtenegger & Lichtenegger, 1978) – show that the retreat was minor between 1952 and 1974 and that the area of the Speke Glacier did not decrease up to 1977. However, the serac zones in the middle of the glacier seem to be less and less rugged, which is also indicated by a series of vertical air photographs (April 1952, September 1952 and June 1955) published by Osmaston (1961). This gives the impression

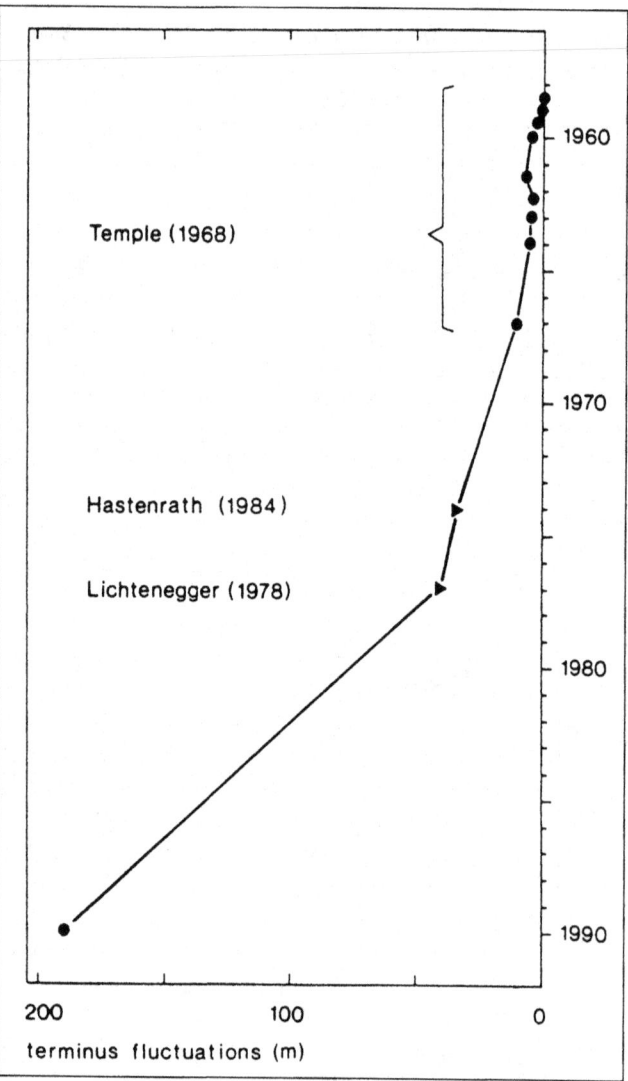

Fig. 6.8.8 Changes of lengths of the tongue of the Speke Glacier from January 1958 to January 1990. Circles: measured values (Temple, 1968; Kaser & Noggler, 1991); triangles: estimated values according photographs taken by Lichtenegger & Lichtenegger (1978) and Hastenrath (1984).

of a decreasing volume. However, the varying thickness of the snow cover can distort the picture.

{10} The latest measurements show an extensive retreat of the tongue of the Speke Glacier (Kaser & Noggler, 1991; Talks, 1993).

Whittow *et al.* (1963) report of the **Elena Glacier**:

{11} The Coronation Glacier lost its connection to the Elena Glacier between 1943 (Firmin, 1945) and 1949 (Menzies, 1951[1]). N.B.: Despite continuing ice falls from Coronation Glacier to Elena Glacier, this had an

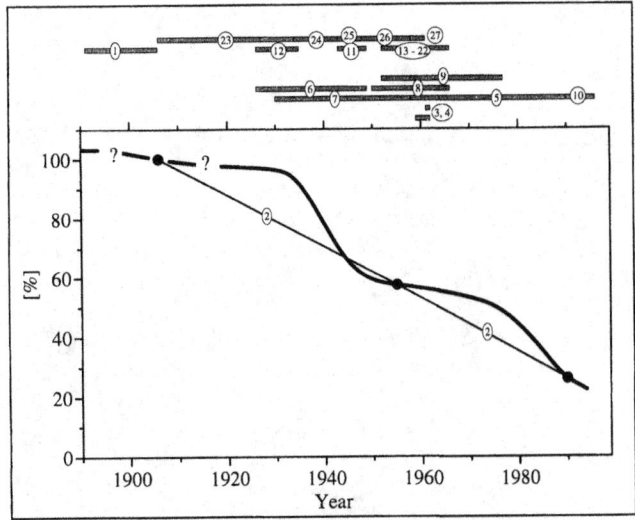

Fig. 6.8.9 Changes of ice extent on the Rwenzori in the twentieth century as a percentage of the 1906 area. The numbers refer to the information in the text, there shown in curly brackets.

influence on the retreat of the tongue of the Elena Glacier (Kaser, 1993).

{12} 'It was after Humphreys' expeditions (1926 and 1927) that the general retreat of the Rwenzori ice became apparent, although no detailed work was undertaken at that time.'

For the **tongue of the Elena Glacier,** the following data are available as well.

{13} August 1952–August 1953 (Osmaston, 1961): 1.9 m advance in the western part, 3.5 m retreat in the eastern part.

{14} 1952–1960 (Whittow *et al.*, 1963): − 52 m (middle) to − 213 m (E).

{15} January 1959–July 1959 (Whittow *et al.*, 1963): +3.7 m.

{16} July 1959–January 1960 (Whittow *et al.*, 1963): +4.5 m.

{17} January 1960–July 1961 (Temple, 1968): +6.8 m.

{18} July 1961–March 1962 (Temple, 1968): +0.9 m.

{19} March 1962–December 1962 (Temple, 1968): −0.6 m.

{20} December 1962–December 1963 (Temple, 1968): −1.5 m.

{21} December 1963–December 1966 (Temple, 1968): 0 m.

{22} Lateral retreat of the Elena Glacier: 1952–1960 (Whittow *et al.*, 1963): −8 m to −55 m.

Whittow *et al.* (1963) summarize the behaviour of the **Savoia Glacier:**

{23} From the comparison of photographs: 1906 (De Filippi, 1909) – 1927–1931 (Humphreys, 1927, 1933) – 1934 (Synge, 1937) slight retreat.

{24} 1934–1943 (Firmin, 1945): rapid retreat.

{25} 1943–1948 (photograph P. Jenkins, archives Mountain Club of Uganda): disintegration of the middle part of the tongue.

{26} 1948–1958: no changes.

{27} 1958–1961: slight retreat.

Morton (1968) reports:

'The Speke, Elena and Savoia Glaciers have all acted in a similar manner. Retreat up to 1961 was followed by a well marked advance in 1961–62. From 1962 to 1966 there was retreat again . . . Finally the period December 1966 to June 1968 has been marked by modest advances (Elena and Savoia) or a period of virtually no change (Speke).'

Whittow *et al.* (1963) again give a summarizing overview for the **Moore Glacier:**

'An examination of the ice in 1958 left little doubt that the Moore Glacier is hardly more than a thin tongue of stagnant ice. There are no crevasses or surface moraines which might be indicative of movement ∴ .

The ultimate cause of the rapid recession, however, is probably the gradually increasing elevation of the firn line in the Rwenzori since the last ice advance, possibly in the mid nineteenth century. At present the firn line on the Uganda slopes stands at an elevation of almost 4573 m, which means that the upper basin of the Moore Glacier lies entirely below it. Since there is no longer a positive budget, and since the ice thickness has been so greatly diminished during the last 50 years, it would appear that if present trends continue the Moore Glacier will cease to exist within the next 20 to 30 years.'

Due to its special situation, details about the Moore Glacier's recession are not used for the reconstruction of the glacier fluctuations on the Rwenzori.

The fact that the Moore Glacier did not disappear by the beginning of the 1990s is due less to the climatic conditions than to the fact that its remains have survived in the shadow of the rock wall below Wollaston and Moore peaks.

6.8.1 Summary

Before 1906, the glaciers of the Rwenzori had retreated from an advance which occurred not long ago. Up to the end of the 1930s, the total retreat was small. However, nothing is known of its course during that time. Between the 1930s and the 1950s, the glaciers of the Rwenzori underwent a generally drastic retreat, and at the beginning of the 1960s, some portions of the glaciers advanced slightly (Whittow *et al.*, 1963; Temple, 1968). Up to the end of the 1970s, a further retreat took place, which only caused slight area changes, as the few photographs indicate. Since then, the glaciers have melted

Fig. 6.8.10 Alexandra (left) and Margherita in 1906.
Photo: V. Sella (De Filippi, 1909).

Fig. 6.8.11 Alexandra (left) and Margherita in 1990.
Photo: G. Kaser.

heavily. The period between 1960 and 1993, however, is mainly based on photographs of the Speke Glacier, the tongue of which lay in a basin. In Fig. 6.8.9, the change of the glaciers on the Rwenzori is depicted schematically.

The picture series in section 6.7 and in Fig. 6.7.5 clearly shows that the recession of the glaciers was not limited to the tongues and to lower elevations of the glacier. A further impression of mass loss in higher elevations is given by a series of pictures which shows drastic changes in the area of the highest peaks, Margherita (5109 m a.s.l.), Alexandra (5091 m a.s.l.), and Albert (5087 m a.s.l.), between 1906 and 1990 (Figs. 6.8.10–6.8.19; see also Figs. 1–3 in Osmaston *et al.*, 1998).

The amount of the mass losses as a direct consequence of climatic influences cannot be reconstructed quantitatively from the information available.

Fig. 6.8.12 Alexandra (left) and Margherita in 1906. Photo: V. Sella (De Filippi, 1909).

Fig. 6.8.13 Alexandra (left) and Margherita in 1990. Photo: G. Kaser.

Fig. 6.8.14 The ascent to Alexandra Peak in 1906. Photo: V. Sella (De Filippi, 1909).

Fig. 6.8.15 The ascent to Alexandra Peak in 1990. Photo: G. Kaser.

Fig. 6.8.16 Margherita Peak from Alexandra Peak in 1906.
Photo: V. Sella (De Filippi, 1909).

Fig. 6.8.17 Margherita Peak from Alexandra Peak in 1990.
Photo: G. Kaser.

Fig. 6.8.18 Albert Peak from Alexandra Peak in 1906.
Photo: V. Sella (De Filippi, 1909).

Fig. 6.8.19 Albert Peak from Alexandra Peak in 1990.
Photo: G. Kaser.

7 Modern glacier fluctuations in the Cordillera Blanca

7.1 THE CORDILLERA BLANCA

In the Cordillera Blanca 722 individual glaciers covered a total area of 723 km^2 (Fig. 7.1.1) at the time of the last photogrammetrical evaluation of aerial photographs in 1970 (Ames *et al.*, 1989; WGMS, 1989). Consequently, this mountain range is by far the most glacierized in the tropical region (Kaser *et al.*, 1996[2]; chapter 3, Table 3.1). As part of the South American Andes, the Cordillera Blanca stretches over about 180 km from 8°30′ S to 10° S, 27 peaks reach elevations of over 6000 m a.s.l., and over 200 peaks overtop 5000 m a.s.l. The highest peak, the Nevado Huascarán Sur, has an elevation of 6768 m a.s.l. (Fig. 7.1.2).

The eastern side of the Cordillera Blanca drains into the Rio Marañon, and consequently into the Atlantic. The western side drains into the Rio Santa which flows into the Pacific. The non-glacierized Cordillera Negra delimits the Rio Santa catchment area, the Callejon de Huaylas, at its western side. The coastal Cordillera is missing in this area.

The geological structure, comprising granodiorite bodies which were lifted tectonically, freed from more easily erodible sediments and, today, form the range of highest peaks (Kinzl, 1935[3]), leads to a peculiarity that can be seen in Fig. 7.1.1 and which is depicted in a schematic cross-section in Fig. 7.1.3. The range of the highest peaks lies to the west of the watershed and, therefore, within the catchment area of the Rio Santa. Fliri (1968) and Kaser *et al.* (1990) have referred to the consequences this has for the climate and water budget, which will be discussed in the following section (Fig. 7.2.2).

The majority of the glaciers are thin, with heavily fissured slopes that are fed from steep sides (Figs. 5.3.3, 5.5.10, and 7.1.4). They left behind indistinct lateral and thin frontal moraines after their retreat from one of the last advances. The individual valley glaciers (Ames *et al.*, 1989 have classified 61 as such) are mostly heavily covered with debris (Glaciar Kinzl is the largest with 9 km^2; see Fig. 5.3.1) and have left behind big moraines after their retreat. The often catastrophic consequences have already been mentioned in the Prologue.

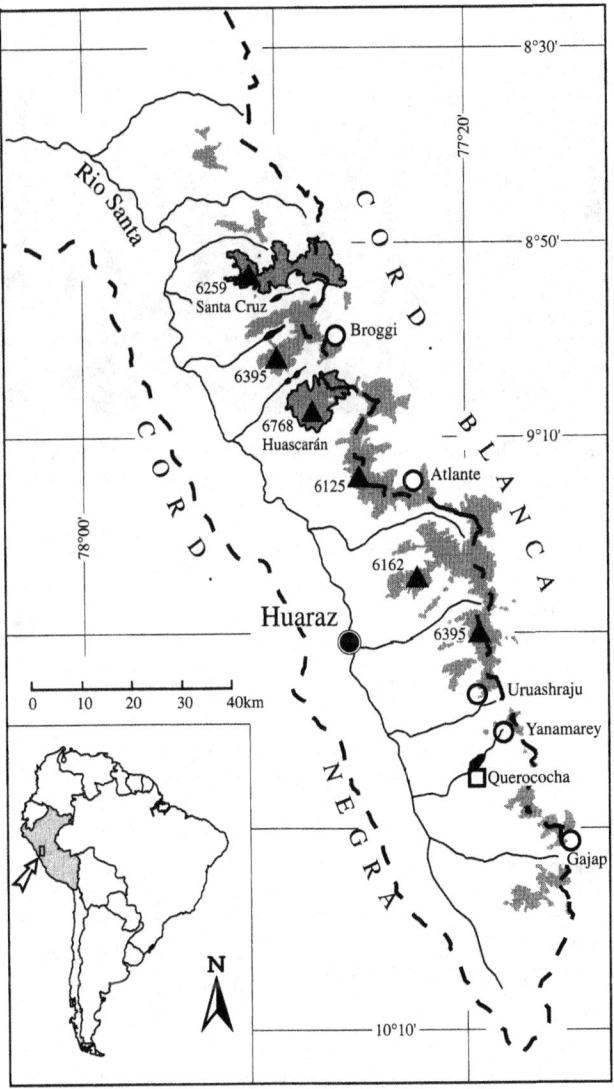

Fig. 7.1.1 The Cordillera Blanca and the catchment area of the Rio Santa. The glacier cover (grey) shows the extent around 1970. The open circles designate individual glaciers mentioned in this chapter, and the square represents the Querococha weather station. The triangles indicate the range of the highest peaks. The broken line outlines the catchment area of the Rio Santa.

Fig. 7.1.2 The Huascarán Norte (6655 m) (left) and the Huascarán Sur (6768 m) from the west in 1995. Photo: G. Kaser.

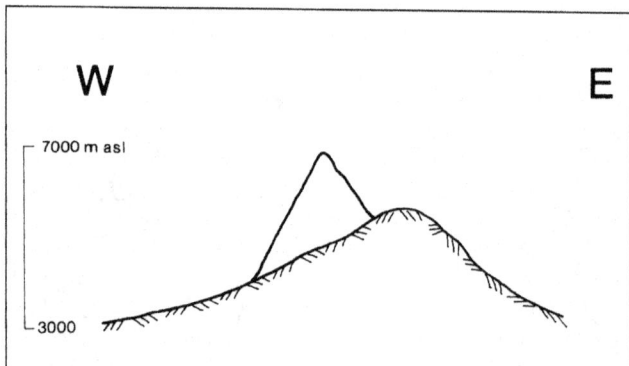

Fig. 7.1.3 A schematic east–west cross-section of the Cordillera Blanca. The range of the highest peaks lies to the west of the watershed.

The tectonic fracture zone of the 'Cordillera Blanca Fault' (Fig. 7.1.5) runs from north to south through the mountain range, where there have already been sudden vertical shifts, sometimes of a few metres, in the geologically recent past (Schwartz, 1983). Frequent and sometimes large earthquakes confirm the constant tectonic activity. The last catastrophic

earthquake on May 31, 1970, caused the deaths of about 70 000 people, due especially to its fatal combination with a dangerous glacier situation (Stadelmann, 1985; Prologue of this book).

7.2 ESSENTIAL FEATURES OF THE CLIMATE IN THE CORDILLERA BLANCA

Although the Cordillera Blanca is situated close to the equator at about 9° S, the weak oscillation of the ITCZ over the Pacific leads to the prevalence of a typically outer tropical climate (Fig. 7.2.1).

Fig. 7.2.2 shows the Cordillera Blanca during the two extreme positions of the ITCZ and, in combination with the special topography, the consequences for the climate. While the adjoining coastal desert is under the influence of the Humboldt current and subtropical Pacific high pressure cells, the Cordillera Blanca is under the influence of wet air masses from the wet Amazon basin. In the humid season, wet air masses are brought in by the ITCZ, and during the dry season trade winds dominate. Both originate from the southeast. As is typical in the tropics and especially within the ITCZ,

Fig. 7.1.4 The Nevado Santa Cruz (6259 m) from the west.
Photo: J. Alean, 1980.

Fig. 7.1.5 The 'Cordillera Blanca Fault' at Catac in the southern
Callejon de Huaylas. Photo: G. Kaser, 1988.

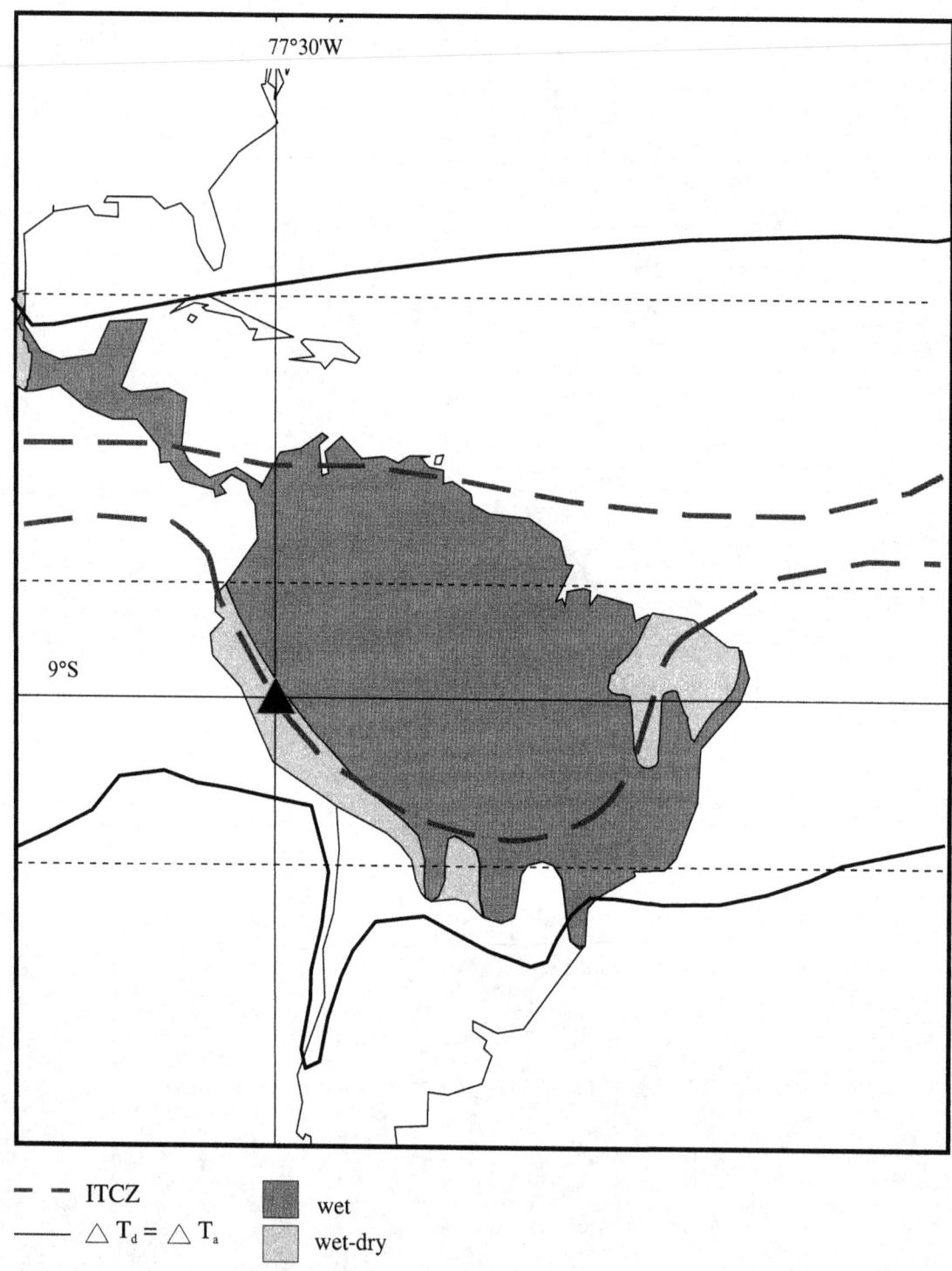

77°30'W

9°S

— — ITCZ   ▨ wet
——— △ T_d = △ T_a   ▨ wet-dry

Fig. 7.2.1 The position of the Cordillera Blanca (▲) within the
tropical regions.
ΔT_d, diurnal temperature range;
ΔT_a, annual temperature range.
(Sources: Lauer, 1975; Liljequist and Cehak, 1984; Paffen, 1967.)

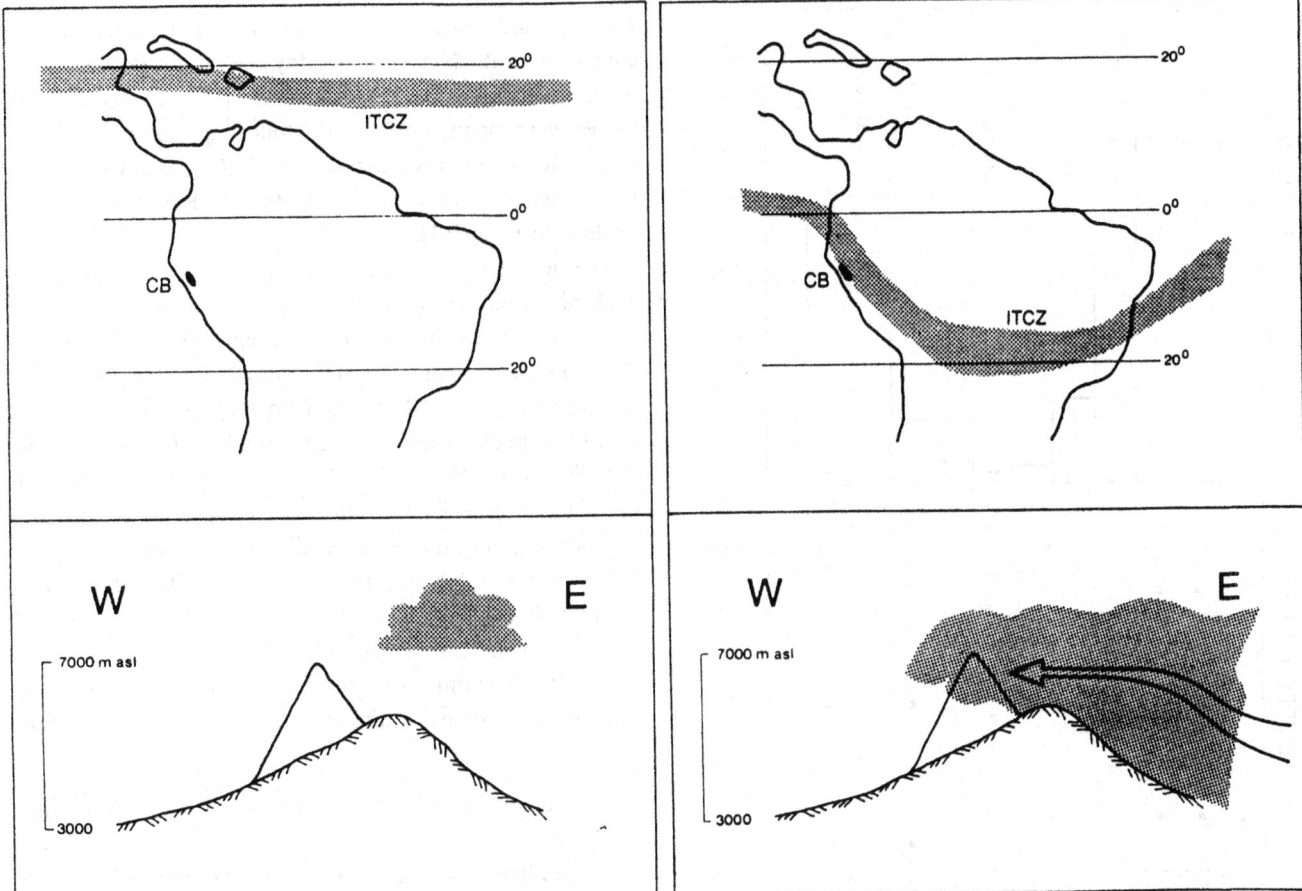

Fig. 7.2.2 (Above) The position of the Intertropical Convergence Zone (ITCZ) over South America in July and August (left) and during the southern winter (according to Graf, 1986). (Below) The weather in the Cordillera Blanca (CB) is dependent on the respective position of the ITCZ and the peculiarities of the relief (Kaser *et al.*, 1990).

convective cells within the extensive advection currents lead to precipitation. Besides these flow patterns, the denser vegetation and above all the lower elevation of the glacier tongues on the eastern side of the Cordillera Blanca (Kinzl, 1942) indicate that there is more precipitation than in the west. This corresponds well with the extensive east–west precipitation gradient which was reported by Johnson (1976). However, measurements for the east of the Cordillera Blanca are missing.

The records from the Querococha station (3955 m; Fig. 7.1.1) offer a clear picture of the climate (Fig. 7.2.3). The mean monthly total precipitation shows a wet season from October to April and a distinct dry season from May to September. Of the annual precipitation 70–80% falls during the wet months (Johnson, 1976; Niedertscheider, 1990). The mean temperature, calculated from the monthly means of maximum and minimum temperature, shows no annual variation. The difference between maximum and minimum is naturally slightly higher during the dry period than under wet conditions.

Niedertscheider (1990) analysed the precipitation ratios in the Callejon de Huaylas in detail. The dependence of precipitation on the altitude is depicted in Fig. 5.1.7 which shows the mean annual totals that rise from 250 mm on the northern valley floor to about 1000 mm at 4500 m a.s.l. Along the valley floor the totals increase up the valley from north to south. A detailed climatology of the large area is offered by Johnson (1976), among others.

7.3 THE GLACIOLOGY OF THE CORDILLERA BLANCA

Numerous archaeological discoveries from the pre-Columbian and also from the pre-Inca periods show that even high regions in the Cordillera Blanca have been settled for a long time (Kinzl, 1935[4]), and that the nearby glaciers have probably been a threat to the native people since that time. The first Spaniards that came to Peru, however, also reported on the Inca irrigation

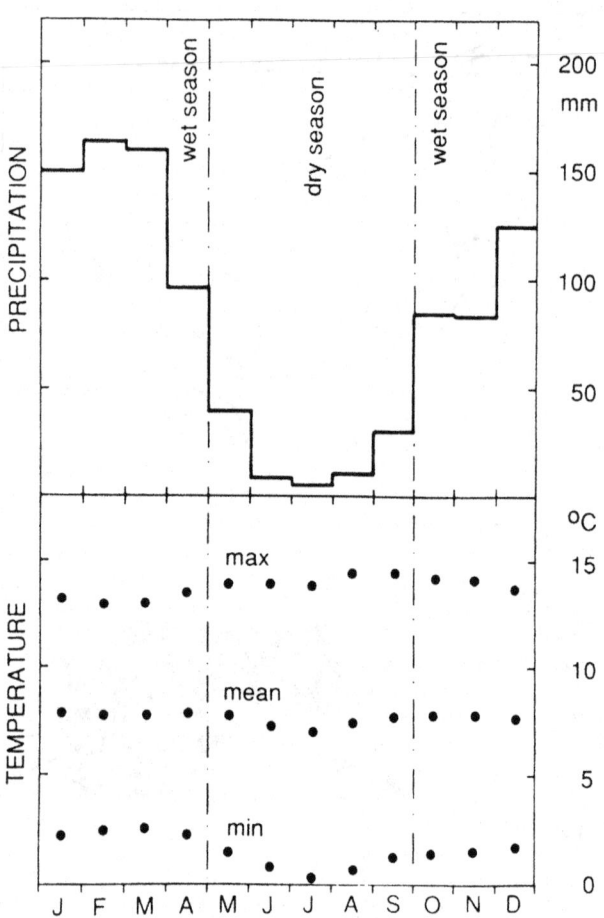

Fig. 7.2.3 Station Querococha (3955 m): mean monthly total precipitation (1954–1987) as well as maximum, minimum and mean values of the mean monthly temperatures (1965–1987) (Kaser *et al.*, 1990).

In 1764, Cosme Bueno reported, in his *Descripción de las Provincias del Arzobispado de Lima,* of an ice avalanche which, coming from the Nevado Huandoy, had totally destroyed the town Ancash and its 1500 inhabitants on January 6, 1725 (Ames & Francou, 1995). In the following century, travellers and explorers often wandered through the country. However, the glaciers were hardly noticed (Raimondi, 1876; Middendorf, 1895; Haënke, 1901).

The actual scientific and consequently also glaciological exploration on the Cordillera Blanca did not start until 1932, when the first expedition of the German and Austrian Alpine Club (Deutscher und Österreichischer Alpenverein) visited the mountain range (Borchers, 1935[1]). Hans Kinzl accompanied the expedition and later led a number of other scientific expeditions (1936, 1939, 1940, 1954) into the Peruvian Cordilleras, especially to the Cordillera Blanca. A large number of publications, especially on glaciology (Borchers, 1935[1,2]; Kinzl, 1935[1,2,3,4], 1937[1,2], 1940[1,2,3], 1941, 1942, 1943, 1944, 1949, 1950, 1954, 1955[1,2], 1965, 1970[1,2]; Kinzl & Wagner, 1938; Kinzl *et al.*, 1942; Kinzl *et al.*, 1964), were issued, the most impressive results of which are the following topographical maps by the cartography department of the Alpenverein:

- Cordillera Blanca 1:100000 (Borchers, 1935[2]; Kinzl, 1942),
- Cordillera Blanca, parte sur 1:100000 (Deutscher Alpenverein, 1945; Kinzl *et al.*, 1964),
- Cordillera Blanca 1:200000 (Alpenverein, year unknown), and
- Cordillera Huayhuash 1:50000 (Deutscher Alpenverein, 1939; Kinzl *et al.*, 1942).

Furthermore, other scientists dealt with former glacier extensions (Broggi, 1943; Oppenheim, 1945; Oppenheim & Spann, 1946; Clapperton, 1972, 1993; Lliboutry *et al.*, 1977[1]; Mercer, 1979; Seltzer, 1990; Rodbell, 1992[1,2]). Notably, catastrophes caused by glaciers or glacier lakes are repeatedly at the centre of interest (Cosme Buenos report; Kinzl, 1940[1]; Spann, 1947; Trask, 1953; Fernández, 1957; Morales, 1966, 1969, 1979; Morales *et al.*, 1979; Kinzl, 1970[1]; Welsch & Kinzl, 1970; Lliboutry, 1975, 1977; Lliboutry *et al.*, 1977[1,2]; Patzelt, 1983 – three maps of the Nevado Huascarán 1:25000, of the ice-rock avalanche in 1962 1:15000, and of the one in 1970 at the scale of 1:25000 are enclosed in this work; Reynolds *et al.*, 1988; see also the Prologue). Additionally, this special situation was the reason for the foundation of the Comisión del Control de las Lagunas de la Cordillera Blanca (see Prologue), which also started measuring glacier tongues, in addition to the observation and the inventory of dangerous lakes (Fernández, 1957). At the same time, the collection of climate records began and

systems (Hagen, 1959; Kinzl, 1965) and it can be assumed that glacier water had also been used in the Callejon de Huaylas and on the eastern sides of the Cordillera Blanca.

Hernando Pizarro and his companions were probably the first Europeans to have seen the Cordillera Blanca. On their expedition along the Pacific coast from Cajamarca to Pachacamac, slightly north of present day Lima, they crossed a snow-covered pass between Corongo and Huaylas on the western side of the Cordillera Rosko, the northern part of the Cordillera Blanca, on January 14, 1533. On their way back, they probably crossed the mountain range in the area of the Cordillera Raura in order to reach the Inca road at Jauja. This time the Spaniards moved along the eastern side of the Cordillera Blanca on their way to Cajamarca. Miguel de Estete, the chronicler of the expedition, mentioned the crossing of a high and cold, snow-covered pass (Xerez, 1534).

aerial photographs for photogrammetric evaluation were taken:

1948/1950: SAN project 2524, about 1:20 000; flight altitude: 27 000 ft.; the black-and-white pictures are of high quality.

1962/63: USAF-AST-9; about 1:30 000, flight altitude: 30 000 ft.; the black-and-white pictures do not provide the quality of those taken in 1948/50 and are also badly exposed in the area of the glaciers. Considerable parts of the glaciated areas are covered by clouds as well.

July 13–18, 1970: SAN project 176–70-A, NASA, about 1:30 000, flight altitude: 30 000 ft.; the infrared images were taken immediately after the earthquake in 1970.

The first lake register was compiled from the aerial photographs of 1948 and 1950 (Fernández, 1957), and the maps by the Dirección General de Reforma Agraria y Asentamiento Rurál 1:25 000 (1972) were drawn from those taken in 1962/63, with improvements made using the pictures from 1970. They were used as a basis for the Carta Nacional 1:100 000 (IGM, 1970; IGN, 1986), but also as a basis for the glacier inventory of Peru together with the aerial photographs of 1970 (Ames *et al.*, 1989; WGMS, 1989).

Fliri (1968) made the first analysis of the glaciological, hydrological and climatological data from the Cordillera Blanca, and Niedertscheider (1990) worked on the precipitation and runoff data from the Callejon de Huaylas. Kaser *et al.* (1990) analysed the glaciological and climatological data and discussed possible reasons for the glacier fluctuations in the Cordillera Blanca in the twentieth century. Francou *et al.* (1995[2]) examined the data in connection with El Niño occurrences. Hastenrath & Ames (1995[1,2]) modelled the often measured Glaciar Yanamarey and Ames & Francou (1995) offered a detailed summary of the history of glacier research in the Cordillera Blanca. Thompson (1995) and Thompson *et al.* (1995) have gained new knowledge from ice cores in the Garganta on the Huascarán. Ames (1998) has compiled a paper about the modern retreat of individual glacier tongues and the consequent origin of lakes. The papers by Kaser *et al.* (1996[1]), Georges (1996), and Kaser & Georges (1997) are the basis for the following section about the modern glacier fluctuations in the Cordillera Blanca.

7.4 MODERN GLACIER FLUCTUATIONS

As with the Rwenzori, an attempt to reconstruct the changes of glacierization of the Cordillera Blanca since the maximum advance of the 'Little Ice Age' in the nineteenth century follows. Two areas were studied using aerial photographs:

- The **Huascarán–Chopicalqui** Massif in the middle of the Cordillera Blanca and
- the range that stretches from east to west from the **Nevado Santa Cruz** over the **Nevado Quitaraju** and the **Nevado Alpamayo** to the **Pucahirca Group** in the north of the Cordillera Blanca (Fig. 7.1.1).

Additionally, earlier reports, recent measurements of changes of the lengths of a few tongues, and studies of the Glaciar Yanamarey (Hastenrath & Ames, 1995[1,2]) are considered.

The different sets of aerial photographs and the working maps of the Dirección General de Reforma Agraria y Asentamiento Rurál at a scale of 1:25 000 (1972) offered the possibility to determine several successive glacier extents and thus to record the changes of the glacier extent in the twentieth century. The different ice extents were drawn onto the working maps by hand, using the aerial photographs, and then digitized. At this point, it is important to mention that a more detailed and thus more precise resolution was striven for than that given in the more broadly based Peruvian glacier inventory (Ames *et al.*, 1989).

The glaciers in the aerial pictures of 1948 and 1950 had clearly withdrawn behind the youngest moraines which are still not overgrown with vegetation and seem very distinct. These fresh moraines lie within the thick moraine walls which are covered with slightly developed and low vegetation on their inner side but are already overgrown with bushes on their outer side. Kinzl (1942) had already described this sequence of moraines and attributed the thick moraines to a maximum glacier extension over several centuries which retreated only towards the end of the nineteenth century. He found the most recent fresh moraines in many locations on top of these older moraines and attributed them to an advance that lasted only a short time and that had taken place shortly before the Alpenverein expeditions. The date of this advance can be deduced as a result of an observation on the eastern side of the Cordillera Blanca:

'According to corresponding reports of miners [of the Mina Atlante], the [Atlante] glacier advanced 12 to 15 m in 1923/24 and pushed a boulder in front of it. . . . After a period of slight changes, the glacier started to retreat again in 1927. In 1932, it was already 45 m from the mark of the advance.' Kinzl (1942).

By 1939, the glacier had retreated a further 60 m. At the time of the advance of the Glaciar Atlante, miners in other Peruvian mines also had problems with advancing glacier tongues (Oppenheim, 1945).

The area around the Mina Atlante had already been the centre of observation of glacier behaviour:

'I first reached a glacier which I call A; in 1909, it ended at the Socabon tunnel, that goes into the mountain below the miners'

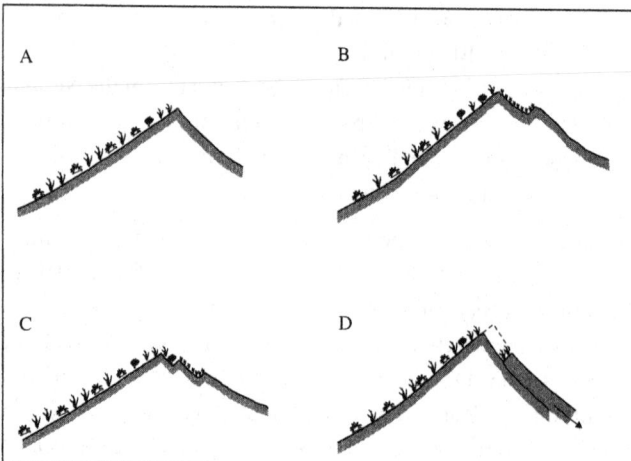

Fig. 7.4.1 The position of the youngest moraine (about 1930) in relation to the next older one (19th century) in the Cordillera Blanca, as they can be seen on the aerial photos of 1948/50.

houses. . . . This glacier descends steeply for about 50 m and has a crevasse from top to bottom. I was told that it had retreated during the previous decade, and that the firn limit had moved about 50 m upwards since 1895, within 14 years, according to the information given by Mr. José Olivieri . . . The ground moraine of the glacier can clearly be seen and consists of granite. A second glacier, B, is moving downwards from the big semicircle of névé . . . whose tongue is better developed than the one of glacier A, as well. . . . Today's glacier terminus is heavily thinned and apparently about to retreat. Additionally, it can be seen that the glacier had previously advanced about 600 m, as the ground moraine, consisting of black slate, lies bare. Next to it is a thick side moraine. . . . The area of the Mina Atlante is remarkable for this fact that real glacier tongues appear that are otherwise rare in the Cordillera. The reason for this lies in the shape of the source area which offers very favourable conditions for the development of glacier tongues.' (Sievers, 1914)

Sievers also observed the glacier retreat in the Cordillera Blanca in other areas during his journey in Peru and Equador in 1909:

'The Yanganuco Pass.
. . . At 4100 m, one reaches the start of the glaciated areas only recently exposed and, at the same time, the end of two frontal moraines of the glacier coming down from the pass. . . . the retreat of the firn limit [is] everywhere without a doubt, being about 150–200 m . . . The light grey zone left recently is followed downwards by a zone of older glaciation, with polished and cut rocks covering a length of 200–250 m, and finally the zone of the moraine covered with quenal [trees] and so on.' Sievers (1914).

The start of this retreat is difficult to determine. Broggi (1943) quotes information by A. Raimondi, according to which the

retreat in the Cordillera Blanca had started in 1862. It can therefore be assumed that the glaciers of the Cordillera Blanca, similar to the ones in the Alps, started their retreat from the maximum extent of the 'Little Ice Age', just after the middle of the century. Towards the end of the century it probably accelerated. In the middle of the 1920s there was a short, distinct advance which was already finished towards the end of the 1920s. A survey by Broggi (1943) of snowline altitudes (*linea de nieve*) on the Huascarán supports that image. However, these individual observations cannot be reliably considered as climatological means.

Fig. 7.4.1 shows four alternative schematic types of moraine sequences as they appear again and again in the aerial photographs of 1948/50.

A. During a younger advance, the glaciers reached the same extent as that of the 'Little Ice Age'.
B. The younger advance stayed within the 'Little Ice Age' moraines.
C. Like B, but a second younger advance failed to reach the earlier one.
D. The younger advance reached the 'Little Ice Age' extent, but the moraine subsided over long stretches after the retreat of the ice.

The A and B cases are by far the most frequent, whereas only on a few tongues were the 'Little Ice Age' extensions significantly larger. The thick frontal moraine of the 'Little Ice Age' was often broken through by ruptures of lakes, and thus the extension of the tongue is difficult to reconstruct. Referring to the total area, the extension of the 1920s was about as large as the one around the middle of the nineteenth century, and the difference lies within the error of the evaluation method. Therefore, an evaluation of the older extension was omitted.

The following retreat in the 1930s was drastic and can be derived from the aerial photographs. The following glacier extents could be drawn and evaluated for the **Huascarán–Chopicalqui Massif**:

• The extent at the end of the advance around the middle of the 1920s can be clearly delimited on the aerial pictures of 1948/50.
• The extent which is documented in the series of aerial photographs of 1948 and 1950 will henceforth be termed the extent of 1950.
• One extent can be derived from the aerial pictures of 1962/63.
• The extent of 1970 was derived from the infra-red pictures of SAN/NASA.

In Fig. 7.4.2, the glacier extents of the 1920s, 1950, and 1970 are depicted and the corresponding details of the areas are

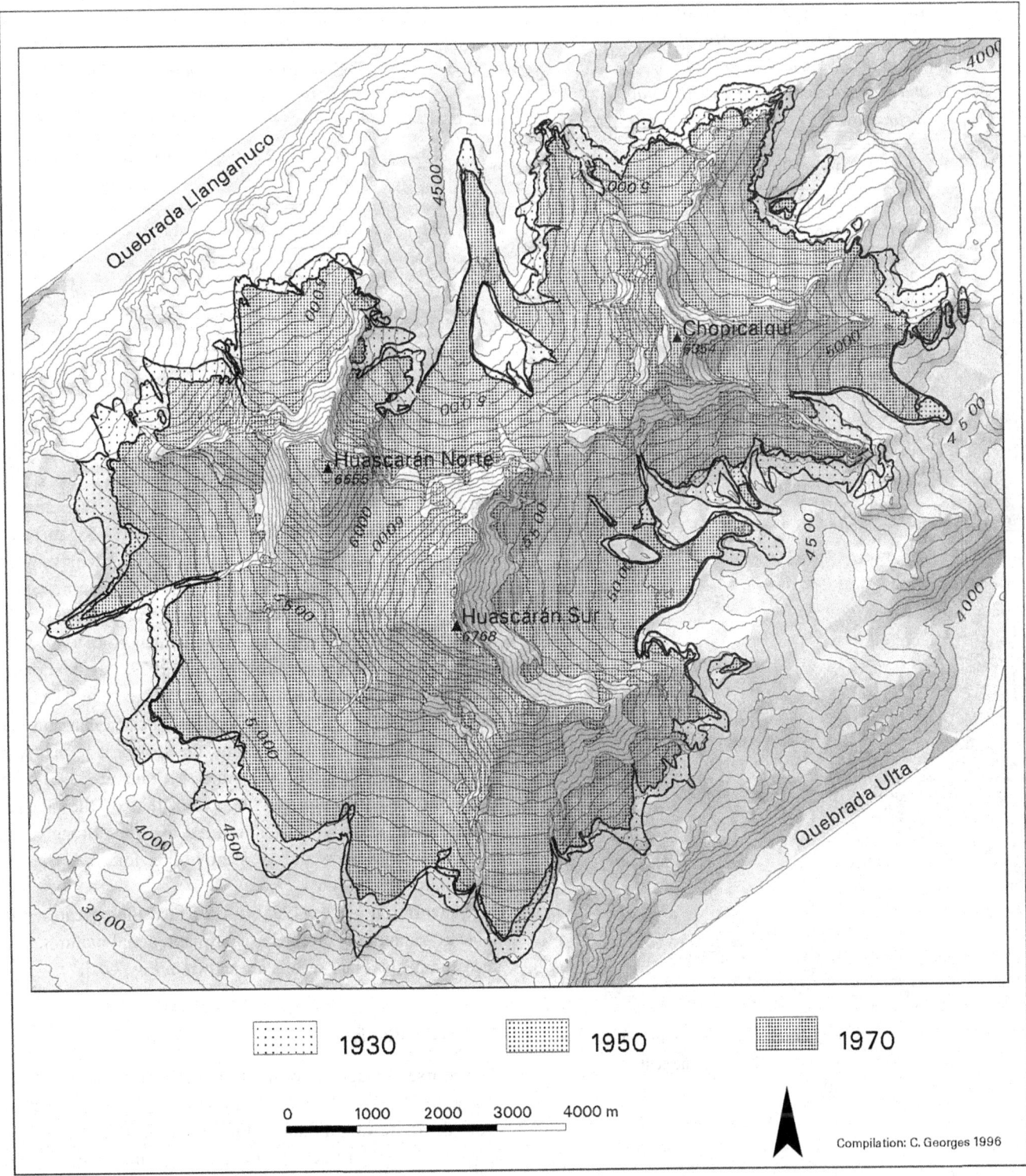

Fig. 7.4.2 The glacier areas on the Huascarán–Chopicalqui Massif.
Base map: Dirección Generál de Reforma Agraria y Asentamiento
Rurál (1972).

Table 7.4.1. *The glacier areas of the Huascarán–Chopicalqui Massif of the 1920s, 1950, and 1970 (in km²), and as a percentage of the 1920s total surface area*

Area	1920s	1950	1970
km²	71.0	59.3	58.2
%	100	84	82

Table 7.4.2. *The glacier areas of the Santa Cruz–Pucahirca group in the 1920s and in 1950 (in km²) and as a percentage of the 1920s total surface area*

Area	1920s	1950
km²	93.7	84.2
%	100	90

Fig. 7.4.3 The Santa Cruz–Quitoraju–Alpamayo–Pucahirca Range from the west. Nevado Alpamayo is shown in the centre. Aerial photo: J. Alean, 1980.

given in Table 7.4.1. The respective areas are depicted as a percentage of the 1920s total surface area in Fig. 7.4.5.

For the evaluation of the glacier extensions in the **Santa Cruz–Pucahirca Range** (Fig. 7.4.3), the photographs of 1962/63 and those of 1970 were of insufficient quality. The glacier extents of the 1920s and 1950 are depicted in Fig. 7.4.4 and Table 7.4.2 lists the corresponding area figures. The

respective areas are depicted as a percentage of the 1920s total surface area in Fig. 7.4.5.

Since the beginning of the 1970s, the changes of the lengths of the tongues have been measured on individual glaciers. The positions of the Gajap (1.20 km²; 1970), Yanamarey (1.35 km²), Uruashraju (2.15 km²), and Broggi (0.58 km²) glaciers can be seen in Fig. 7.1.1. The changes of the lengths were examined by Kaser *et al.* (1990) until 1987 and are represented in Fig. 5.3.6. Older tongue extents were reconstructed from maps. The tongues have retreated moderately since 1948 and, some had a slight advance in 1974 and 1979. A subsequent, greater retreat slowed down again from 1985 to 1986. The least retreat since 1948 was recorded on the Glaciar Gajap on the eastern side of the Cordillera, and the strongest retreat was noticed on the Glaciar Broggi. Ames & Francou (1995) have extended the curves up to 1993, except that of the Glaciar Gajap, and show a faster retreat for the Broggi and Uruashraju glaciers and a steady one for the Glaciar Yanamarey (Fig. 7.4.6).

Hastenrath & Ames (1995[1]) investigated the retreat of the Glaciar Yanamarey on the basis of maps, aerial pictures, and measurements. Their results are shown in Fig. 7.4.7.

7.4.1 A synthesis

The course of glacier cover in the Cordillera Blanca since the 'Little Ice Age' can be graphically summarized and sketched from the quite varied data and information (Fig. 7.4.8). The glaciers of the Cordillera Blanca have retreated since the end of the maximum of the 'Little Ice Age' in the middle of the nineteenth century, which cannot be dated exactly. Since about 1890, the retreat was accelerated. A temporary minimum ice extension, before an advance around the middle of the 1920s, is not datable. Between 1930 and 1950, a striking retreat took place that then became much weaker later on. In the 1970s, a few glaciers advanced only to retreat again faster.

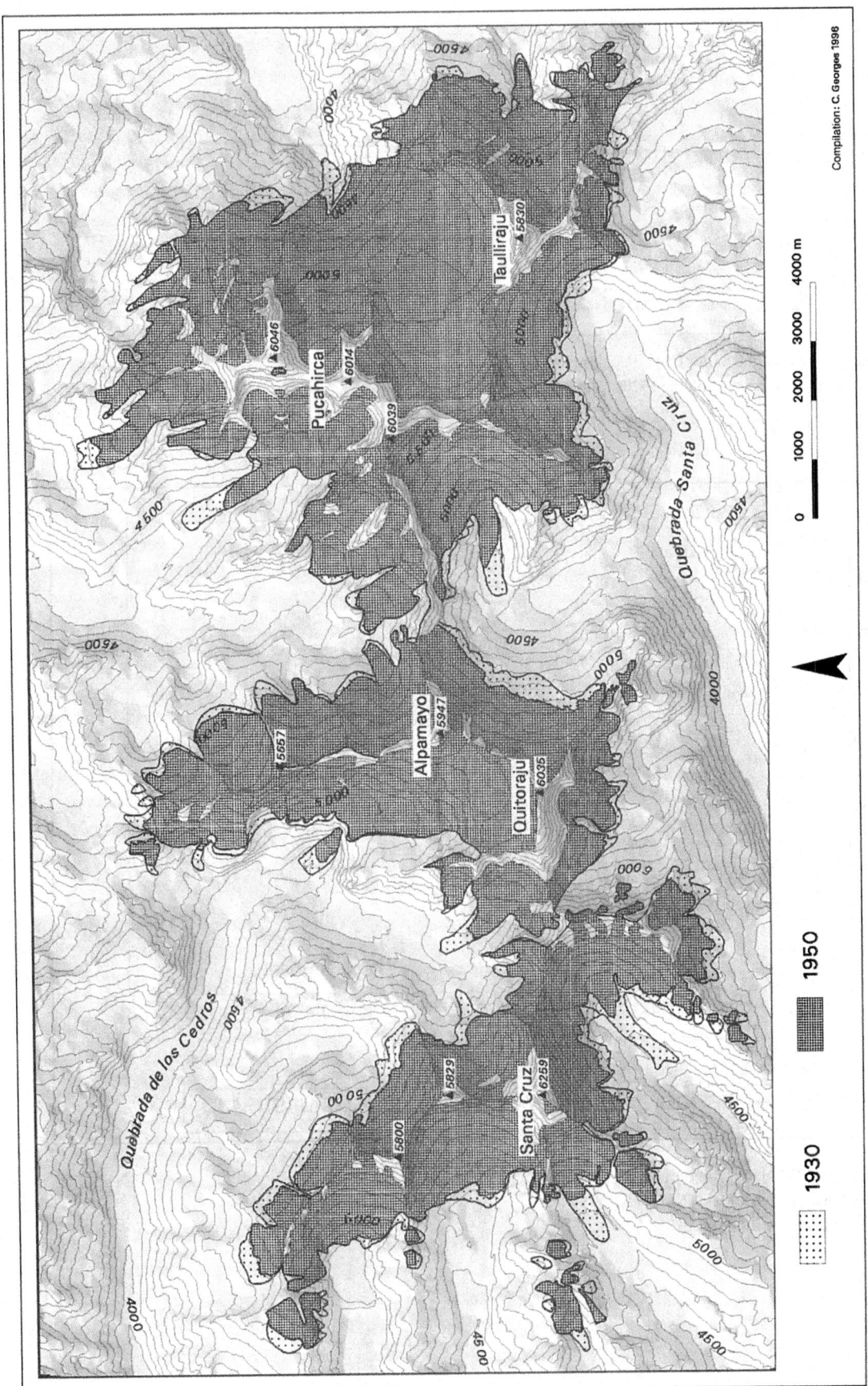

Fig. 7.4.4 The glacier areas in the Santa Cruz–Pucahirca Massif. Base map: Dirección Generál de Reforma Agraria y Asentamiento Rurál (1972).

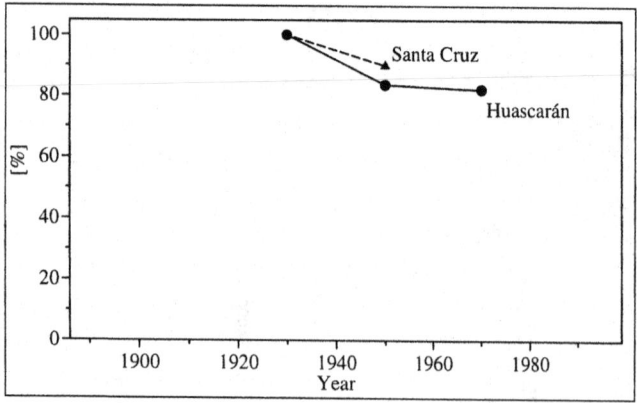

Fig. 7.4.5 The glacier areas of the Santa Cruz–Pucahirca group in the 1920s and in 1950 as a percentage of the 1920s total surface area.

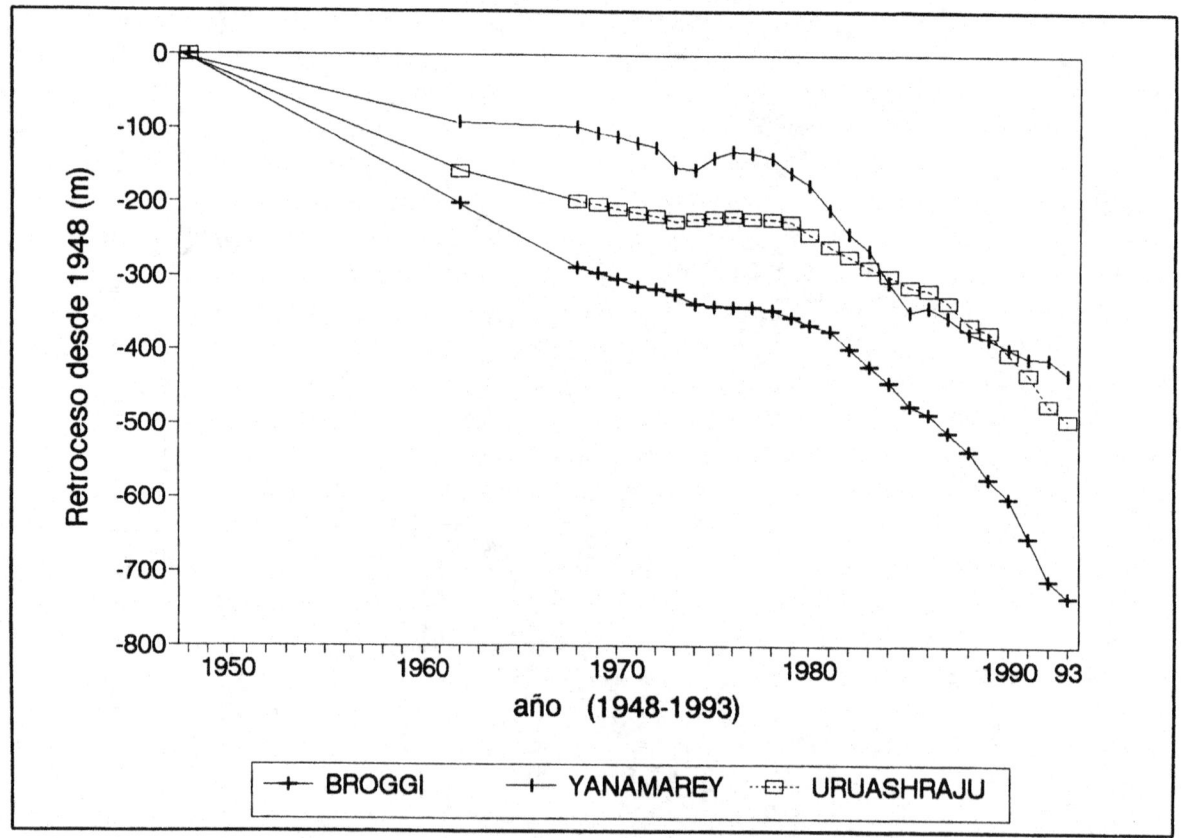

Fig. 7.4.6 The retreat of the glacier tongues on the Glaciar Broggi, Yanamarey, and Uruashraju since 1948 (Kaser *et al.*, 1990, as extended by Ames & Francou, 1995).

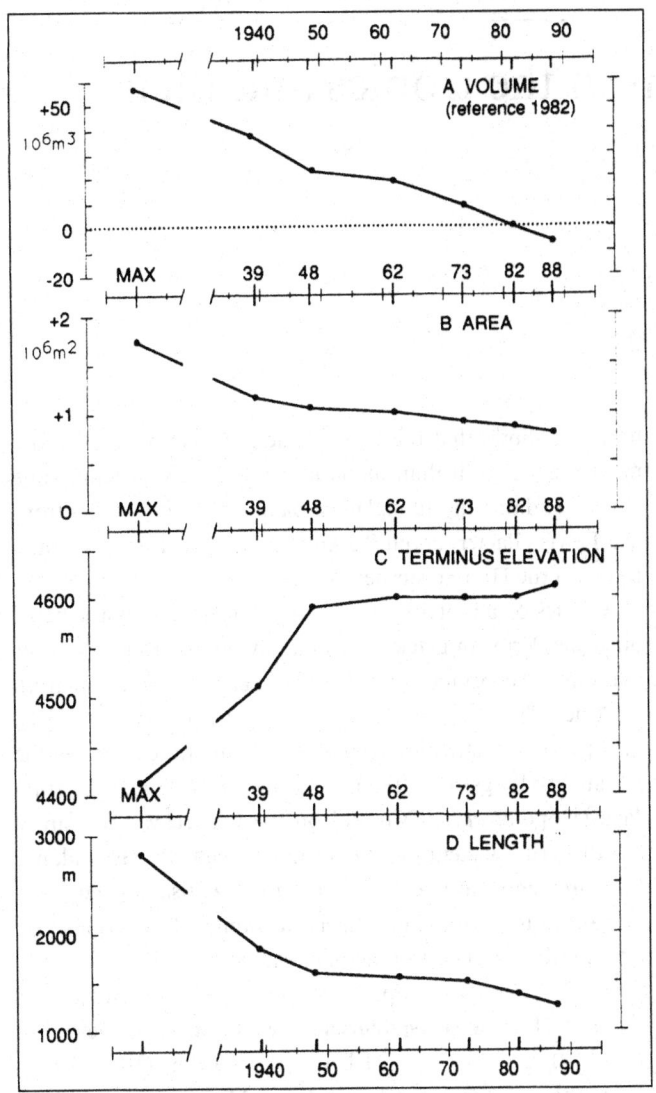

Fig. 7.4.7 The long-term changes of the Glaciar Yanamarey (Hastenrath & Ames, 1995[1]).

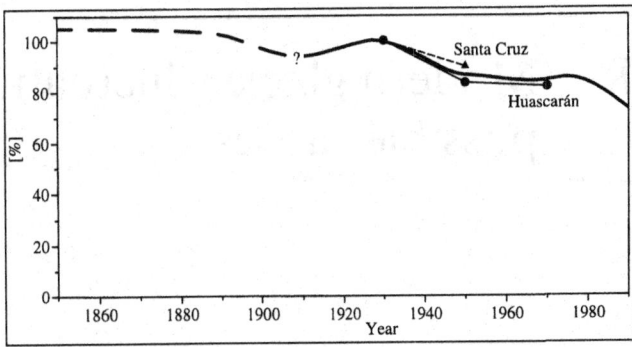

Fig. 7.4.8 Changes of the ice extent in the Cordillera Blanca since the 'Little Ice Age' as a percentage of the area of the highest level at the beginning of the twentieth century (thick line). The thin lines show the changes on the Huascarán–Chopicalqui Massif and in the Santa Cruz–Pucahirca group.

8 Modern glacier fluctuations in the tropics and their possible causes

In this chapter, a synthesis between the above presented investigations and those of other authors, analysing the fluctuation of modern glaciation in tropical high mountains, is attempted.

The glaciological data and observations available vary, such as changes of area and length, but rarely of volume or mass. Additionally, they often deal with different time periods and are often weakened by uncertain dating before the middle of the twentieth century. The attempt to calibrate modern glacier fluctuations in the tropics is, therefore, schematic and comparisons with the behaviour of glaciers in upper latitudes can only be attempted and are selective.

Under these circumstances, it is therefore difficult to investigate quantitatively the possible causes of glacier variations. Additionally, there are only few and incomplete climate records (section 4.2) from tropical high mountains which could assist in the verification of model calculations.

8.1 MODERN GLACIER FLUCTUATIONS IN THE TROPICS – A SYNTHESIS

Mass balance studies on tropical glaciers are extremely rare. There is only one complete long-term series on the Lewis Glacier (Hastenrath, 1984, 1991[2], among others). Table 8.1.1 compares the specific mass balances from the Lewis Glacier with those from the Hintereisferner.

Obvious similarities or systematic differences between the two mass balance series, which are mainly determined by the year to year climate, cannot be expected. The comparison, however, shows that the Lewis Glacier has clearly lost more mass per area unit than the Hintereisferner in the same time period, and that the annual mass balances on the smaller tropical Lewis Glacier (about 0.3 km[2]) vary more drastically than those on the Hintereisferner (about 9 km[2]).

A series of about 10 years of mass balance measurements on a few Peruvian glaciers was mainly limited to the tongues and was not continued recently (Kaser et al., 1990; Hastenrath & Ames, 1995[1,2]; section 5.1.4. (Fig. 5.1.6); section 5.5.1). Short-term investigations (mostly <1 year) were made on the Rwenzori (Bergstrøm, 1955; Whittow et al., 1963), in Irian Jaya (Hope et al., 1976), and on the Quelccaya Ice Cap in South Peru (Hastenrath, 1978). Only recently, mass balance measurements have been started on the Antisana (Ecuador), Artesonraju (Peru), Chacaltaya and Zongo (Bolivia) glaciers (Pouyaud et al., 1995; Francou et al., 1995[1,2]).

In the Austrian Alps the climatologically significant increase of the **mean equilibrium line altitude** since the end of the 'Little Ice Age' was $\Delta ELA_{1850-1969} = 94$ m (Gross, 1987). For the Huascarán–Chopicalqui Massif, a value of $\Delta ELA_{1920s-1970} = 95 \pm 5$ m was determined (Kaser et al., 1996[1], see also section 8.2). As the extent of the glaciers in the Cordillera Blanca was as large in the 1920s as at the end of the 'Little Ice Age', the values can definitely be compared. Both were determined through the accumulation area ratio method, while Allison & Kruss (1977) determined an elevation of the equilibrium line of $\Delta ELA_{1850-1970} = 96$ m for the Carstensz Glacier in Irian Jaya using a numerical model. Finsterwalder's (1987) cartometrically determined increase of

Table 8.1.1. *Annual specific mass balances between 1978/79 and 1990/91 of the Lewis Glacier (L.G.), Mount Kenya, and the Hintereisferner (HEF), their sum (Σ), their mean and their standard deviation (σ) (in kg m^{-2})*

	78–79	79–80	80–81	81–82	82–83	83–84	84–85	85–86	86–87	87–88	88–89	89–90	90–91	Σ	mean	σ
L.G.	−70	−1750	−1210	−370	−720	−900	−950	−680	−770	−2300	770	−1010	−810	−10770	−828	741
HEF	−219	−50	−173	−1240	−581	320	−574	−731	−717	−946	−636	−996	−1325	−7868	−605	476

Data sources: Hastenrath (1984, 1991[2]); Institute for Meteorology and Geophysics, University of Innsbruck.

Table 8.1.2. *Absolute (km²) and relative (%, italic) area changes in the Irian Jaya (I.J.), on Mount Kenya, on the Kibo, on the Rwenzori (Rw.), on the Huascarán–Chopicalqui Massif (H-C) and the Santa Cruz–Pucahirca range (SC-P) in the Cordillera Blanca, on selected glaciers of the Cordillera Real (Bolivia) and on the Pico Bolivar (Venezuela)*

Date	I.J.[a]	Kenya[b]	Kibo[c]	Rw.[d]	H-C[e]	SC-P[f]	CR[g]	PB[h]
c. 1850	**19.3** *(110)*		**20.0** *(110)*					
1899		**1.563** *100*						
1906				**6.509** *100*				
1910								**2.85** *100*
c. 1920					**71.0** *100*	**93.7** *100*	**28.58** *100*	
1936	13.0 *(74)*							
1947		0.874 *56*						
1950					59.3 *84*	84.2 *90*		
c. 1955				3.808 *58*				
1958			6.5 *(38)*					
1963		0.765 *49*					26.63 *93*	
1970			4.9 *(27)*	58.2 *82*				
1972	6.9 *(49)*							0.57 *20*
1975							25.03 *88*	
1987		0.495 *32*						
c. 1990	3.0 *(17)*			1.674 *26*				
1993		0.413 *26*						

Notes:
Bold: reference values; ( ): reference values were determined arbitrarily as ≠100%.
Sources: [a]Allison & Peterson (1976); Peterson & Peterson (1994); [b]Hastenrath *et al.* (1989); Hastenrath (1995); [c]Osmaston (1989[1]); [d]chapter 6; [e]Kaser *et al.* (1996[1]); chapter 7; [f]Georges (1996); Kaser & Georges (1997); chapter 7; [g]Finsterwalder (1987); [h]Schubert (1972); Jordan (1991).

the snow limit of 78 m in the Cordillera Real between 1922 and 1975 shows the same order of magnitude. The few values give rise to the supposition that the climatic causes of the glacier retreat in tropical regions had a similar effect as in the mid-latitudes.

There are also only few comparable values concerning **surface area** changes. Table 8.1.2 shows the changes of the glacier areas on the Rwenzori, on the Huascarán–Chopicalqui Massif and on the Santa Cruz–Pucahirca range compared with those in other tropical high mountains.

The relative changes make the different areas comparable with each other. The extent around 1920 is used as a reference point (bold in the table) for the relative changes, except on the Rwenzori, where the 1906 level is used as a reference. Also since the glacier extent around 1920 was smaller than the one at the end of the 'Little Ice Age' (about 1850), the earlier extent was assumed arbitrarily to be 110% for the Irian Jaya and Kibo (Kilimanjaro) and determined as a reference level. This makes these series more readily comparable with those of other mountain ranges.

In Table 8.1.3, area data of individual glaciers are compared with each other. As reference values for the relative area details, those values at the end of the 'Little Ice Age' (about 1850) are used as there are no values available for 1920. On the Rwenzori, the 1906 values are determined arbitrarily to be 90% of the previous maximum extent.

Due to the different time spans in which there are data available for different mountain ranges, a graphic depiction is omitted. Even though the series of values in the two tables cannot be compared with each other without problems, they still show that tropical glaciers have drastically retreated since the end of the 'Little Ice Age'. As expected, the small glaciers have had the highest relative area losses, whereas the big glacier areas in the Cordillera Blanca and in the Cordillera Real (Table 8.1.2) have had relatively small losses. The dependency of the retreat on the size of the glacier has already been discussed in chapter 5.3; Table 5.3.2 shows that the small glaciers in tropical regions have shrunk more than in the Ötztal Alps.

A further comparison as an overview is offered in Fig. 8.1.1. In this figure, the two reconstructed glacier fluctuations on the Rwenzori (Fig. 6.8.9) and in the Cordillera Blanca (Fig. 7.4.8) are compared with those of other areas.

The non-homogenous data basis makes it difficult to discuss comprehensively the changes in relation to the size of the total area or even the mean glacier size:

- Since the middle of the nineteenth century, the glacier fluctuations have been broadly similar in the Cordillera Blanca and on the Rwenzori. However, the glaciers of the

Table 8.1.3. *Absolute (km^2) and relative (%, italic) areas of individual tropical glaciers*

	LG[a]	Kr[b]	Gr[b]	Da[b]	Ty[b]	Ce[b]	Jo[b]	No[b]	El[c]	Sp[c]	Mo[c]	Sa[c]	Nw[d]	Me[d]	Ca[d]	Wo[d]	Vw[d]	Sw[d]	Ya[e]
c. 1850	**0.69** *100*												**9.1** *100*	**5.1** *100*	**2.5** *100*	**0.5** *100*	**0.6** *100*	**1.0** *100*	**1.75** *100*
1899	0.63 *92*	**0.08** *(90)*	**0.29** *(90)*	**0.09** *(90)*	**0.16** *(90)*	**0.01** *(90)*	**0.06** *(90)*	**0.05** *(90)*											
1906									**0.32** *(90)*	**0.45** *(90)*	**0.17** *(90)*	**0.12** *(90)*							
1913																0.3 *60*	0.2 *33*	0.4 *40*	
c. 1920	0.62 *90*																		
1934	0.50 *72*																		
1936													8.3 *91*	2.8 *55*	1.25 *50*	0.2 *40*	0.15 *25*	0.3 *30*	
1939																			1.2 *69*
1942													5.5 *60*	2.6 *51*	1.1 *44*				
1947		0.04 *(45)*	0.09 *(29)*	0.04 *(40)*	0.10 *(55)*	0.05 *(44)*	0.03 *(49)*	0.04 *(70)*											
1948																			1.08 *62*
1950										0.37 *(74)*									
1951											0.09 *(48)*								
1952									0.20 *(57)*										
1955												0.08 *(57)*							
1956									0.18 *(51)*										
1958	0.38 *55*																		
1960									0.17 *(49)*	0.37 *(73)*									
1962														2.1 *41*	0.95 *38*				1.02 *58*
1963	0.37 *53*	0.04 *(45)*	0.09 *(28)*	0.04 *(42)*	0.09 *(49)*	0.04 *(36)*	0.02 *(36)*	0.03 *(52)*											
1970													3.6 *40*	1.9 *37*	0.9 *36*	0.17 *34*	0.14 *23*	0.2 *20*	
1973																			0.95 *54*
1974	0.32 *47*																		
1978	0.31 *45*																		
1982																			0.87 *50*

Table 8.1.3. (*cont.*)

	LG[a]	Kr[b]	Gr[b]	Da[b]	Ty[b]	Ce[b]	Jo[b]	No[b]	El[c]	Sp[c]	Mo[c]	Sa[c]	Nw[d]	Me[d]	Ca[d]	Wo[d]	Vw[d]	Sw[d]	Ya[e]
1983	**0.28**																		
	41																		
1987	**0.24**	**0.02**	**0.04**	**0.03**	**0.08**	**0.02**	**0.01**	**0.01**											
	35	*(24)*	*(14)*	*(26)*	*(43)*	*(22)*	*(14)*	*(20)*											
1988																			**0.8**
																			46
1990	**0.23**									**0.21**									
	34									*(42)*									
1991									**0.11**		**0.02**	**0.04**							
									(32)		*(9)*	*(32)*							
1993	**0.20**	**0.02**	**0.03**	**0.02**	**0.06**	**0.02**	**0.01**	**0.01**		**0.20**									
	(29)	*(22)*	*(11)*	*(23)*	*(35)*	*(16)*	*(9)*	*(16)*		*(39)*									

Notes:
Lewis (LG), Krapf (Kr), Gregory (Gr), Darwin (Da), Tyndall (Ty), Cesar (Ce), Joseph (Jo), and Northey (No) on Mount Kenya;
Elena (El), Speke (Sp), Moore (Mo), and Savoia (Sa) on the Rwenzori; North Wall Firn (Nw), Meren (Me), Carstensz (Ca),
Wollaston (Wo), Van de Water (Vw), and Southwall Hanging (Sw) in the Irian Jaya; Yanamarey (Ya) in the Cordillera Blanca.
Bold: reference values ( ): reference values were determined arbitrarily as ≠100%.
Sources: [a]Patzelt *et al.* (1984); Hastenrath (1991[2]); [b]Hastenrath (1995); [c]section 6.7; [d]Allison & Peterson (1976); [e]Hastenrath & Ames
(1995[1]).

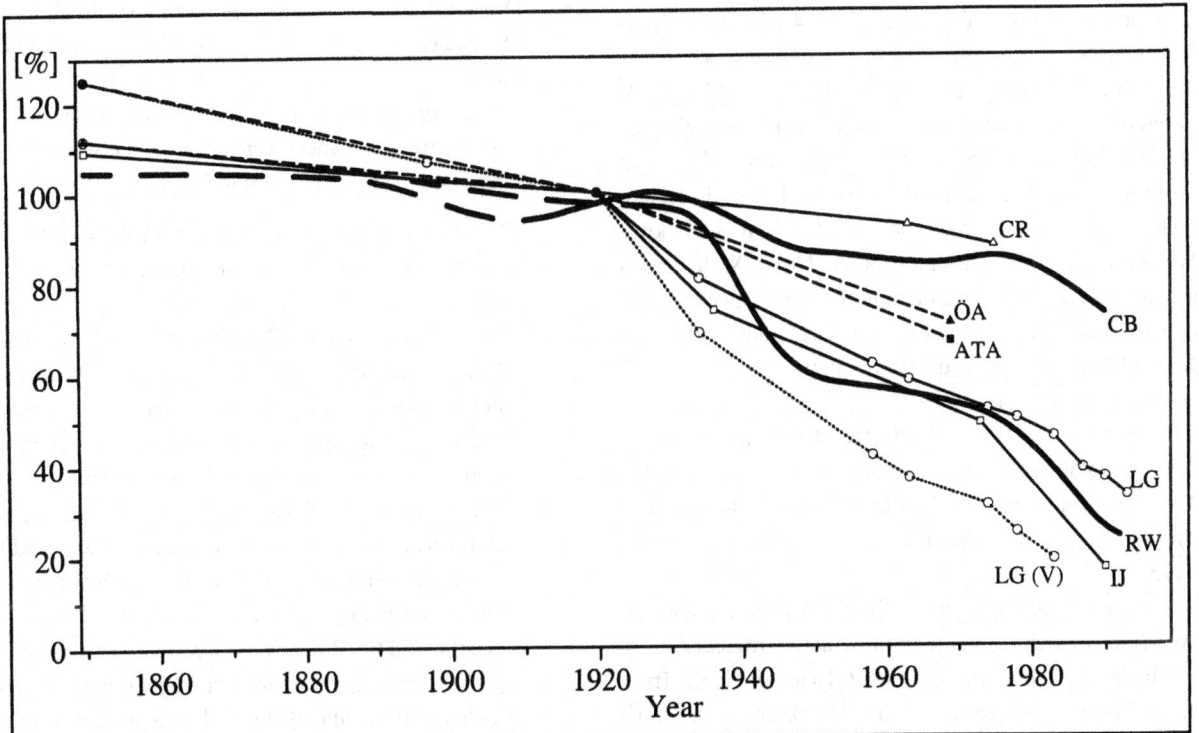

Fig. 8.1.1 The area losses of tropical and Austrian glaciers as a
percentage of the area around 1920. RW: Rwenzori (Fig. 6.8.9); CB:
Cordillera Blanca (Fig. 7.4.8); CR: Cordillera Real (Finsterwalder,
1987); LG: Lewis Glacier (Patzelt *et al.*, 1984; Hastenrath, 1995); IJ:
Irian Jaya (Allison & Peterson, 1976; Peterson & Peterson, 1994);
ÖA: Ötztal Alps; ATA: Austrian Alps (Gross, 1987); LG(V) relative
volume losses of the Lewis Glacier (Patzelt *et al.*, 1984).

Rwenzori (1906: about 6.5 km^2) have clearly experienced a greater relative area loss than those of the Cordillera Blanca [1920s: about 850 km^2 (in Fig. 7.4.8 projected from the extent in 1970 = 723 km^2 according to Ames et al., 1989); the curve, however, is mainly based on the values of the Huascarán–Chopicalqui (71 km^2) and Santa Cruz–Pucahirca areas (93.7 km^2)]. In this case, the mean values of the glaciers can be compared with each other to some extent. If the area in the Cordillera Blanca is divided by the number of glaciers in 1970 (722; Ames et al., 1989), then a mean glacier size of slightly over 1 km^2 is determined; if the glacier area on the Rwenzori is divided by the number of glaciers, which Hastenrath (1984) states as comprising the 'recent' glaciation (44), then the mean glacier has a size of about 0.15 km^2. If the relative losses between about 1920 and about 1970 are applied to the mean glacier areas in the curves of Figs. 6.8.9 and 7.4.8, then the mean Rwenzori glacier has lost 0.07 km^2 and the mean Cordillera Blanca glacier 0.17 km^2.

- The relative area changes of the Lewis Glacier (1920: 0.62 km^2) and those of the total area in the Irian Jaya are similar in size to those on the Rwenzori. The Lewis Glacier lost an area of 0.3 km^2 by 1974. The mean Irian Jaya glacier around 1920 can be calculated to be about 2.5 km^2, according to details given by Allison & Peterson (1976). According to Table 8.1.3, the glaciers around 1970 had about 49% of the area they had around 1920. According to these findings, the mean Irian Jaya glacier would have lost about 1.27 km^2. However, it is worth noting that the North Wall Firn is a long plateau glacier and that the other glaciers also cover small altitude ranges. This gives the mean Irian Jaya glacier a totally different shape compared with those of other areas.

- The relative area loss of selected, large (mean area = 3.8 km^2) glaciers in the Cordillera Real (Finsterwalder, 1987) is similar to that in the Cordillera Blanca. The average glacier lost about 0.46 km^2 between 1922 and 1975.

- The relative area losses in the Ötztal Alps and the entire Austrian Alps are higher than those in the tropical Andes, but lower than those on the Lewis Glacier and in the Irian Jaya. The mean Austrian glacier (1920: about 0.9 km^2) lost about 0.3 km^2 between 1920 and 1969.

- The relative volume loss of the Lewis Glacier was clearly higher than the relative area loss. Similar assumptions can be made for other similarly small glaciers. However, whether this can also be applied to the large glaciers in the tropical Andes remains uncertain.

For a meaningful investigation of the similarities or differences of the retreat values of different tropical glacier areas, the amount and homogeneity of the information is insufficient. Yet, a temporal trend which is widely uniform can be seen:

- A distinct maximum extent, which ended around the middle of the nineteenth century, was recognized on all tropical glaciers that were investigated. Even though it could not be exactly dated in any case, this maximum extent is attributed to the 'Little Ice Age' (Kinzl, 1942; Schubert, 1972; Hope et al., 1976; Patzelt et al., 1984; Osmaston, 1989[1,2]). Hastenrath (1995) proposes that the East African glaciers had only first started their retreat towards the end of the nineteenth century, whereas, in South America and in Irian Jaya, this might have already been the case around the middle of the century.

- The advance during the 1920s observed in the Cordillera Blanca is synchronous with that which was observed in the Alps. It is reported that the Lewis Glacier advanced around 1890 and several times around 1920 (Patzelt et al., 1984). Immediately afterwards, it suffered its highest volume and area losses.

- During this last period (from about 1930 to about 1950), the glaciers also retreated in other tropical high mountain ranges.

- After about 1950, the rate of the retreat was smaller and, in some areas, advances were observed between the beginning of the 1960s and the middle of the 1970s:

  - on the Rwenzori around 1960 (Temple, 1968)
  - in the Cordillera Blanca in the middle of the 1970s (Kaser et al., 1990)
  - on the Lewis Glacier, Patzelt et al. (1984) determined a basic increase of the thickness in the upper area between 1958, 1963, and 1974. It seems that these small, most recent advances can be compared to the processes in the mid-latitudes. Especially in the Alps (but also in Scandinavia, North America and in the Himalayas), a changed trend of mass increases and glacier advances was observed in the course of the 1960s (Patzelt et al., 1984; Letreguilly, 1984; Patzelt, 1985).

- Since the middle of the 1970s, a retreat has occurred in all regions which shows as similarly high area losses, like the ones in the 1930s and 1940s, and it seems that tropical glaciers are clearly out of balance with the current climatic conditions.

8.2 POSSIBLE CAUSES OF MODERN GLACIER FLUCTUATIONS IN THE TROPICS

8.2.1 Basic considerations

If vertical mass balance profiles, surface movements, and different volumes of a glacier are known, then the changed mass balances can be calculated under simplified assumptions which, if they continue over a certain period of time, lead to the observed changes (Budd & Jensen, 1975; Paterson, 1994). Under certain circumstances, changes in the position of a mean equilibrium line (ΔELA), of which the climatic boundary conditions are known (section 5.2), can be reconstructed (section 5.4). If individual key variables of the mass and heat balance (accumulation, air temperature, global radiation, etc.) are treated as if their disturbances were each individually responsible for the required changes in the mass balance, then their amounts could be calculated. It can only be estimated which combination of these individual key variables has really caused the observed change of the glacier. Additional information (long-term climate records, fluctuation of lake levels, runoffs, etc.) from that region is needed for more evidence. However, the uncertainty about the chronology of glacier fluctuations compared with climatic disturbances still remains (section 5.3). Detailed analyses can only be made if long-term records of measurements of individual climatic elements and glacier fluctuations are available, which rarely is the case in the tropics.

The most common approach to the understanding of glacier–climate interaction considers the changes of air temperature (δT_a) and precipitation (δP). Several series of measurements of these climate variables exist, but from stations which are far away from the glaciers. The temperature is used to represent the whole energy balance and consequently the ablation. Accumulation is represented by precipitation. This approach can be useful in areas with accumulation and ablation periods which can clearly be delimited from each other. For example, in the mid-latitudes an increase of precipitation can compensate for the effects of an increase in summer temperature. An increase in temperature will hardly influence the accumulation. Yet, in tropical regions, an increase of the temperature changes the precipitation into rain above a certain altitude and withdraws it from the accumulation. Under these circumstances, when looking for the causes of glacier fluctuations, the individual components of energy and mass balance and their interaction must be investigated differently in the tropics than in the mid-latitudes.

Fig. 8.1.1 shows that the long-term glacier fluctuations in the tropics have a global character. Consequently, the causes must be searched for in general changes of the climatic conditions. However, corresponding climatic changes can, even if of a large-scale character, continental or even global, have spatially differentiated consequences differing in degree.

It can be assumed that the air temperature changes homogeneously in the tropics over large areas (section 1.2). This is true for the horizontal extension as well as for possible changes of the vertical temperature gradient. The temperature of a tropical air mass will not change considerably during advection over vast distances. However, this varies with the amount of humidity in the atmosphere. The air humidity not only has larger spatial differences than the temperature field, but a change in the water vapour can also have locally different consequent effects (windward/lee, convective activities, precipitation, cloud cover, global radiation).

If changes in the air temperature and the air humidity are used as primary climatic key variables on a large scale when attempting to explain long-term glacier retreat, several combinations of different consequent effects for the mass balance of glaciers become apparent.

The symbols for the variables are as follows:

T_a	air temperature
e_a	vapour pressure in the atmosphere (absolute humidity)
T_s	temperature of the glacier surface
$RH_a(T_a, e_a)$	relative humidity in the atmosphere as a function of T_a and e_a
$clc.(RH_a)$	cloud cover as a function of RH_a
$G(clc.)$	global radiation as a function of cloud cover
$E(T_s)$	outgoing longwave radiation from the earth surface as a function of T_s
$A(T_a, e_a)$	incoming longwave radiation as a function of T_a and e_a
$Q_S(T_s, T_a)$	sensible heat flux as a function of T_s and T_a
$Q_L(T_s, e_a)$	latent heat flux assuming predominance of sublimation into the atmosphere as a function of T_s and e_a
$P(RH_a)$	precipitation as a function of relative humidity
$c(P)$	accumulation as a function of precipitation
δ	the respective disturbances or changes of the variables
$\downarrow$	the heat flux is directed towards the surface
$\uparrow$	the heat flux is directed away from the surface

1. $\delta T_a > 0$; $\delta e_a = 0 \rightarrow$

   $\delta Q_S \downarrow > 0$
   $\delta A(T_a, e_a) > 0$
   $\delta RH_a < 0 \leftarrow \delta clc. < 0 \leftarrow \delta G \downarrow > 0$
   δablation > 0

$\delta c(\delta T_a) < 0$

$\delta RH_a < 0 \leftarrow \delta P < 0 \leftarrow \delta c(\delta RH_a) < 0$

$\delta accumulation < 0$

While the direct influences of the temperature disturbances on the heat budget (δQ_S, $\delta A(T_a, e_a)$) and accumulation ($\delta c(\delta T_a)$) are spatially homogeneous, those that are caused by a change of the relative humidity (δG; $\delta c(\delta RH_a)$) can be spatially different. The clarity of the outcome of the spatially variable effects depends on the initial values of the relative humidity. In very wet (inner tropics) and in very dry areas (subtropics), the consequences of a change of the relative humidity are small and, therefore, the total effect on the glaciers is spatially uniform. All consequent effects will prompt a glacier retreat in this case.

2. $\delta T_a = 0$; $\delta e_a < 0 \rightarrow$

$\delta Q_S \downarrow = 0$

$\delta(A) \downarrow < 0$

$\delta RH_a < 0 \leftarrow \delta clc. < 0 \leftarrow \delta G \downarrow > 0$

$\delta Q_L \uparrow > 0$

$\delta ablation \lesseqgtr 0$

$\delta RH_a < 0 \leftarrow \delta P < 0 \leftarrow \delta c(\delta RH_a) < 0$

$\delta accumulation < 0$

None of the changes is spatially homogeneous. The changes of global radiation and accumulation contribute to the retreat of the glaciers. The change of the longwave radiation balance and the change of the latent heat flux have a reversed effect. The extent and possibly also the signs of the resulting effect on the mass balance will depend on the initial values, and especially on the relative humidity. In wet mountains, the changes of global radiation and accumulation will predominate, a decrease of the air humidity will, therefore, lead to a glacier retreat. In dry areas, where the latent heat flux, which takes energy from the surface into the atmosphere due to evaporation and sublimation, plays a dominant role in the energy and mass balance, it can be assumed that a decrease of the air humidity is an advantage for the glacier. The effects are not definite in this case and depend on the extent of the disturbance as well as on the initial conditions.

3. $\delta T_a = 0$; $\delta e_a > 0 \rightarrow$

$\delta Q_S = 0$

$\delta(A - E) \downarrow > 0$

$\delta RH_a > 0 \leftarrow \delta clc. > 0 \leftarrow \delta G < 0$

$\delta Q_L \uparrow < 0$

$\delta ablation \lesseqgtr 0$

$\delta RH_a > 0 \leftarrow \delta P > 0 \leftarrow \delta c(\delta RH_a) > 0$

$\delta accumulation > 0$

In this case, too, none of the changes is spatially homogeneous and processes which increase glacier retreat are again opposed to others that are an advantage for the glacier. Which processes predominate depends on the extent of the disturbance and on the initial conditions. When δe_a is sufficient to clearly increase precipitation, increasing accumulation will predominate. If this is not the case, an increase of the air humidity can possibly lead to net mass losses. This can be particularly expected in very dry areas where the sublimation is very important. In humid areas, however, an increase in humidity is more likely to be favourable for glaciers.

4. $\delta T_a > 0$; $\delta e_a < 0 \rightarrow$

$\delta Q_S \downarrow > 0$

$\delta RH_a(\delta T_a) < 0 \leftarrow \delta clc.(\delta T_a) < 0 \leftarrow \delta G \downarrow(\delta T_a) > 0$

$\delta RH_a(\delta e_a) < 0 \leftarrow \delta clc.(\delta e_a) < 0 \leftarrow \delta G \downarrow(\delta e_a) > 0$

$\delta(A - E) \downarrow < 0$

$\delta Q_L \uparrow > 0$

$\delta ablation \lesseqgtr 0$

$\delta c(\delta T_a) < 0$

$\delta RH_a(\delta T_a) < 0 \leftarrow \delta P(\delta T_a) < 0 \leftarrow \delta c(\delta RH_a(\delta T_a)) < 0$

$\delta RH_a(\delta e_a) < 0 \leftarrow \delta P(\delta e_a) < 0 \leftarrow \delta c(\delta RH_a(\delta e_a)) < 0$

$\delta accumulation < 0$

This combination has spatially homogeneous components as well as several components that have different effects. The effects that protect the glacier, such as higher evaporation and stronger outgoing longwave radiation (E) at night, are faced with a number of negative effects on the mass balance which, in tropical regions in any case, might be clearly predominant. Again, this could be different in very dry areas.

5. $\delta T_a > 0$; $\delta e_a > 0 \rightarrow$

$\delta Q_S \downarrow > 0$

$(\delta RH_a(\delta T_a) < 0 \leftarrow \delta clc. < 0 \leftarrow \delta G \downarrow > 0)$

$(\delta RH_a(\delta e_a) > 0 \leftarrow \delta clc. > 0 \leftarrow \delta G \downarrow < 0)$

$\delta(A - E) \downarrow > 0$

$\delta Q_L \uparrow < 0$

$\delta ablation (\lessgtr) 0$

$\delta c(\delta T_a) < 0$

$(\delta RH_a(\delta T_a) < 0 \leftarrow \delta P < 0 \leftarrow \delta c(\delta RH_a) < 0)$

$(\delta RH_a(\delta e_a) > 0 \leftarrow \delta P > 0 \leftarrow \delta c(\delta RH_a) > 0)$

$\delta accumulation \gtrless 0$

This combination has spatially homogeneous as well as non-homogeneous components. The changes in ablation and accumulation can have both signs. The effects of the two input key variables on the relative humidity have opposite effects. If the temperature and the absolute humidity increase at a similar rate, their effects on the relative humidity and consequently on cloud cover, global radiation, and precipitation will be minimal. Consequently, these key variables which, on the one hand, increase ablation and, on the other hand, decrease accumulation, combine to create clearly unfavourable conditions for the glacier.

As global radiation plays a predominant part in the heat balance of the glacier surface, a slight change of albedo can have considerable consequences. This was not included in the considerations mentioned above, as its causes and consequences are more complex than those that have already been discussed. The following reasons could lead to a change in the albedo in a long-term mean:

- A change in the frequency of precipitation (but with possible equal sums of precipitation) and, consequently, in the duration of the new snow cover in the accumulation area. This could be particularly effective in dry areas (subtropics) and during dry seasons (outer tropics).
- A glacier surface, characterized by strong evaporation/sublimation, has less liquid water in the pores, and consequently has a higher reflectivity than one that is characterized by predominant melting or even condensation. When visiting the Glaciar Zongo in the Cordillera Real in Bolivia on June 17, 1995, the snow surface in the area of the mean equilibrium line was characterized by *penitentes* up to 30 cm high, which took many days or even weeks to develop and gave the surface a high reflectivity. At 11:30 a.m., the surface temperature was $-1.3\,°C$, and the saturation vapour pressure was consequently 5.48 hPa. The vapour pressure in the air was 3.34 hPa. Thus, the vapour pressure gradient was favourable for pronounced sublimation and for the further growth of the *penitentes*. At noon, individual cumuli and fractocumuli had gathered from the east and an audible melting process had started in the depressions of the *penitentes*. In small snow lysimeters, sublimation had taken place since 11:30 a.m. At 1 p.m. the vapour pressure in the air was 5.40 hPa over a melted surface with a 6.1 hPa saturation vapour pressure. The gradient had clearly decreased, and sublimation had also diminished. At 2 p.m., the measurement site was covered in fog, the atmospheric vapour pressure had risen to 5.74 hPa with a still melting surface and the lysimeter had gained in weight. The *penitentes* had collapsed and melted to a large extent, and the surface was saturated with water and seemed much

darker than before. A period of humid air from the eastern lowland suddenly led to perfect melting conditions and melted the *penitentes* that had formed over a long period of time within a very short duration. An increasing frequency of such events could considerably change the mass balance of outer tropical and subtropical glaciers.

- Increased or decreased influence of dust. Thompson *et al.* (1995) infer clearly stronger influences of dust from the Amazon basin during the Ice Age from their drillings in the Garganta (Cordillera Blanca). However, it cannot be assumed for modern glacier fluctuations that such an effect extends over large areas or even globally. It is not known to what extent a broadening of the desert especially to the south of the Sahara influences the glaciers of East Africa.
- Allison & Kruss (1977) also consider possible accumulation of cryo-vegetation and consequently changes in the albedo.

None of the possible reasons for a change in the albedo can, however, be assumed to have had a global impact within modern climate fluctuations and, consequently, be able to explain the general changes of glaciers. Kruss (1983) has also referred to this issue.

8.2.2 Possible causes on individual tropical high mountains

Glacier fluctuations can be numerically modelled for the Lewis Glacier on Mount Kenya (Hastenrath & Kruss, 1982; Kruss, 1983, 1984; Hastenrath, 1989), the two Irian Jaya glaciers, Meren and Carstensz (Allison & Kruss, 1977), as well as the Yanamarey (Hastenrath & Ames, 1995[1,2]) and the Uruashraju glaciers in the Cordillera Blanca (Ames & Hastenrath, 1996), of which surface movements and mass balance values, in addition to different extents and tongue positions, were known. The above named authors attributed the calculated mass changes to different climatic causes:

- Allison & Kruss (1977) have modelled different glacier extensions characterized by the moraines of the Meren and Carstensz glaciers in the Irian Jaya. They calculate a necessary rise of the equilibrium line as being 96 m for the glacier retreat from the maximum extent between around 1850 and around 1970. Based on temperature analyses by Callendar (1961) and Mitchell (1961), the retreat was attributed to a rise in temperature, although it was considered that other climatic variables could have an influence as well. A temperature increase of $0.6\,°C$ per century could explain the retreat if no other variables are taken into consideration.

- There are different assumptions about the beginning of the retreat of the Lewis Glacier on Mount Kenya since its modern maximum extent. While Kruss (1983), and Hastenrath (various publications), assume that the glacier only first retreated towards the end of the nineteenth and at the beginning of the twentieth century, Patzelt *et al.* (1984) assume that a retreat had already taken place earlier, and that several advances and possibly even an increase in thickness of the glacier took place between 1890 and 1920. In a more recent paper, Hastenrath (1995) writes about the stable condition of the glaciers on Mount Kenya at the beginning of the twentieth century. In any case, both groups attribute the retreat to a decrease of precipitation and base their assumptions on data from Kraus (1955) and Berger (then unpublished; 1989). Kruss (1983) assumes a decrease of precipitation by 15% or 160 ± 70 mm a^{-1}. Kruss & Hastenrath (1987) attribute the retreat of the glaciers on Mount Kenya not only to decreasing precipitation, but also to a simultaneous decrease in the cloud cover and the consequently increasing global radiation. According to Kruss (1983), a rise in temperature of $0.35 \pm 0.2\,°C$ during the first half of the twentieth century can be included as a further cause (section 8.2.1, case 4).

- In a 1992 paper, Hastenrath & Kruss investigate the causes of the dramatic retreat of the Mount Kenya glaciers between 1963 and 1987 and, through the method of exclusion, arrived at the assumption that, in contrast to the first strong retreat at the beginning of the century, in this instance a combination of increases of the air temperature and absolute humidity was the cause (section 8.2.1, case 5). Precipitation deficits could not be determined from the measurements taken at the foot of Mount Kenya and the uniformity of the recession ratio excludes changes in the area of the shortwave radiation budget. A combination of a $0.2\,°C$ rise in temperature and a 0.3 g kg^{-1} increase of the specific humidity would explain the observed mass loss.

- Hastenrath (1995) gives a summary of the investigations of possible causes of the retreat of the glaciers on Mount Kenya.

- Hastenrath & Ames (1995[2]) and Ames & Hastenrath (1996) only give a sensitivity analysis for the Glaciar Yanamarey and the Glaciar Uruashraju of those climate variables which could stop the present mass loss of the glaciers.

In several other publications, different authors have formulated the following possible causes of the observed modern glacier fluctuations on the basis of observations on the investigated mountains or in their surroundings.

Temple (1968) assumes that temporary high precipitation led to a limited advance on the **Rwenzori** at the beginning of the 1960s. Kaser & Noggler (1991) confirmed this assumption. A comparison of photographs in Figs. 6.8.10 to 6.8.19, as well as the drawing in Fig. 6.7.5, shows that particularly the peaks, ridges, and crests on the Rwenzori have become exposed over vast stretches. This must have been caused by decreased precipitation. The different retreat of the two lobes of the tongue of the Speke Glacier, especially in the second half of the twentieth century (section 6.7.2), indicates that global radiation has increased. This can be attributed to decreased relative air humidity and a consequently decreased cloud cover (Kaser & Noggler, 1991; Kaser, 1993). The determination of the shifting of the mean equilibrium line for the glaciers of the Rwenzori from the area analyses in chapter 6 is not advisable due to the following reasons (see section 5.4):

- None of the analysed glacier extents can be considered steady-state or quasi-steady-state.
- The retreat was so pronounced that the glaciers have considerably changed their relief.
- Therefore, the 1955 contour lines cannot be transferred to other extents within the context of inaccuracies that could be omitted.

A speculation may be permitted for the glacierization of **Kibo** (Kilimanjaro). The ice remnants in and around the crater have shapes arising from dramatic, distinct, and selective ablation (Fig. 8.2.1), as if they were developing on glaciers in the most extreme outer tropics, the subtropics (e.g. in northern Chile) or in extremely continental mountains in the mid-latitudes (e.g. in the Tianshan), in very dry seasonal or year-round conditions.

These typical ablation and surface shapes (*penitentes*) have often been described and discussed (e.g. Workman, 1914; Klute, 1915; Troll, 1942, 1949; Lliboutry, 1954; Kotlyakov & Lebeveda, 1974). Hofmann (1963, 1965) and Kraus (1966, 1972) theoretically proved selective ablation to be a cause. The glaciers on Kibo show this type of ablation shapes, although the mountain is situated at a geographical latitude of 3° S in the middle of the tropics. This can be explained by the altitude and the shape of the mountain. On the one hand, Kibo might be mostly above the trade wind inversion due to its height of nearly 6000 m and, on the other hand, the convex shape of the peak prevents essential convective precipitation. Convective cells cannot, as for example on Mount Kenya, meet and arrange themselves annularly around the peak (Fig. 8.2.2).

The hypothesis states that, besides the shape of the peak, the ice cover particularly at the edge of the crater also stops the

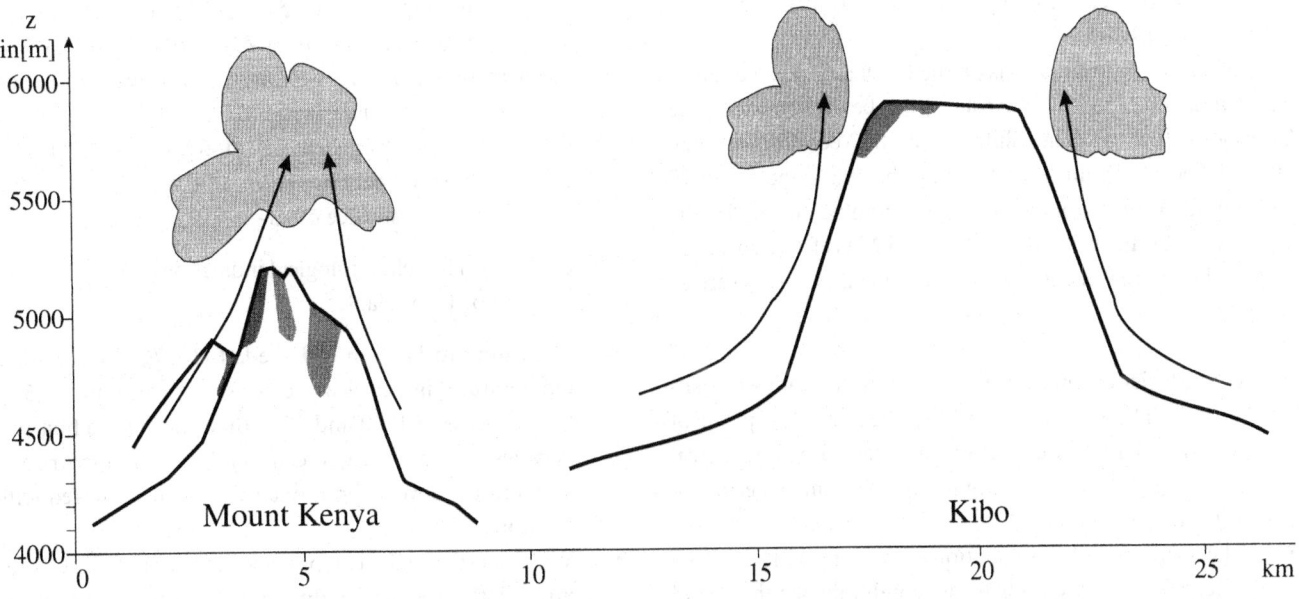

Fig. 8.2.1 Selective ablation forms on the glacier remnants in the crater of Kibo. Photo: G. Kaser, March 1988.

Fig. 8.2.2 Different convective patterns on East African mountains.

development of convective cloud cover that causes precipitation. In contrast, it can be expected that the convective cells above Kibo could develop better and produce more precipitation if there was no ice cover. If ice cover distinctly develops, the convection would again lift off and the glacier would 'starve'. This cycle would be superimposed upon the large-scale long-term climate fluctuations. This hypothesis lacks, however, model calculations as well as respective analyses of glacier fluctuations.

In the **Cordillera Blanca**, Kaser *et al.* (1990) have shown that the equilibrium lines of the Glaciar Yanamarey and the Glaciar Uruashraju show a connection with the maximum, the minimum and the mean temperature of the Querococha station at the end of the rainy season between 1977/78 and 1986/87, but not with the measured precipitation quantities. During the dry season, ablation values on the tongues were mainly lower than during the rainy season. This indicates that the glaciers are sensitive to temperature fluctuations during the rainy season and to fluctuations of the air humidity (see above) during the dry season. Francou *et al.* (1995²) also show that temperature fluctuations in Querococha (Fig. 7.1.1) display a high negative correlation with the El Niño – Southern Oscillation Index (S.O.I.), while precipitation show no clear connection with it (Fig. 8.2.3).

The question of whether an accumulation of negative S.O.I. values is responsible for accelerated recessional periods has not yet been investigated. The increased frequency of positive indices in the 1970s could be connected with the stagnation and the advances of individual tongues in the Cordillera Blanca (Fig. 5.3.6). They were preceded by slightly higher precipitation and were accompanied by lower temperatures (Kaser *et al.*, 1990).

Contrary to the evaluations on the Rwenzori, mean equilibrium lines, and especially their shifts, can be determined in the Cordillera Blanca. The shifting of the equilibrium line between the maximum extent during the 1920s and the 1970 level could be determined from hypsographic curves for the glaciers of the Huascarán–Chopicalqui Massif (Kaser *et al.*, 1996¹). The requirements (section 5.4) for this curve were as follows:

- Both glacier extents correspond to quasi-steady-state situations. The moraines from the 1920s mark a point of return and 1970 was the end of a period of low retreat values, just before the beginning of a small period of advance.
- The area–altitude distribution of the total glacier area (Fig. 8.2.4) corresponds to the conditions for the use of an area ratio model. This condition would not be met for single glaciers.

- The relief of the glaciers was not fundamentally changed by the retreat (Fig. 8.2.4).

Both area–altitude distributions were evaluated on the basis of the 1970 contour. This leads to a certain inaccuracy, above all in the area of the tongues. This inaccuracy can be ignored in the area of possible equilibrium lines due to the relatively very slight changes.

The hypsographic curves of both situations are almost linear in the entire middle portion and mainly parallel. As a consequence, the shifting of the equilibrium line $\Delta ELA_{1920s-1970}$ is constant if one remains within the realistic accumulation area ratios (AAR) (Fig. 8.2.5). The shift is 95 ± 5 m. The absolute altitudes of the two equilibrium lines, however, depend on the accumulation area ratio. The equilibrium lines $ELA_{1920s} = 4900-5150$ m a.s.l. and $ELA_{1970} = 5000-5250$ m a.s.l. lie within the margin of the AAR $= 0.75 - 0.5$ (Fig. 8.2.5).

In order to calculate the possible climatic causes of the rise of the equilibrium line, in accordance with Fig. 5.2.1, it was assumed that the mean equilibrium line in the Cordillera Blanca lies clearly above the 0°C level. With this in mind, equations 5.2.4 to 5.2.9 can be used for the calculation of the possible climatic disturbances. The vertical gradients of accumulation and air temperature were assumed to be $\partial c/\partial z = 0.30$ kg m^{-2} m^{-1} (according to Fig. 5.1.8) and $\partial T_a/\partial z = -0.0065$ K m^{-1} (according to Prohaska, 1973 and Osmaston, 1989² – also see Fig. 6.2.5). The following would be necessary under these boundary conditions for the 95 m rise of the equilibrium line:

- a rise in air temperature of $\delta T_a = 0.50$ K or
- a decrease of accumulation of $\delta c = -1041$ kg m^{-2} a^{-1} or
- an increase of the energy from the shortwave radiation balance of $\delta[G(1 - r)] = 0.952$ MJ m^{-2} d^{-1} or
- a decrease of the latent heat flux of $\delta Q_L = -1.080$ MJ m^{-2} d^{-1}, which would correspond to a decrease of the sublimation from the surface of $\delta s = -139$ kg m^{-2} a^{-1}.

There are only a few climatological indications for the possible combination of the variables:

- According to Hansen & Lebedeff (1987), the mean air temperature in tropical regions was around 0.15 °C higher between 1920 and 1970 than during the first two decades of the twentieth century. This observed rise in temperature covers about one third of the required temperature.
- With an estimated mean precipitation sum of about 1200 mm for the present conditions in the area of the mean equilibrium line, cited from Niedertscheider (1990), it would have had to have been about double the amount

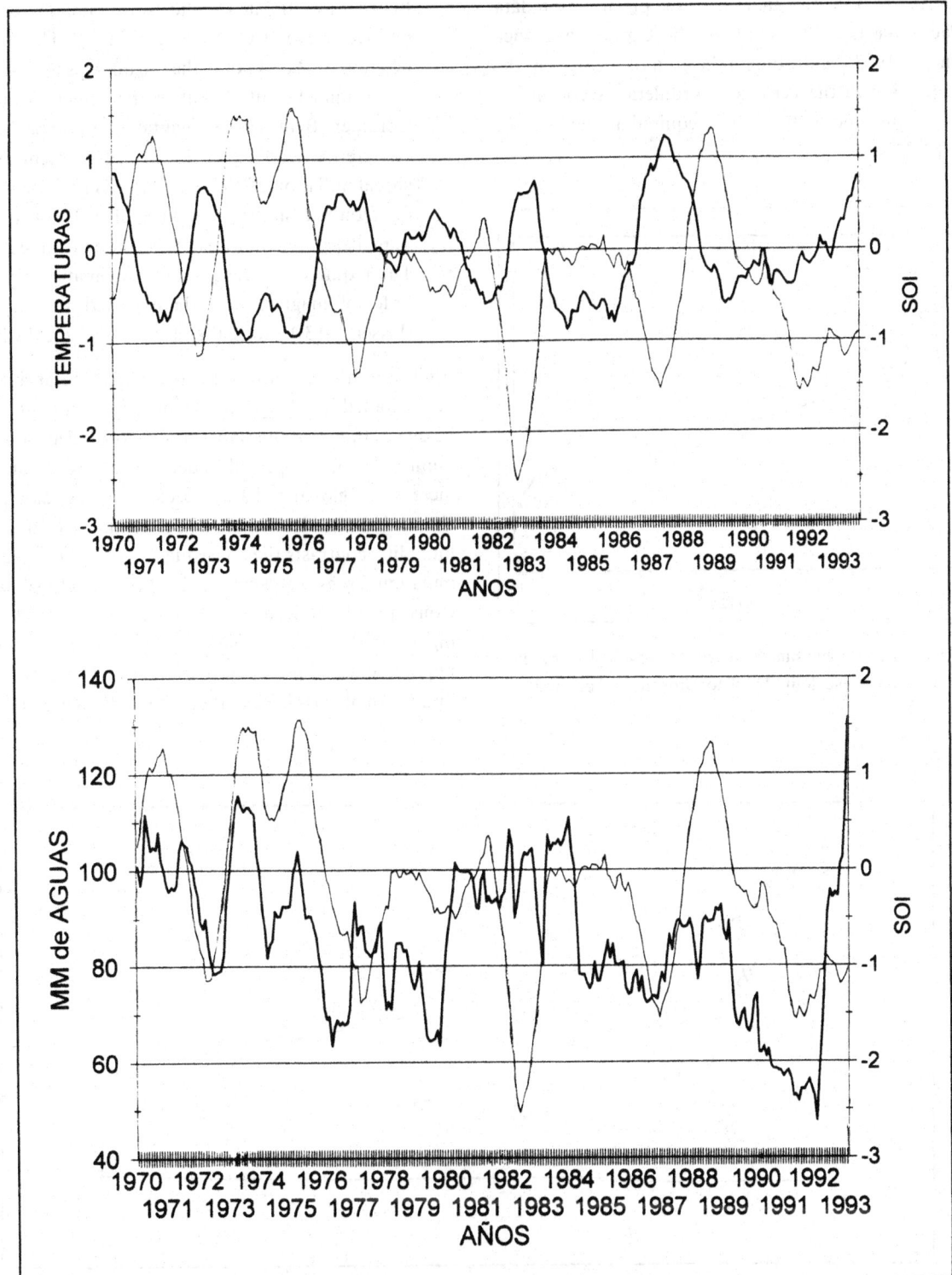

Fig. 8.2.3 Querococha station (3955 m a.s.l.). Cordillera Blanca:
Top: standardized temperature deviations (bold) and Southern
Oscillation Index (thin). Bottom: 12 months running mean
precipitation sums (bold) and Southern Oscillation Index (thin).
(From: Francou *et al.*, 1995[2].)

before 1920. Francou's analysis of the precipitation data in Pachachaca, to the south of the Cordillera Blanca (Francou, 1983), is the only indication of changes in the accumulation in the Peruvian Cordillera. According to his data, only about 10% of the required amount can be explained.

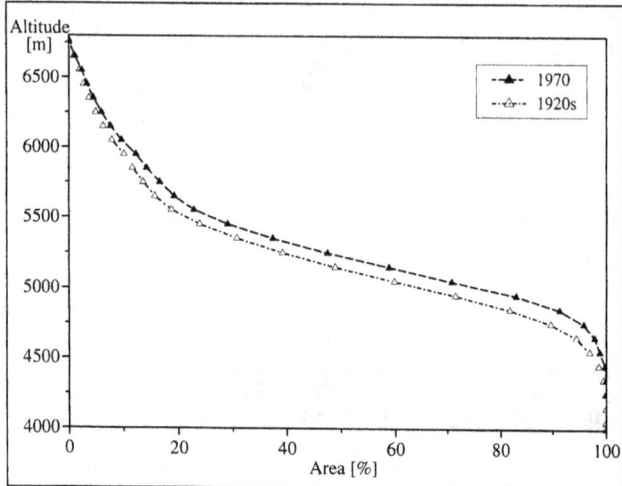

Fig. 8.2.4 The hypsographic sum curves over the total glacier area of the Huascarán–Chopicalqui Massif towards the end of the 1920s and in 1970.

- In order to estimate the shortwave radiation balance, the mid-latitudes must be again considered. The fact that the position of the sun over the Ötztal Alps (47° N) in July is at a similar height to that of the annual mean over the Cordillera Blanca (9° S) facilitates the comparison. At the Hintereisferner, Wagner (1979) measured a mean effective global radiation of $G(1 - r) = 15$ MJ m^{-2} d^{-1} in July. The full required amount would be only 6.3% of that value so a smaller proportion than this could be quite realistic.
- The required decrease of the sublimation is within the order of magnitude of the expected absolute amounts (Kaser, 1983[2]) and is, therefore, unrealistically high.

While around one third of the rise of the equilibrium line can be attributed to changes in the air temperature, the rest must be due to those variables that result from a decrease of the air humidity: that is, especially a decrease of precipitation and an increase of global radiation because of less cloud cover. A change in the latent heat flux, as a result of a change in the air humidity, contrasts the consequences of the changes in accumulation, sensible heat and effective global radiation. Consequently, it would have to be compensated with even higher amounts of the last variables in order to reach the required effects on the equilibrium line. For simplification, a discussion of possible changes in the sublimation is omitted at

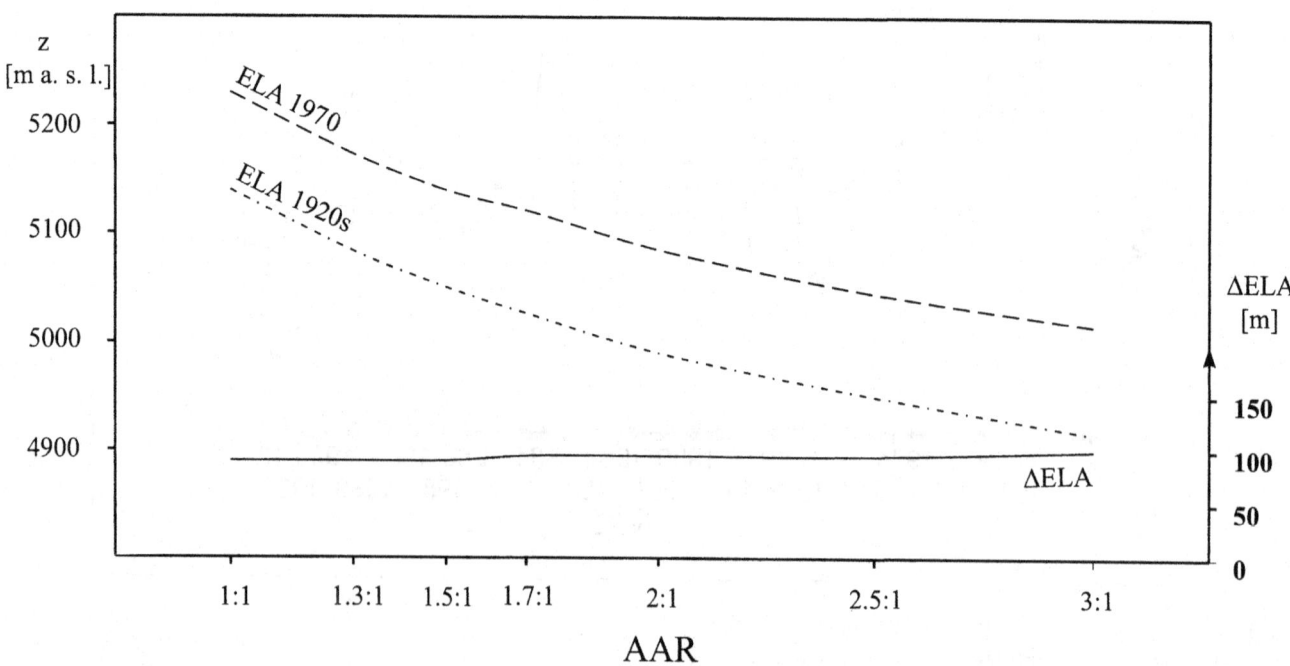

Fig. 8.2.5 The altitudes of the equilibrium lines (ELA) in the 1920s and in 1970 as well as their difference $\Delta ELA_{1920s-1970}$ for different accumulation area ratios AAR.

this point and in further considerations. For a more exact analysis on the basis of detailed climate data (possibly measured in the future), however, such a discussion cannot be neglected, particularly in the partly dry outer tropics (e.g. Cordillera Blanca).

The fact that changes in the air humidity and their consequences have spatial differences and, therefore, cause spatially different glacier behaviour, offers opportunity for their study. Despite the favourable position as a barrier in the generally eastern current, the Huascarán–Chopicalqui Massif was not very useful for such an analysis. No glacier areas could be combined which, on the one hand, fulfill the conditions for using an accumulation area ratio method and, on the other hand, have a common exposure character.

These requirements were fulfilled in the **Santa Cruz–Quitaraju–Alpamayo–Pucahirca** chain in the north of the Cordillera Blanca (Figs. 7.1.1, 7.4.3). The partial massif with the furthest west–east extension in the Cordillera Blanca is characterized by three ridges towards the north: one ridge starting from the Nevado Santa Cruz (6259 m a.s.l.) and those starting from the Nevados Alpamayo (5947 m a.s.l.) and the Pucahirca (6039 m a.s.l.). Georges carried out the corresponding analyses in his doctorate thesis (Georges, 1996). A summary of the results is given by Kaser & Georges (1997). However, it was not possible to reconstruct the changes in that group up to 1970 (section 7.4). The glacier retreat that could be discussed took place between the short extension of the 1920s and the period around 1950, which was at the beginning of a phase of slight changes (Figs. 7.4.5 and 7.4.8). Both conditions can again be considered quasi-steady-state.

Fig. 8.2.6 shows the subdivision of the area of investigation into six different portions, of which the glacier areas were again, as with the investigations on the Huascarán–Chopicalqui, analysed as a whole and not as a mean of single glaciers. In Table 8.2.1, the areas for the situations in the 1920s and in 1950 are compiled.

In Fig. 8.2.7, the hypsographic curves for the individual areas and the two glacier extents, as well as an accumulation area ratio of AAR = 0.75, are depicted.

The hypsographic curves are again linear in their middle portion but not as parallel as they were on the Huascarán. According to the considerations in section 5.4, an accumulation area ratio of AAR = 0.75 was chosen as the most likely to occur. The resulting altitudes of the equilibrium lines and their rises are compiled and graphically depicted for the three north–south chains and for the six glacier areas in Table 8.2.2 and in Fig. 8.2.8.

On the one hand, it is shown that the equilibrium lines on the eastern side of the Cordillera Blanca, as assumed by Kinzl (1942), lie clearly lower than on their western flank. When the areas of investigation are looked at individually, this trend, however, is not regular from east to west. The east–west trend covering the whole mountain range is eclipsed by an opposite trend which moves the equilibrium lines higher on the eastern sides of the Santa Cruz and the Alpamayo than on their western side. The lowest equilibrium lines are in the area of the Pucahirca east, the highest in the area of the Santa Cruz east. This pattern is the same for both situations.

The causes of the distribution pattern of the altitudes of the equilibrium lines can be explained as follows:

A. The humid air is almost exclusively advected from the east. In the Cordillera, this causes both the humidity and the convective processes to decrease towards the western side (Fig. 8.2.9). Consequently, both precipitation and accumulation decrease from east to west and, finally, are the main reason that the equilibrium lines lie clearly lower in the east of the mountains than in the west. Measurements are still required from within the Cordillera Blanca and from its eastern side, but differences in the precipitation over vast areas indicate this particular precipitation gradient. The difference in the mean annual precipitation between Tingo Maria, at the foot of the eastern slope of the Andes ($\varphi = 09°08'$ S; $\lambda = 75°57'$ W; 665 m a.s.l.), and Chiclayo, close to the Pacific coast ($\varphi = 06°47'$ S; $\lambda = 79°50'$ W), is more than 3000 mm (Johnson, 1976).

B. This system is superimposed on a diurnally convective circulation pattern (Fig. 8.2.10), which Troll (1942), Troll & Wien (1949) and afterwards Hastenrath (1991[1]) describe as the typical process on tropical mountains. The fact that the convective cloud cover develops in the course of the day leads to an asymmetry in the radiation balance and, hence, in the ablation conditions on tropical mountains. The eastern flanks receive more direct sun radiation than the western flanks. As a result, the equilibrium lines are higher in the east.

The influence of (A) decreases from east to west and allows the daily circulation pattern and its consequences to have more and more influence. While (A) explains the general rise of the equilibrium lines from east to west as well as the one on the most eastern Pucahirca Ridge, (B) is responsible for the opposite patterns on the Alpamayo and the Santa Cruz.

In this case, climatological data are also lacking, as with the investigations of the Huascarán–Chopicalqui Massif, in order to verify the hypothesis of the overlapping circulation patterns. It is again possible to estimate the individual climate variables with the help of the equations from section 5.2, not to calculate the temporal differences of the altitude of the equilibrium line in this case, but rather the spatial pattern. The ELA is assumed again to lie above the 0 °C level. It is also

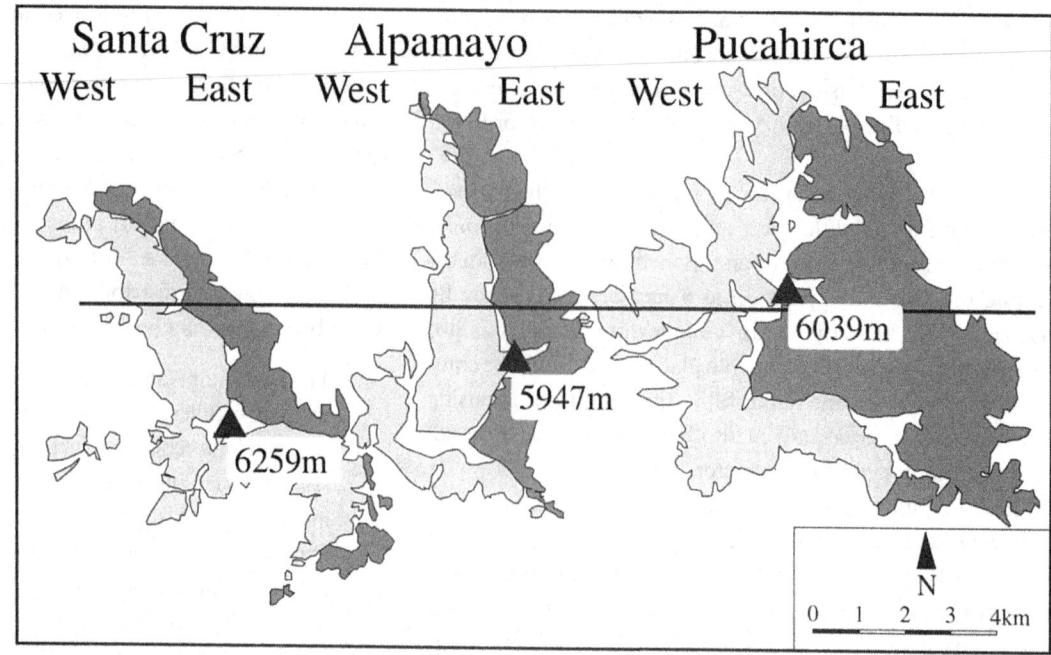

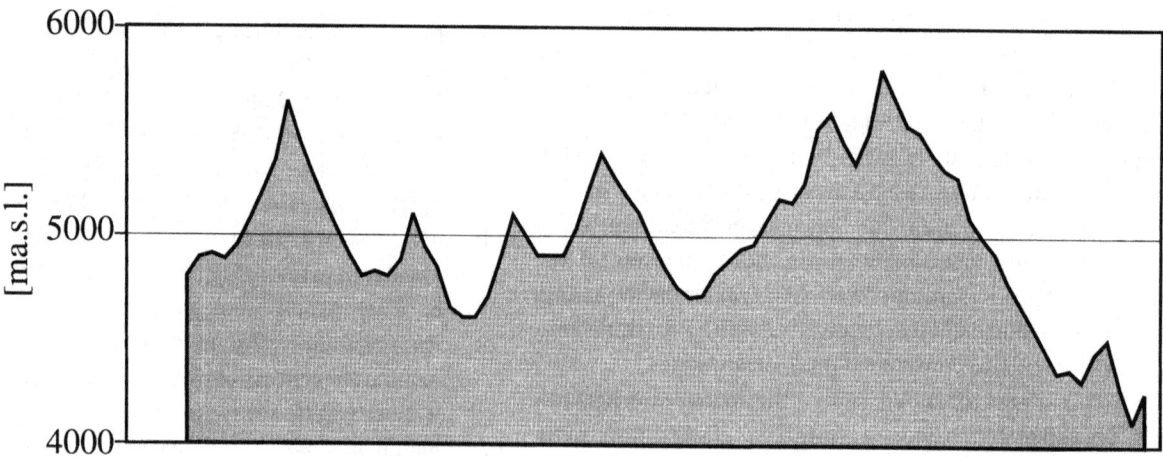

Fig. 8.2.6 The six glacier areas of the Santa Cruz–Pucahirca chain (above) and their altitudes along the marked east–west section line (below). Darkly shaded areas of ice face the east, lightly shaded areas face west. There is a strong E–W gradient in humidity which decreases to the west.

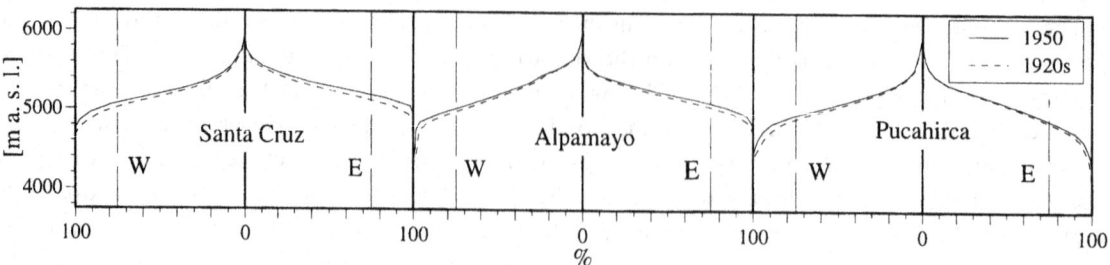

Fig. 8.2.7 The hypsographic curves of the six glacier areas for the situations in the 1920s and in 1950. The vertical broken lines mark an accumulation area ratio of AAR = 0.75.

Table 8.2.1. *Extents of the six investigated glacier areas in the Santa Cruz–Pucahirca chain in the 1920s and in 1950 (km²)*

Date	Santa Cruz West	Santa Cruz East	Alpamayo West	Alpamayo East	Pucahirca West	Pucahirca East	Total
1920s	16.0	8.7	12.1	11.0	20.3	25.8	93.9
1950	13.5	6.8	11.2	9.5	18.4	24.8	84.2

Table 8.2.2. *The altitudes of the equilibrium lines (ELA) in the 1920s and in 1950 and their elevation $\Delta ELA_{1920s-1950}$ for the three north–south chains and the six investigated glacier areas*

	Nevada Santa Cruz	Nevada Alpamayo	Nevada Pucahirca
ELA_{1920s}	5051	5015	4894
ELA_{1950}	5109	5056	4928
$\Delta ELA_{1920s-1950}$	58	41	34

	West	East	West	East	West	East
ELA_{1920s}	5014	5118	4981	5053	4911	4882
ELA_{1950}	5068	5189	5019	5099	4958	4905
$\Delta ELA_{1920s-1950}$	54	71	38	46	47	23

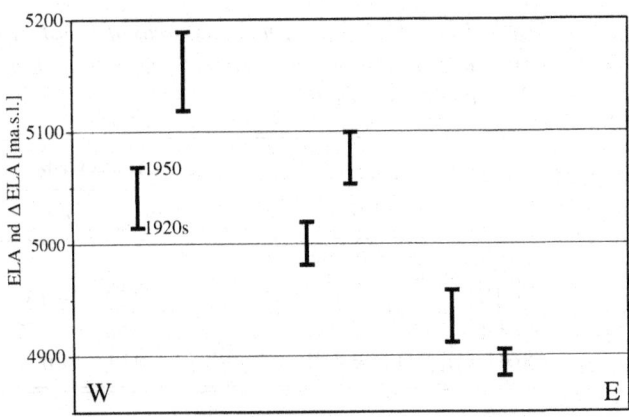

Fig. 8.2.8 The equilibrium lines (ELA) in the 1920s and in 1950 and their differences $\Delta ELA_{1920s-1950}$ in the individual areas of investigation from west to east.

cover, are possible. In relation to the amount of $G(1-r)$ = 15 MJ m⁻² d⁻¹ on the Hintereisferner (see above), the maximum amount required is 18%.

The combination of the two differences, depending on humidity along the Santa Cruz–Pucahirca chain, can very well make the different positions of the equilibrium lines plausible.

The pattern observed in the spatial distribution of the equilibrium lines also repeats itself in the rise of the equilibrium line $\Delta ELA_{1920s-1950}$. The largest rises are where the equilibrium lines are at the highest altitudes and vice versa (Fig. 8.2.8.). A rise in temperature, under the assumption that it occurs spatially uniformly, can explain at best the basic sum of $\Delta ELA_{1920s-1950}$, which was observed in the entire area. This is the shortest distance the equilibrium line rose on the eastern side of the Pucahirca Ridge. The amounts of possible climatic disturbances, calculated with the help of the equations in section 5.2, are compiled in Table 8.2.4.

The given increase of the mean temperature in the tropical atmosphere after about 1920 of $\delta T_a = 0.15$ K, according to Hansen & Lebedeff (1987), corresponds in good approximation to the value that is required as the basic value of $\Delta ELA_{1920s-1950} = 23$ m ($\delta T_a = 0.12$ K) (Table 8.2.4). Hence, 33%

presumed that the temperature is not the cause of the spatial differences (see above and chapter 4). The results for the 1920s are compiled in Table 8.2.3. The ratios around 1950 are similar.

- Different accumulation amounts are definitely part of the cause. The required difference of an accumulation of almost 3000 kg m⁻² a⁻¹, however, is a value that is unattainable, considering that about 1200 mm of precipitation fall on the western flanks of the Cordillera Blanca (Niedertscheider, 1990 and Fig. 5.1.8) and 3072 mm at one of the stations with the most precipitation on the eastern slope, in Tingo Maria (Johnson, 1976).

- As an additional cause, different heat fluxes from the shortwave radiation balance, which depend on the cloud

Table 8.2.3. *The spatial differences of the altitude of the equilibrium line ($dELA_{1920s}$), based on the values on the Pucahirca east side and its climatic causes under the assumption that either a change of the accumulation δc or the shortwave radiation balance $\delta[G(1-r)]$ were each solely responsible*

	Santa Cruz West	Santa Cruz East	Alpamayo West	Alpamayo East	Pucahirca West	Pucahirca East
$dELA_{(1920s)}$ [m]	132	236	99	171	29	0
δc [kg m⁻² a⁻¹]	−1634	−2921	−1225	−2116	−359	0
$\delta[G(1-r)]$ [MJ m⁻² d⁻¹]	+1.49	+2.67	+1.12	+1.94	+0.33	0

Table 8.2.4. *ELA*$_{1920s-1950}$ *in the six areas of investigation and their climatic causes under the assumption that either a change in the temperature* T$_a$*, the accumulation* c *or the shortwave radiation balance* G*(1−r) are each individually responsible*

| | Santa Cruz | | Alpamayo | | Pucahirca | | |
	West	East	West	East	West	East	Measured
ΔELA$_{(1920s-1950)}$ [m]	54	71	38	46	47	23	
δT_a [K]	+0.29	+0.38	+0.20	+0.25	+0.25	+0.12	+0.15[a]
δc [kg m^{-2} a^{-1}]	−668	−879	−470	−269	−582	−285	
$\delta[G(1-r)]$ [MJ m^{-2} d^{-1}]	+0.61	+0.80	+0.43	+0.52	+0.53	+0.26	

Note:
[a]Hansen & Lebedeff (1987).

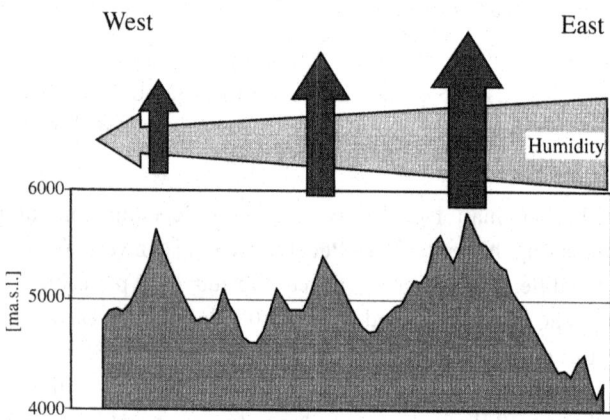

Fig. 8.2.9 A schematic depiction of the combination of advection (light grey) and convection (dark grey) in an east–west profile over the Santa Cruz–Pucahirca group.

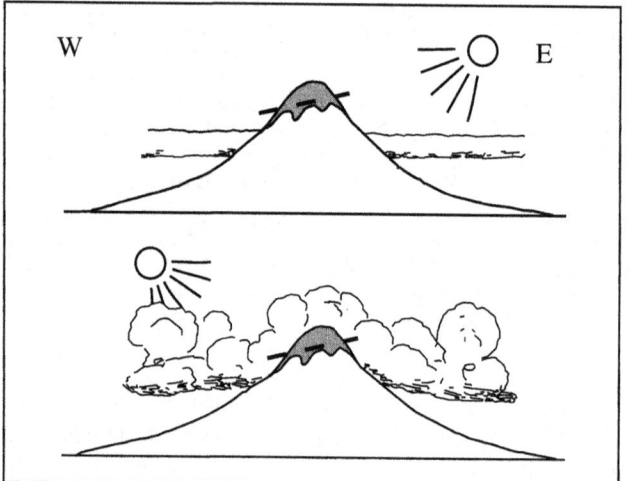

Fig. 8.2.10 The diurnal course of the convective cloud cover and the resulting asymmetry of the equilibrium lines on tropical mountains. (According to Troll & Wien, 1949.)

of the maximum rise on the eastern side of the Santa Cruz Ridge is explainable. The cause of the spatial differences of the rise of the equilibrium line are again found in a combination of those climatic variables which follow a change in air humidity. A combination of effective global radiation of $\delta[G(1-r)] =$ +0.4 MJ m^{-2} d^{-1} (2.7% of the mean value on the Hintereisferner in July) and accumulation of $\delta c = -155$ kg m^{-2} a^{-1} (about 13% of the amount on the western flank of the Cordillera Blanca) would be sufficient for the maximum difference between the Pucahirca East and the Santa Cruz East.

As an explanation, the model of a combination of a large-scale advection–convection system (Fig. 8.2.9) and daily local circulation (Fig. 8.2.10) can again be used. The changes within the model which are caused by changes in the air humidity are thus included. It can be assumed that these changes are lower under humid conditions, on the humid side of the mountains and/or during the wet season, than under dry conditions, on the dry side and/or during the dry season (section 8.2.1).

8.2.3 Summary of section 8.2

The individual investigations, using partly different methods and approaches, indicate that the strong retreat between the 1920s and about 1950 was mainly caused by a decrease of the air humidity and the corresponding consequences within the tropical atmosphere and, for only a small part, by a rise in temperature (case 2 or 4 in section 8.2.1). This can also be presumed for the first phase of retreat after the 'Little Ice Age'. The causes of the advances or the reduced rates of retreat at the beginning of the twentieth century cannot be clearly reconstructed due to the short duration of this period. The small advances in the 1960s and 1970s could be traced back to increased accumulation. In an estimation by Hastenrath & Kruss (1992), a rise in temperature and humidity is responsible for the most recent high rates of retreat (case 5 in section 8.2.1).

III

Former Quaternary tropical glaciers

9 The nature, extents and climates of former Quaternary tropical glaciers, with special reference to the East African mountains*

9.1 INTRODUCTION

From the Witwatersrand Glaciation (2650 Ma BP) onwards the earth has been repeatedly glaciated, sometimes with multiple repetitions in each period. In the late Precambrian the whole earth was ice-covered, both cooling and rewarming being due to major changes in atmospheric CO_2 emitted by volcanoes. Subsequent glaciations, e.g. in the Permo-Carboniferous and Quaternary, have been less extensive, perhaps partly limited by the stabilizing effects of photosynthetic and marine fixation of CO_2, and failing to reach the tropics except on the highest mountain-tops. These events and their causes are relevant to the current debate about the greenhouse effect and climatic change, but the study of pre-Quaternary glaciations is difficult because of their fragmentary remains and because of substantial displacements of the continental plates involved.

During the Quaternary, however, the lateral movements of continents have been slight and the uplift of mountains is independently quantifiable, so it is possible to concentrate on other factors. Although the overriding impetus for Quaternary glaciation is still a matter of debate, it is generally accepted that the sequence of glacial advances and retreats is linked to the Milankovitch astronomic cycles, with low atmospheric pCO_2 as a contributory cause in at least the last glacial. That theory, however, implies that events in the northern and southern hemispheres should be generally out of phase, though with a slight lag. Inconveniently, even the earliest dating results failed to confirm this on the scale of major events. The most recent comparison (Clapperton, 1998) indicates rather a synchrony between major and minor glacial advances during the last 50 ka over the length of the Andes and with cold periods inferred from northern hemisphere ice and ocean cores. Williams (1975) claimed a similar synchroneity of late Pleistocene arid periods in the north and south tropics. In contrast, Gillespie & Molnar (1995) concluded from a wider study including northern, central and southern America, Europe, Asia and Hawaii that

although there was some global uniformity in the occurrence of glacial advances, there was variability in the dates of the maximum advances. In particular they concluded that in many mountain ranges including tropical ones the glaciers reached their maximum extents one or two stades before the last maximum of the continental ice-sheets (20000–18000 BP), and so suggested that mountain glaciers reacted more quickly to climatic events and were more subject to local climatic influences as well as global ones. However, this study did not include the equatorial glaciers of Africa, Southeast Asia and Australasia where the limited evidence mostly supports a late maximum advance.

Thus tropical glaciers and especially equatorial ones occupy an ambivalent but significant position in relation to theory. Moreover, for several reasons it is possible to obtain more detailed information about their past history and environment than for many extra-tropical ones. Many of the mountains involved are isolated and relatively steep sided, with glacial extents clearly demarcated by moraines and tills, so that estimates of former ELAs (equilibrium line altitudes) can be made more reliably than from the large trunk glaciers of mid- and high-latitudes and particularly from large ice-sheets of uncertain thickness which generate their own weather. Many tropical mountains are active volcanoes so that moraines may be protected by and datable from overlying and underlying tephra or the soils developed in them.

However, former ELAs are of limited interest in themselves; their main use is as indicators of former climates. Because of the steepness of the mountain sides, altitudinal shifts of vegetation zones due to climatic changes are particularly emphasized, and are often well recorded from nearby and closely dated pollen

---

* Acronyms used in this section: LGM, Last Glacial Maximum; ELA, equilibrium line altitude; BP, Before Present; THAR, toe–headwall altitude ratio; AAR, accumulation area ratio; BR, balance ratio, the ratio of the mean mass balance gradient below the ELA to the mean gradient above the ELA; AABR, area–altitude balance ratio; AHA, area–height–accumulation method.

profiles, which are sometimes very long and span more than one glacial advance. Similarly because the mountains are usually in tectonically active areas, there are often nearby lakes in closed drainage basins where datable fossil beaches indicate former lake highstands. Both these sources provide proxy climatic data which can be compared with ELA estimates to yield conclusions about the interacting effects of temperature and precipitation changes.

To provide a full review of Quaternary tropical glaciations would require more than a whole book. However, the most important questions about them are 'When did they occur? How extensive were they? How can we draw valid conclusions about their nature and the contemporary climate? What are these conclusions?' The evidence from which to answer the last three questions is incomparably better for the last major advance than for earlier ones, the traces of which are only fragmentary. Therefore this chapter has four restricted aims. First, a brief account of the main characteristics of tropical glaciers and how ELAs may be estimated from both present and past ones by indirect methods. Second, a condensed summary of the evidence for several Quaternary glaciations on tropical mountains world-wide, concentrating on those major (but not necessarily maximum) glacial advances which appear to have coincided more or less with the Last Glacial Maximum (LGM, 20000–18000 BP) elsewhere, though including with these those areas (especially the Andes) where an advance at *c.* 30000 BP appears to have been the greater or the only one. Third, a more detailed review of glaciations and ELA estimates on the equatorial East African mountains. Fourth, a discussion of what climatic inferences can be drawn from evidence of ELA changes in the tropics. The concentration on East Africa and particularly the Rwenzori Mountains is partly to provide a context for a major part of the studies earlier in this book (the tropical Andes are too extensive and variable to cover concisely) and to complement the map which accompanies this book; partly also because there are especially detailed data available which demonstrate very clearly some of the problems and their solutions.

All the currently glacierized tropical mountains discussed in the earlier chapters of this book carry clear traces of former much greater extensions of their glaciers during the Quaternary. In addition others, not quite high enough to have glaciers now or recently, carry substantial evidence of former ones. The most conspicuous evidence is usually in the form of glaciated valleys with large lateral or terminal moraines. On the Rwenzori, for example, they are up to 150 m high and 5 km long and record the extent of 75 separate glaciers or glacier tongues, covering an area of 260 km², while on the Carstensz Mountains (Irian Jaya) similar moraines delimit an area of over 900 km². On some more flat-topped mountains (usually stratovolcanoes), however, such as Mount Giluwe (PNG) and the Bale Mountains (Ethiopia), there was an extensive ice-cap with rather small outflow glaciers, which, however, still left conspicuous moraines. The extent of this last glaciation on each tropical massif has now in most cases been well identified and mapped, often partly or mainly from air photographs, on which the moraines show distinctly even in forested country (Figs. 9.1.1, 9.1.2).

On each mountain, judging by their similar and slight degree of degradation, those moraines apparently date from a common advance episode. Their altitudes and their positions relative to each other and the existing glaciers (if any) give general support to this, although they vary greatly with the size of the glacier, but more significantly, when they are used statistically to infer the contemporary ELAs, these show a great consistency. For similar reasons and from a growing number of radiometric dates it appears that these episodes may have been contemporaneous throughout the tropics, terminating shortly before 15000 BP. However, many of the dates are poorly constrained, and on a few mountains there is evidence that this advance was small or absent, being preceded by a similar one up to 15 ka earlier (and see section 9.4.2). The majority were probably contemporaneous with the Last Glacial Maximum (LGM) in mid-latitudes (Devensian/ Weichselian/Würm/Wisconsin) and were presumably caused by related climatic changes. On each mountain this major glacial advance has often been given appropriate local names, but when we wish to refer to these generally we shall also term it the LGM (though it was not necessarily the most extensive glaciation and the assumption of contemporaneity both within the tropics and globally is provisional).

In addition to this well-attested LGM advance, there is on some of the mountains fragmentary evidence of previous major advances, ranging from one to three, mostly in the form of moraines and tills on interfluves or under later tephra, and thus protected from subsequent fluvial and glacial erosion active in the valleys. The evidence comprises mostly those which were more extensive than the LGM (in area, if not in lowest altitude reached) and so may reflect lower ELAs, but most have not been studied in detail. However, most tropical mountains are in tectonically active areas so that the possible effects of uplift or subsidence must be considered. Clapperton (1991) considered that in parts of the central Andes the similar extents of the penultimate glaciation and the LGM glaciations did not indicate similar ELAs, but that after making allowance for uplift the earlier ELA would have been lower. Conversely in Hawaii, Porter (1979[2]) and Gillespie & Molnar (1995) concluded that, though the moraines of the oldest glaciation (300 ka BP) now extend 550 m lower than those of the LGM, the ELA using a THAR (toe–headwall altitude ratio) of 0.5

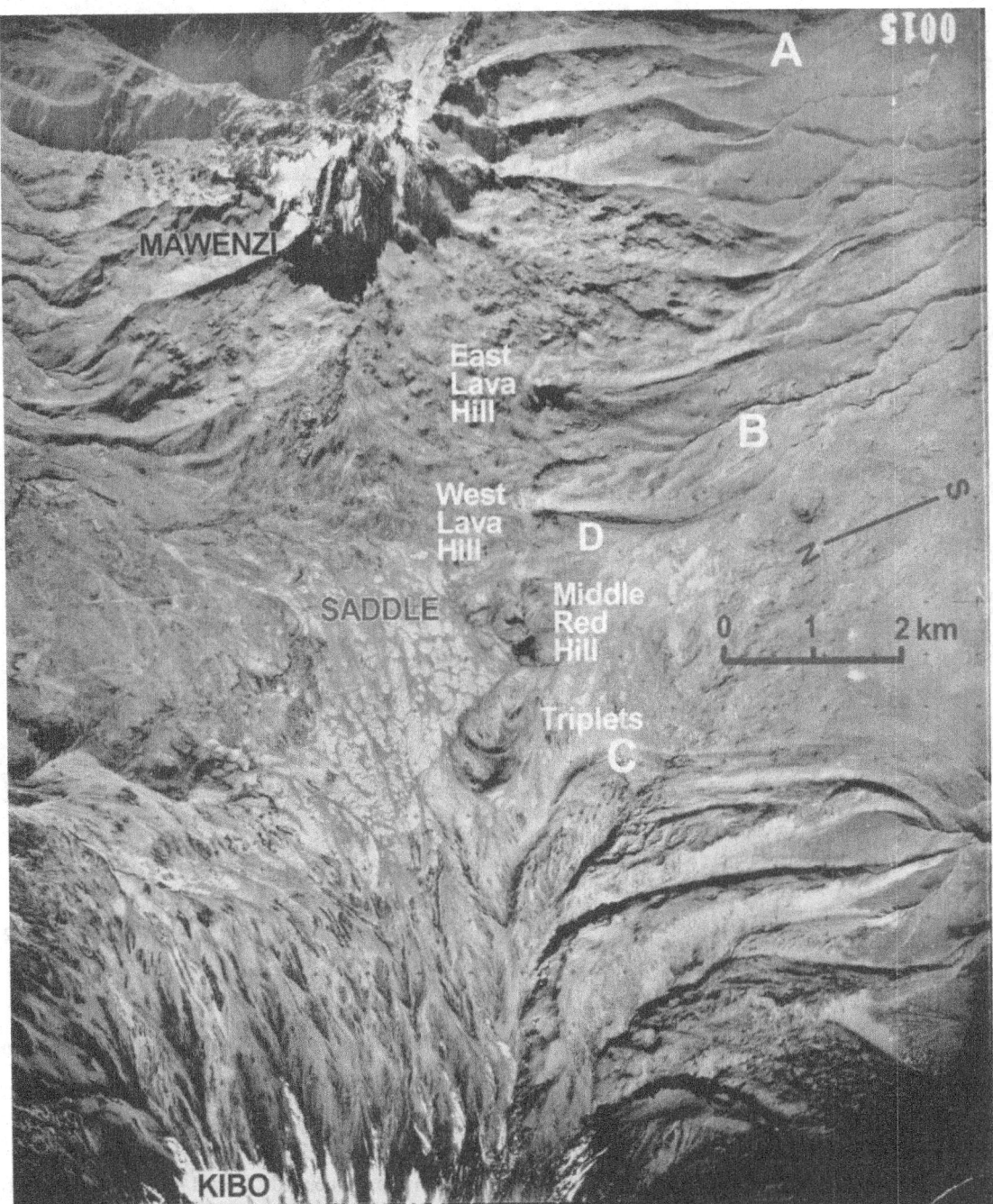

Fig. 9.1.1 Kibo (5895 m) and Mawenzi (5147 m) peaks and the Saddle on Kilimanjaro. East-southeast is towards the top of the photo, Mawenzi is at the top edge, Kibo at the bottom. The conspicuous moraines are those of the Main (= Fourth) Glaciation, dated by ^{36}Cl at positions at **A** and **B** at 20 000 BP and 17 000 BP; retreat or minor readvance stages **B** and **C** are dated to 16 000 BP and 14 000 BP respectively (Shanahan & Zreda, 2000). Large glaciers on the west face of Mawenzi flowed towards the Saddle (4300 m), but then diverged to north and south. However, note the only slight development of this moraine stage on the east (centre and left-hand) slopes of Kibo descending to the Saddle. Nor are moraines visible on or descending southwards from the centre of the Saddle, as would be expected if it had been completely ice-covered in the main glaciation, particularly below Middle Red Hill and the Triplets on the south (right), where moraines would have formed like those below East and West Lava Hills close to the east of them. A single erratic boulder at **D** dated at 34 000 BP supports this view, but would accord with there having been ice on the Saddle in an earlier glaciation. Older relic moraines, just downslope from **A** but not visible on the photos, have an age of over 360 000 BP. The surface is mostly scree or bare rock, with semi-arid montane moorland on the lower slopes, comprising only sparse tussock plants. (Air photo adapted from RAF V/13A/RAF/688 No.0015, f = 6 in, alt. = 42 000 ft, 2/3/1958.)

Fig. 9.1.2 Rwenzori Mountains, Uganda and Dem. Rep. of Congo. **LM**, main glaciation moraines (Lake Mahoma Stage) in a valley at 3000–3800 m, clearly visible despite a thick cover of upper montane forest.
RR, tills of the preceding Rwamya–Rwanoli Stage on interfluves at 3650–4100 m altitude, clearly distinguished by moorland of *Helichrysum* scrub (darker grey) and shorter *Alchemilla* (pale grey). (Air photo adapted from DOS/Huntings 15.UG 33 No.025, f = 6in, alt. = 24000ft, October 1955.)

was only 140 m lower after correcting for subsidence and sea-level change (but if they had used a more appropriate THAR value of 0.2 to 0.3 for a conical ice-cap the inferred ELAs would have been nearly the same).

There are also small valley-bottom moraines attesting to some minor advances since the LGM described above, some of which may be correlated with the Younger Dryas of northern latitudes.

9.2 THE NATURE OF TROPICAL QUATERNARY GLACIERS AND ESTIMATION METHODS FOR THEIR ELAS

9.2.1 The morphology of the glaciers and their moraines

At the LGM the different climate caused a great extension of tropical glaciers, but it is not yet clear what other interactions there may have been. Observation, maps and air photos show that on most glaciated tropical mountains (other than some recently active volcanoes) the LGM glaciers were valley glaciers resembling typical modern ones on other steep massifs such as the Himalaya. They had long tongues which usually left substantial lateral and converging terminal moraines, particularly in the main trunk valleys even if they were of low gradient but had been widened in previous glaciations. Where the glacier emerged on to a piedmont or altiplano its widening lobe usually left smaller frontal moraines. In this they differ from most present and recent tropical glaciers, which often have clean surfaces and are leaving little in the way of moraines even at standstill positions. Was this just a question of glacier size and a long period with a stationary terminus?

Or are other factors involved which increased the load? In chapter 5 it was shown (Fig. 5.3.4) and supported on theoretical grounds that *modern* tropical glaciers in various regions have shorter tongues (ablation areas) relative to their accumulation area than those in mid-latitudes. Consequently with increasing glacier size the absolute vertical extent of the ablation zone increases more slowly than that of the accumulation zone. Fig. 9.2.1 illustrates similar data for three regional groups of LGM glaciers on the Rwenzori Mountains, and shows that for a wide range of glacier sizes the vertical limits

of top and bottom are approximately mirror images; indeed, a few of the longest glaciers have especially long tongues. For this purpose it is essential to group glaciers together thus by environmental similarity and so expected similarity of their ELAs, otherwise differences in the latter may mask the pattern. In each of these groups with a wide range of sizes, the ELA is narrowly constrained for the smallest ones, where it must lie roughly midway between top and bottom, and in Fig. 9.2.1 it is indicated that the same ELA is applicable to the others. Though it could be argued that the ELA was lower on the larger glaciers, resulting in a smaller ablation area, this is in fact contrary to the evidence, for the equilibrium lines of the largest glaciers lay nearer the centre of the range where the ELA surface is higher (see Fig. 9.4.8) than on the small peripheral glaciers, so increasing the effect shown here. Thus it appears that in this respect the glaciers of the LGM on the Rwenzori differed from these modern near-stationary tropical ones; the very small and rapidly retreating present glaciers on the Rwenzori are not comparable, being far from stationary. There is a lag in the response of the terminus to a rising ELA, so that in these circumstances the ablation area is relatively larger and longer.

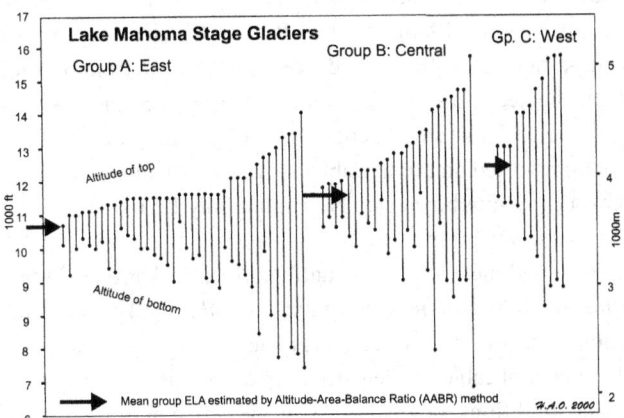

Fig. 9.2.1 Rwenzori Mountains, Uganda and Dem. Rep. of Congo. Lake Mahoma Stage glaciers. Altitudes of top and bottom, with mean ELAs, for three contrasted regional groups. 'Top' is the highest altitude of the headwall.

A constraint on the use of some methods of estimating ELAs is that the hypsographic (area–height distribution) curve should resemble those of current Alpine glaciers and be at a smooth maximum near the equilibrium line if similar ratios are to be used. Fig. 9.2.2 depicts curves for three of the largest former glaciers on the Rwenzori, showing that these

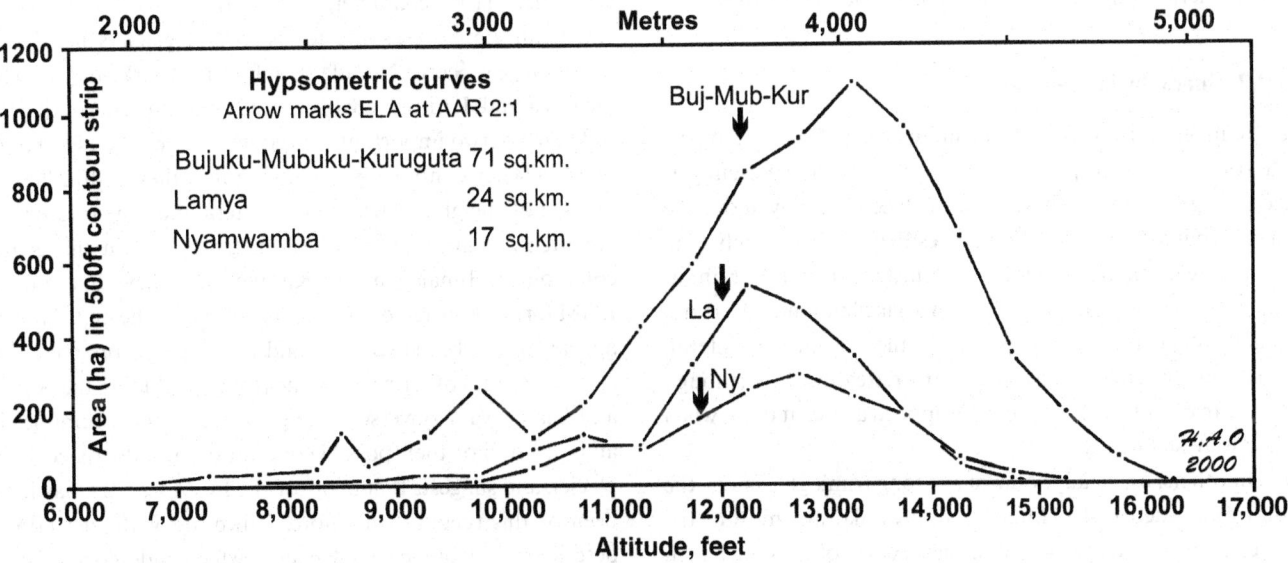

Fig. 9.2.2 Rwenzori Mountains, Uganda and Dem. Rep. of Congo. Hypsographic curves of three large glaciers of the Lake Mahoma Stage. Areas are for present valley floors, not reconstructed glacier surfaces.

comply adequately with this constraint as do most of the others. However, this is not true of ice-caps on domed or conical volcanoes (see below).

The large size and 'textbook' form of most LGM converging lateral moraines indicate either a long period of near still-stand of the glacier terminus or a very large volume of debris being brought by the glacier, probably a combination of both. It is remarkable that, on the Rwenzori at least, even the smallest former glaciers down to 50 ha in size had similar moraines. Many occupied sites just below the crests of long ridges where on many mid-latitude mountains a cirque would instead have formed, with often a small lake in a rock basin or dammed by a frontal moraine. For example, 34 of such forms are recorded in the English Lake District (Sissons, 1980; Evans & Cox, 1995) but Osmaston has found just one (hitherto unrecorded) with a pair of moraines similar to the Rwenzori ones, at the head of the Seathwaite valley.

Tricart (1971) observed very big Quaternary moraines with sharp crests, 50–100 m high, in Venezuela and Colombia and proposed that they were attributable to a year-round diurnal melting regime on the glaciers producing a less competent runoff and hence less effective removal of moraine material than the seasonal regime of glaciers elsewhere. He proposed that they characterized an *Equatorial* type of glacier. This factor may complement the effect of a particularly heavy load of moraine on the glacier surface. Also at moderately high altitudes on moist equatorial mountains where relic moraines occur, the high total rainfalls are achieved mainly from frequent and prolonged rain, rather than the very heavy storms that sometimes cause serious erosion on the lower slopes.

9.2.2 Supraglacial moraine

In many glacierized regions, including the polar and temperate zones and the tropical mountains, many of the present glaciers, large and small, have clean surfaces not only above the firnline but also below it. Sometimes they are completely bare of surface moraine, sometimes with just a stripe of moraine at the side or below a spur where two glaciers join. These are usually depositing little moraine either frontal or lateral, partly at present because many are retreating fast (though there are exceptions) but the same applied at recent times when they were stationary.

On others, especially some of the long trunk glaciers of the subtropics such as the Himalaya and Karakoram, most of the glacier surface below the firnline has a cover of moraine. Such glaciers with long debris-covered tongues are commonly associated with large, long, lateral moraines similar to those described above. These reflect the combination of suitable geology, valley form and climate to produce the debris. Until

recently the dynamics of these have been ignored as being too complicated, as the process is often confused by pseudo-karst melt hollows and ice pyramids, but they are now the subject of intensive study (e.g. Nakawo *et al.*, 1999).

The debris is often quite thin (*c.* 1 m) but has a major effect in absorbing radiation. However, if it is thick enough, it insulates the ice from the effects of radiation and hence greatly reduces the ablation rate. Comparative measurements of this at particular sites have often been made: on the Terong Glacier in the Karakoram, ice under a cover of up to 1 cm of grit melted twice as fast as clean ice which lost 3–4 cm d^{-1}, whereas under a cover 5 cm thick it melted at only half the rate of clean ice (Osmaston, 1986); on the Batura Glacier also in the Karakoram, up to 18 m of ice a year were melted from bare ice near the terminus whereas under moraine cover this was reduced to 5 m (Shi Yafeng *et al.*, 1979). Inoue & Yoshida (1980) reported an average ablation rate of 20 mm d^{-1} on bare ice on the Khumbu glacier, Nepal, but only 1.5 mm d^{-1} under a debris cover 1.2 m thick.

Proposed methods of estimating it (e.g. Nakawo & Young, 1982; Bozhinsky *et al.*, 1986; Nakawo *et al.*, 1999) depend on a knowledge of current meteorological variables so are not directly applicable to Pleistocene ones. Nor is it clear exactly what conditions predispose a glacier to a moraine cover. Factors which may be important are the lithology around the accumulation zone, and rapid incision with steep valley sides resulting in frequent rockfalls on to the glacier surface. Adjoining glaciers, otherwise rather similar, may be clean or debris covered (e.g. fig. 12 in Lliboutry, 1998). Although they are particularly characteristic of the Asian and South American subtropics, some debris-covered glaciers also occur elsewhere, e.g. Sierra Nevada (California) (Clark *et al.*, 1994), New Zealand (Kirkbride & Brazier, 1998), and even Antarctica.

There are two important consequences. First, if the tongue of the glacier is not fully occupying its valley width, lateral melting of the glacier will deposit lateral moraine ridges that are often very large and many kilometres long. Not only is this common in Himalayan and Karakoram valleys but similar LGM forms occurred on the Rwenzori, where the main valleys appear to have been excavated and widened by a previous glaciation, besides other parts of the tropics including the Andes and Irian Jaya. Conversely, the presence of such moraines is an indication of their possible deposition by a debris-covered glacier, and suggests that many of the LGM glaciers may have been of this type. Furthermore, Kirkbride & Brazier (1998) describe how debris-covered glaciers which undergo a series of minor readvances build up large lateral moraines by successive superposition of material. This too could be relevant to large tropical LGM moraines.

Second, in order to dispose of the same discharge of ice

despite the lower ablation rate, the area of a debris-covered ablation zone will be greatly extended, perhaps by a factor of 2× or more compared with a clean one. Kirkbride & Warren (1999) describe debris cover as the normal state of the Tasman Glacier, New Zealand, regardless of historic mass balance changes which has led to the preservation of a long ice tongue at low gradient. The expression of this in the mass balance curve will be to reduce its gradient in the ablation area, compared with that of a bare glacier. On many other present glaciers, and especially tropical ones, the mass balance gradient below the ELA is steeper than above it, but on such debris-covered glaciers this relationship is commonly reversed. This has major effects on morphometric estimates of ELAs (sections 5.4, 9.2.4). Kirkbride (1989) reports on three clean glaciers in New Zealand with a mean AAR (accumulation area ratio) of 0.78 while ten debris-covered ones had a mean AAR of 0.46. Grove (1988) reported that debris-covered glaciers in the Everest area had a mean AAR of 0.41. Clark et al. (1994) reported that in the Sierra Nevada debris-covered glaciers had AARs ranging from 0.2 to <0.1, and THARs (toe–headwall altitude ratio) of 0.6–0.8, though they were small and there is controversy over their distinction from rock glaciers.

Accordingly it cannot be assumed that LGM glaciers had the same AARs as modern clean glaciers, nor the higher values proposed in chapter 5. Furthermore the moraines suggest that many of the glaciers of the LGM operated in a substantially different climate and regime to present tropical ones, and that one resembling the present subtropics (though cooler) should be considered. But it is dangerous to generalize. Just as the few lowest reaching tongues of glaciers in the Cordillera Blanca (Fig. 5.3.4) are heavily debris covered today and so do not conform to the general modern tropical model, so at the LGM the converse may have applied, with many glaciers debris-covered though in regions that did not become significantly drier and colder. Klein et al. (1999) comment about the LGM glaciers of the central Andes that 'One conspicuous feature of these former glaciers are their long narrow tongues, which are not general features of the region's modern glaciers'. The driest southwestern Andean mountains are probably too dry at present for much rock shattering to occur. If these mountains had higher precipitation at the LGM (see section 9.5.3), this could paradoxically have increased rock-shattering there.

Above the ELA, debris falling on the glacier is progressively buried by snowfall, only to be re-exposed below the ELA. Thus lateral moraines can only be formed along the sides of the glacier in the ablation zone. The upper limit of such moraines is therefore a lower limit for the possible elevation of the ELA, though this may be a variable amount higher. In general the highest lateral moraines of former glaciers on both the Rwenzori and Kilimanjaro are at or slightly below the altitude of the ELA estimated by quantitative morphometric methods. For example, the upper limits of moraines of the LGM on south Mawenzi (Kilimanjaro) shown on Fig. 9.1.1 are at altitudes between 4000 m and 4300 m, while their mean ELA by the AABR method (see section 9.2.4) is 4300 m. This physical morphological feature thus supports the statistical ones described below (but see section 9.4.3).

9.2.3 ELA estimation – general principles

The ELA of a single glacier is a statistical concept in space and time, which may then be aggregated for a number of glaciers. This rationale and the statistical nature of both the components and the outcome make it possible to apply objective statistical tests to the data, the parameters and the results. This can guide data and parameter selection and acceptance of results (Osmaston, 1975, 1989[2]; Furbish & Andrews, 1984; Burbank, 1991).

For past glaciers ELAs can be estimated indirectly from morphological variables by statistical inference, using parameters derived by direct observation on existing glaciers. However, there remains the problem of whether the reference glaciers are strictly comparable with the study set (c.f. Clark et al., 1994), which is particularly acute when it is expected that the result of the study will indicate a substantial climatic difference. Indeed, the methodology usually employed in the ELA approach and its climatic interpretation is subject to several major assumptions and uncertainties.

An indirectly estimated ELA of a single glacier is unreliable, so normally several are aggregated, often simply by taking the mean of a group that is presumed or demonstrated to be acceptably homogeneous. However, the variability between glaciers reflects environmental factors at five scales: global, regional, the mountain, the group and the individual. At the mountain scale, such factors as the dominant wind direction and humidity, overall relief and temperature are major factors. At the individual scale the aspect, steepness, valley form, debris cover, windblow, etc. are important. The group (usually a particular aspect of the mountain) represents a compromise which may be convenient if data are poor but inevitably discards some information. In most studies it is the mountain scale variation which is desired. If the number and distribution of the data points are adequate it is better to derive from them a trend line or trend surface which reflects mountain scale variation while the deviations are 'noise'. Since trends can be made to reflect various degrees of detail, the choice is to some extent subjective, but should at least correspond to major relief; hence linear trends are usually inadequate.

Especially where there was a large range of glacier sizes so that termini were at different altitudes, it can be difficult to distinguish those of each stade. A statistical criterion based on the conformity of their ELAs may provide a useful means of testing the coherence of the data. Selection of the most appropriate parameters for the studied population should result in group means or mountain scale trends from which the individual deviations are a minimum; it is therefore possible to test a range of parameters to find the optimum. Finally, the results can be accompanied by their confidence limits which may be important when differences between trends are sought.

9.2.4 ELA estimation – methodology

Three main quantitative methods have been used by various authors to infer the former position of the equilibrium line on tropical and other glaciers from the morphology of glaciated valleys and associated moraines, depending on the quality of the available information and the effort available. All have a longer history than is usually realized. However, users have often not taken account of the constraints of each method and have seldom tested their precision in use. The first (THAR, toe–headwall altitude ratio) takes no account of factors other than the altitudes of the top and terminus (toe) of the glacier; the second (AAR, accumulation area ratio) also considers its form but not its hypsometry; the third (AABR, area–altitude balance ratio) considers both of these and also the mass balance gradient. The parameters usually adopted for the first two methods are those appropriate to glaciers of 'standard alpine valley form', e.g. a THAR of 0.4. However, both Osmaston (1965, 1975) and Furbish & Andrews (1984) have analysed the effects of common 'non-standard' shapes (e.g. piedmont glaciers, ice-caps) and demonstrated that different parameters should be used. This is particularly relevant to tropical glaciers descending on to an altiplano and to ice-caps on undissected volcanic domes where THARs of 0.5 and 0.3 respectively are more appropriate.

The simplest and most frequently used method (requiring least data and least effort) simply takes the altitude of the top and toe of the glacier and estimates the ELA as at some given ratio between top and bottom. This was originally applied by Höfer (1879) to existing glaciers in the Alps, using a simple 1:1 ratio and taking the top as the mean altitude of the mountain-tops around it. The more recent version, THAR, is commonly used for past glaciers with a ratio of 2:3, or of 0.4 (for toe to ELA/toe to headwall), derived from observations on present glaciers, and this value has been recommended by Meierding (1982) from criteria of minimum variance for estimates on former valley glaciers in Colorado. However, a graphical version of THAR analysis of three topographically distinct groups of LGM glaciers on the Rwenzori indicated (see Fig.

9.4.6) that ratios of 0.46, 0.50 and 0.57 gave best fit results (Osmaston, 1965, 1975) and indicated mean ELAs of 3780 m, 4040 m and 3600 m respectively. Failure to group the glaciers or the use of an arbitrary ratio would have given seriously misleading results.

When it is possible to reconstruct the outline of the former glacier and some altitude information is available at least along its centre line, the AAR method is frequently used, dividing the glacier into accumulation and ablation areas in a ratio based on glaciological observations on groups of current Alpine glaciers. Originating with Brückner (1886) this has been frequently studied, criticized and used (e.g. Drygalski & Machatschek, 1942), a ratio of 2:1 (now conventionally expressed as the AAR, the ratio of the accumulation area to the whole, i.e. 0.67) often being considered appropriate. Support for this was provided by Gross et al. (1977), who studied six Austrian glaciers for a decade or two: zero mass balance was obtained for AARs between 0.60 and 0.69 (average 0.64). Again Meierding's work on former glaciers (1982) confirmed this value. Workers on former glaciers elsewhere, lacking locally and temporally relevant information about this ratio, have commonly adopted the same value, but seldom considered the uncertainty thus introduced nor the further constraints discussed below. However, Sutherland (1984) warned that the appropriate ratio varies with climate and with glacier shape (area–altitude distribution), and may be between 0.34 and 0.77 with an average of 0.55.

When a contoured map of the valley floor is available, or better if the former ice surface can be reconstructed, a more rigorous approach is possible which accommodates divergencies from the standard glacier shape and in an advanced form also accommodates non-linear vertical mass balance gradients. Simple ratio methods such as THAR and AAR make no allowance for the hypsometry of a glacier, but treat all areas above the ELA as equivalent and similarly all those below it. However, areas which are just above or below the ELA may not contribute much to the mass balance, while ones which are near the headwall or terminus may have high accumulation or suffer very heavy losses.

This method depends not on mean morphological ratios but on weighting the **area** of each altitude contour band of the glacier surface by its **altitude** above or below the ELA. Developed originally by Kurowski (1891), it was ignored for over half a century until it was again used on the East African mountains (Osmaston, 1965, 1975, 1989[1,2]), then independently rediscovered by Sissons (1974, 1980) and used by him on Late Quaternary glaciers in the English Lake District. In this original simple form it assumes a linear mass balance gradient throughout, and can be estimated very simply from the following formula where

s = the area of successive contour bands of the glacier surface

a = the mean altitude of each contour band

then
$$\text{ELA} = \frac{\Sigma sa}{\Sigma s}$$

However, on some glaciers the gradient is steeper below the ELA than above. It was therefore improved by Osmaston to apply different **ratios** of the mass balance gradients (BR) below and above the ELA by the use of a simple iterative computer programme (Osmaston, 1975; see also 1965; this is currently being up-dated by him) for a group of glaciers, and termed by him the area–height accumulation (AHA) method. Because the necessary contour data were not always available and the method was not widely known, this has seldom been used, but it was revived and thoroughly tested on existing (though not necessarily stationary) glaciers by Furbish & Andrews (1984) who concluded that it gave very precise results, the estimated ELAs falling within a range of only ±50 m for some groups. A convenient spreadsheet programme for the calculations was devised by Benn & Gemmell (1997) terming it the 'balance ratio' method, providing for the use of an a priori selected ratio on a single glacier. To match better the other acronyms in use and be more explicit, the name area–altitude balance ratio (AABR) is now proposed, indicating that each of these three variables is involved, and this is used in this chapter. Besides its major use for former glaciers, it is applicable to present glaciers when other data are not available.

The mass balance curves of modern glaciers (from measurements on actual glacier surfaces) vary greatly in detail, but many can be generalized into a steep slope segment below the ELA and a gentle one above it, commonly with balance ratios of the order of 2:1. Chapter 5 has emphasized the prevalence of this in modern tropical glaciers. A further example is provided by the Carstensz glaciers in Irian Jaya, where the slope is 3 m water equivalent per 1000 m below the ELA but 1 m or less above it, a balance ratio of 3:1 (Hope et al., 1976).

With a linear gradient of mass balance (BR=1) each of a group of glaciers has an ELA (individually estimated by AABR), and the group will have a mean ELA and variance. If the mass balance gradient below the ELA has 2× the gradient (BR=2) compared with above it, the ELAs must be shifted down glacier to achieve equilibrium. The new ELA (again estimated by AABR) for each of the group of glaciers will have a fresh deviation from the new group mean ELA. The new variance (or standard deviation) may be more or less than the previous variance. If the new variance is more, the linear gradient is a better choice. If the new variance is less, then the BR=2 gradient is to be preferred and its associated estimate of the

ELA. Further values of BR can be tested. However, Osmaston (1965, 1989[2]) found that for LGM glaciers on the Rwenzori a balance ratio of 2:1 was not superior to a factor of 1:1 judged by a minimum variance criterion. For modern debris-covered glaciers ratios of less than 1:1 would be applicable.

Where there were a sufficient number of suitably distributed glaciers, an improvement on the group method can be achieved by plotting a best-fit elevational profile or trend surface for the ELA by manual or multiple regression techniques. The individual glacier estimates can be derived by THAR, AAR or AABR methods using a range of parameter values but in all cases the variance from the profile or surface should be used as a criterion of the most appropriate parameter value. The results of such methods give more detailed information about the relationship between the ELA and such factors as mountain relief, aspect and the prevailing snow-bearing wind.

9.2.5 Constraints

The applicability of these methods is subject to certain constraints:

(a) The ratios are statistical parameters which should not be used to draw general inferences from one or a few glaciers but only from a sufficiently large group to mask individual differences. This is not always possible. However, it is also important that the groups should be glaciologically homogeneous, so it is permissible and desirable to define regional groups with consistent ELAs by preliminary assessment of the ELAs of individual glaciers, using a priori parameters.

(b) The glaciers should be in a near stationary condition. This constraint is usually satisfied when studying past glacial advances, since a large terminal moraine may reasonably be thought to represent a stationary condition, a position reached repetitively, or at least a transition from positive to negative balance, but is seldom correct for modern glaciers.

(c) For THAR and AAR, the hypsographic (area–altitude distribution) curves of the glaciers should resemble those of current Alpine glaciers and be at a smooth maximum near the equilibrium line. This constraint has seldom been considered and numerous former glaciers clearly did not conform to it, particularly those extending into a piedmont lobe or forming a conical ice-cap (Table 9.2.1; Osmaston, 1975; Furbish & Andrews, 1984).

(d) Their mean vertical mass balance gradient should conform to that of the reference group. This is rarely considered but has been critically examined in section 5.1.

Table 9.2.1 *Approximate morphometric parameters for various glacier types*

	Valley	Conical ice-cap or piedmont	Debris covered
THAR	0.4–0.5	0.2–0.3	0.6–0.8
AAR	(0.5) 0.6–0.7 (0.8)	0.4–0.5	0.1–0.5

(e) It is desirable to use the reconstructed ice surface of each former glacier for morphometric analysis (e.g. Sissons, 1974) rather than its bed, the present valley floor. However, this is difficult, somewhat arbitrary and time consuming. It is easier and often sufficiently accurate to add a figure for ice thickness at the estimated ELAs, more for big glaciers than small. Later in this chapter, Fig. 9.4.7 shows unadjusted data for present valley floor measurements, while Fig. 9.4.8 has been corrected for estimated ice thickness.

(f) Under different climatic conditions the mean atmospheric temperature at the ELA varies considerably. In moist regions the ELA is commonly below or only just above (e.g. Irian Jaya; Hope *et al.*, 1976) the altitude of the mean freezing point, 0°C, but in dry regions the ELA is often at −5 °C to −10°C. Not only does that make it unreliable to compare apparent ELA changes in different climatic regions, but it also affects the inferences to be drawn from ELA changes in the same region, when these may have been due to major changes in humidity regime, not just temperature.

(g) On mountains with dome-shaped or conical ice-caps, the derivation of useful ELA profiles or trend surfaces is usually impossible as the individual ELA estimates just form a single ring round the summit, though probably with systematically varying altitudes on different aspects. Since the equilibrium lines of different stages are horizontally displaced, the simple differences between their altitudes are not reliable estimates of ELA change for the reasons discussed below; usually they will be excessive.

(h) Comparisons are often made between inferred ELAs of former glaciers and observed or inferred ELAs of present ones, particularly to yield a figure for the change in ELA. For such comparisons it is valid to use an ELA obtained by direct measurements on a present stationary glacier covering a period of several years, but on a receding glacier the extent and form lag the mass balance and rising ELA. Most tropical glaciers are retreating now, and are in a far from stationary state. Hence morphometric methods will give ELA values that are too low, and

can only be used to give minimum estimates of the present ELA and of its difference from the LGM ELA. Conversely their ELAs directly measured by observation are not appropriate for deriving THAR or AAR ratios to apply elsewhere or to past glaciers. For example on the Lewis Glacier, Mount Kenya, from detailed observations the mean AAR over the period 1978–1982 was about 0.1 (est. from Hastenrath, 1984, map 5.4.3:5) with a mean net balance of *c.* −1 m; after continued warming and ELA rise, this glacier probably now has an AAR of zero. Hope *et al.* (1976, p. 45) record that in Irian Jaya in 1973 with an ELA of 4580 m, the AAR of the Carstensz Glacier was 0.66 and of the Meren Glacier 0.59, but neither of these glaciers was stationary, both having large mass balance deficits. If their ELAs remain constant while the glaciers shrink to equilibrium extents, these two ratios will be much larger as proposed by Kaser above.

9.2.6 Estimation of ELA changes on tropical mountains

The differing ELAs of present and former glaciers are the most important indications we have of the possible contemporary climatic changes. However, it is essential to ensure that ELAs estimated by indirect methods are as accurate as possible and that comparisons between them are also soundly based.

It is important to distinguish between global, regional and local variations of ELAs (and similarly glacierization levels and snowlines). Many authors have presented numerical data and prepared maps and vertical profiles of these at various scales, and at various dates, but have not always appreciated that it is essential to use comparable models of the data to make valid comparisons of ELAs at different dates. The ELA of a single glacier has both altitude and position and a change in extent changes both; consequently the mean ELA of a homogeneous group of glaciers on the same aspect of a mountain may also change significantly in position when their ELAs change.

Numerical data have often been presented for the mean ELAs on various tropical mountain ranges at various glacial stages, and the arithmetic differences between stages then used to indicate the change in ELA (and hence of climate). Although such data are sometimes presented and discussed as if they represent directly comparable point values or values that are uniform over a considerable area, in fact they are always more or less generalized and inadequate substitutes for complex trend surfaces, defined by data points which may lie at different positions at different times.

From detailed data presented by Nogami (1976) on Pleistocene ELAs in the North and South American

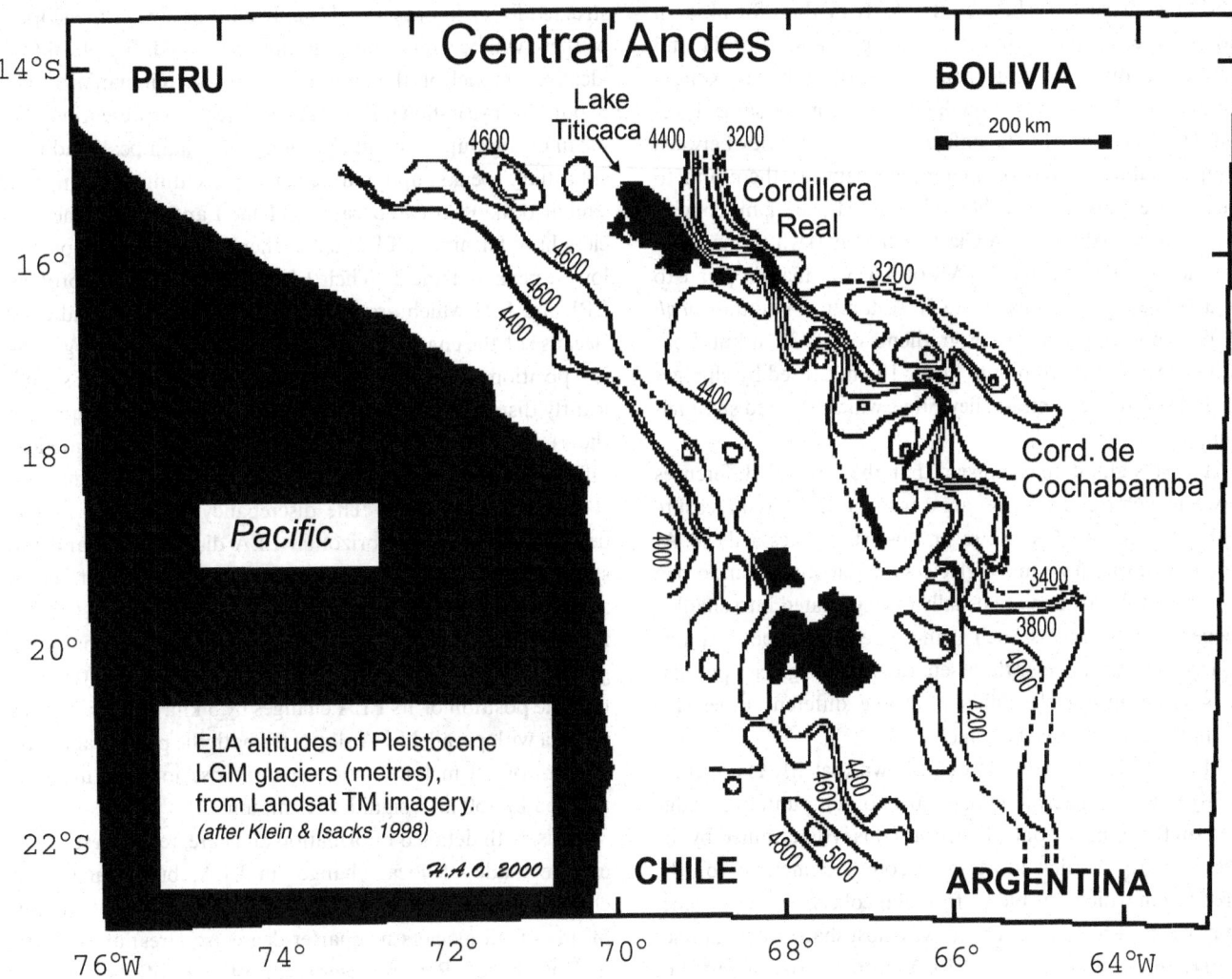

Fig. 9.2.3 ELAs of LGM glaciers in the Central Andes.
(After Klein & Isacks, 1998 by permission.)

cordilleras, derived from thousands of individual cirques at 450 locations, one can show that at a global scale the north–south gradient in ELA between the equator and 24° S is a rise of about 0.6 m km⁻¹. Over shorter regional distances of 50–200 km covering several massifs the gradient is about ten times this, 5–10 m km⁻¹, but it is not possible to infer local gradients on single massifs.

Much steeper regional gradients at right-angles to the main cordilleras have been mapped by Klein & Isacks (1998; Klein *et al.*, 1999) who used air photos and Landsat TM imagery to map present snowlines and LGM glaciers (from their moraines) in the Central Andes, to estimate the regional ELAs at the LGM by the THAR (0.5) method, and to draw smoothed trend surfaces for snowlines and ELAs. That for the present glaciers was derived from average values for the snowline in a moving 0.25° square, while the LGM one used the

lowest ELA estimate in a moving 0.25° square. Thus they are closely but not exactly comparable. Although the general form of the two surfaces is similar, the maximum slope of the present ELA surface is about 18 m km⁻¹ and of the LGM one about 31 m km⁻¹ (Fig. 9.2.3). They also plotted a trend surface of the vertical differences between the two surfaces, which ranged in amount from 500 m to 1200 m and had slopes of up to 6 m km⁻¹. This difference surface is quite complex and demonstrates the variability of the factors which determine ELAs even on a regional basis, and it is likely that on a local basis of individual ranges or mountains the trend surface slopes would be greater by a factor of two or more. In East Africa, Osmaston (1989[1,2]) found slopes much steeper than these on individual mountains or ranges (see below).

Outside the tropics similar slopes have been reported. Evans (1990) mapped a linear trend surface of glaciation level in west Canada sloping at 6 m km⁻¹ but noted local anomalies sloping at 17–25 m km⁻¹ over distances of 11–18 km between windward and leeward sides of mountain ranges. Burbank

(1991) reconstructed ELAs by the AAR method for modern and Pleistocene glaciers in the Sierra Nevada, California, and showed that on a local scale their trend surfaces had maximum slopes of 12–15 m km^{-1}, though not always in the same direction. He also claimed that conformity or discrepancy between groups could be used to differentiate between them and to counter previous views of their relations. However, his conclusions have been disputed by Clark et al. (1994) who considered that the 'standard clean ice' AAR of 0.65 was inappropriate to the debris-covered glaciers there. In Antarctica, Fountain et al. (1999) reported general ELA gradients of 18–30 m km^{-1} but with a steep rise of 70 m km^{-1} over 10 km caused by changes in wind speed over a mid-valley ridge, which affected sublimation.

At first sight it may appear that the vertical differences between such trend surfaces are a true and unbiased measure of the changes in ELA. However, this is not necessarily so. At a local scale of a few tens of glaciers the statistically smoothed regional ELA surfaces are usually disaggregated into smaller, steeper slope segments, such as those described above. It may be necessary to consider these separately if the glacier populations are significantly different at the different times, for example because some have melted.

At such a local scale it is well known that ELAs are commonly lower on glaciers on one aspect of a mountain or ridge than on the opposite one. Sometimes this is determined by the direction from which most snow comes, sometimes by the direction in which it is blown from flat collecting areas, sometimes by the direction which receives most shade from summer insolation. Consequently the ELA profile across a series of such ridges has a sawtooth shape in detail; usually there is also an overriding general gradient over a number of 'teeth' related to the direction of the snow-bearing winds. This is well demonstrated in fig. 14.12 of Clapperton (1993) which shows that right across the Andes LGM glaciers extended further on the east sides of successive ranges (facing moisture-bearing winds from the Atlantic–Amazon domain) than on the drier west sides. Detailed and precise comparisons of ELA change must ideally be restricted to glaciers on the same side of the same 'tooth' (see Table 8.2.2). Regional profiles smooth such local variations and have lower gradients.

When ELAs of a group of glaciers are being compared at different occasions, and especially when many of the glaciers have greatly retreated or melted over the interval, a careful approach is essential. At each occasion there will have been a general ELA gradient across the group, and if there are outlying peaks, then those that are upwind may have much lower ELAs than the main group. If there are only a few glaciers their ELAs can be generalized as linear profiles across the area; if there are more, then ELA trend surfaces can be con-

structed for each occasion. Usually but not always the slopes of these will be similar but vertically separated. Fig. 9.2.4 (an idealized model of the situations on both Kilimanjaro and Mount Kenya) shows some glaciers (each representing the mean of a group of similar glaciers) on a main peak and two subsidiary peaks up-wind (each representing a group of similar peaks). At two occasions, Time 1 and Time 2, the glaciers have termini at T1 and T2 (there are no glaciers on the lowest peak at Time 2). Their ELAs lie on local ELA profiles LP1 and LP2 which are assumed to have parallel gradients. Because of the change in extent of the glaciers the topographical position of ELA1 and ELA2 of the main glacier is significantly displaced between Time 1 and Time 2. Consequently the true change in ELA is not given by the simple arithmetical difference **A** between these ELAs but by the vertical difference **B** between the profiles. The discrepancy between **A** and **B** depends on both the horizontal ELA displacement and the slope(s) of LP1 and LP2. With very small glaciers or gently sloping ELA profiles the difference between these estimates may be less than their errors and so be insignificant, but for a glacier (or a group of glaciers) which retreats by say 10 km so that the position of its ELA changes by 5 km, and which is in an area with parallel ELA slopes for both the present and past glaciers of 30 m km^{-1}, the apparent rise in ELA must be reduced by 150 m, a significant amount.

Thus, with detailed information and care, reliable estimates can be made of local changes in ELA, but estimates of changes in regional ELAs from more generalized data (e.g. means of all glaciers in quarter-degree squares) are fraught with problems. RP1 represents an estimate of the regional ELA (a small part of the regional profile) at Time 1, derived from the mean of ELA1, ELA1* and ELA1**. At Time 2 only the higher peaks will be glacierized, so that RP2 will be the mean of ELA2 and ELA2*. Therefore RP2 is derived from a restricted population of higher peaks from which the smaller ones are excluded, so the arithmetic difference **R** between RP1 and RP2 is an excessive estimate for the rise in ELA.

We have the paradox that, while R1 and R2 may be considered correct representations of the regional ELA at Times 1 and 2, the difference between them is not a correct figure for the ELA change. Valid regional changes must be obtained via detailed local ones.

9.2.7 Deglacierization dates

Soon cosmogenic dating may supply reliable ages for actual LGM moraines and older ones. After writing the previous sentence and just before this book went to press the first cosmogenic ^{36}Cl dates for tropical moraines were published for Kilimanjaro and Mount Kenya in East Africa (Shanahan &

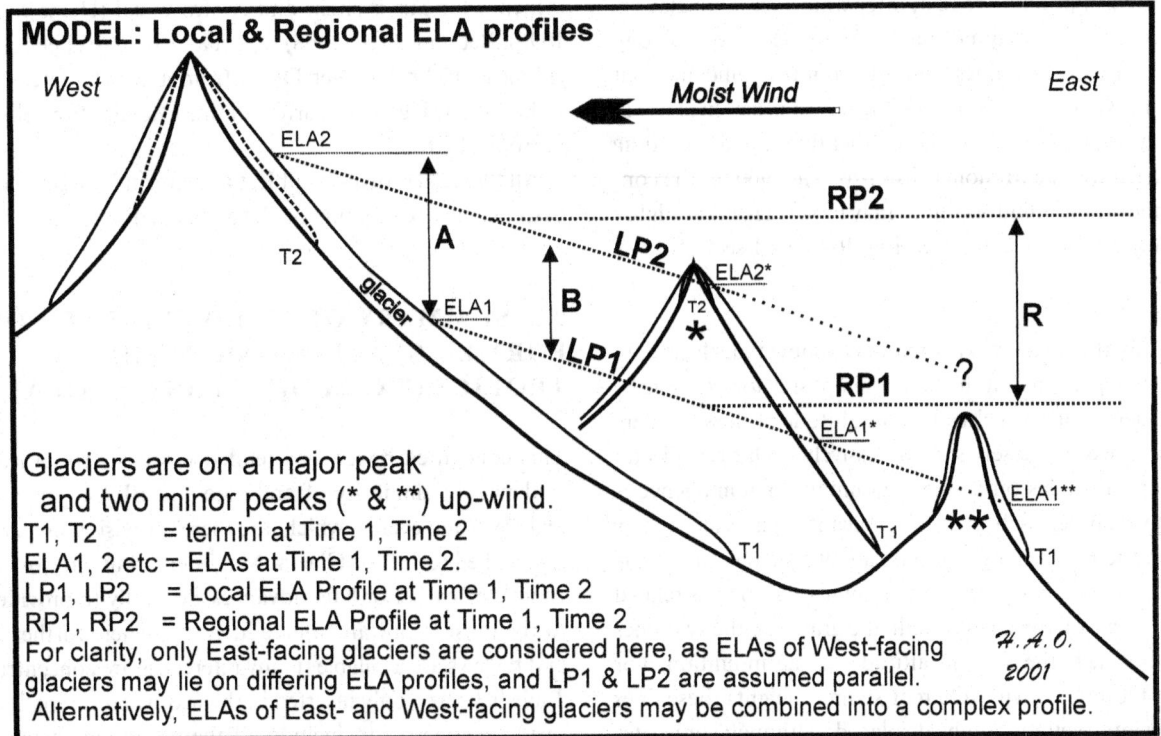

Fig. 9.2.4 Modelled relationships between glacier ELAs at Time 1 and Time 2. On the main peak and on the lower peaks each glacier represents the mean of a group of glaciers in approximately these positions. The local ELA change between Time 1 and Time 2 is not correctly given by A, the difference between ELA1 and ELA2 (nor by the difference between ELA1* and ELA2*), but by B the difference between ELA profiles. Nor is the regional change in ELA correctly given by R, the difference between RP1 (mean of ELA1, ELA1*, ELA1**) and RP2 (mean of ELA2, ELA2*), but by the mean of B and other neighbouring local ELA changes.

Zreda, 2000). These confirm previously accepted relative ages for Kilimanjaro but radically change those for Mount Kenya – see discussion later. At present this is an expensive process but in due course must be applied more widely, and may be expected to modify current ideas considerably in some cases.

In the meantime authors publish and quote *'minimum dates for deglaciation'* (i.e. *'deglacierization'*).* At present this is mostly dependent on ^{14}C dates from organic deposits that are assumed to have been formed shortly after the retreat of the ice from the sites of those deposits, and so provide a minimum age for those events. Numerous papers recording such 'deglaciation dates' are available for mountains throughout the tropics and have been summarized and compared in reviews; however, Gillespie & Molnar (1995) stress how poorly constrained many of these ages are, often being based on a single dated sample originating some time after the start of retreat by the glacier. Moreover, such dates are commonly quoted (as in

Table 9.3.1 below!) without any reference to their relationship to the glacier which preceded them, particularly its toe and head. It is important to distinguish between:

(a) Sites from cirques etc. near the summits of mountains that were only just high enough to have been glacierized. Here melting of the ice corresponded closely with the deglacierization of the mountain as a whole, and with a slight rise of the ELA from just above the site altitude to above (or possibly just below) the summit of the mountain. Within a homogeneous region such events have a precise temporal and altitudinal meaning, which justifies the terminology, but any meaningful altitudinal relationship to similar events in other regions with different climates and different local and regional ELAs is obscure.

(b) Sites at or near the maximum extension of larger glaciers, far below their ELA, which therefore record only the very start of retreat of the glacier, and the relationship of which to the ELA needs individual interpretation, since this is determined as much by the summit height of the mountain and the extent of that particular glacier basin as by the ELA. For these the age is only a minimum one

---

* **Glaciation** is the (usually geomorphic) *effect* of **glacierization**, or the *period* when this happened. Hence **'deglaciation'** should logically mean the opposite – possibly the bulldozing of a moraine? – and is to be deprecated. (*J. Glaciology* **2**, nos. 16, 17, 18.)

for the initiation of retreat, after which actual deglacierization may or may not have taken place, soon or long afterwards, which does have a temporal significance, but has no valid altitudinal relationship to other sites at all. In the extreme case of the frequently quoted 14750 BP date from Lake Mahoma (Rwenzori) at 2960 m, the contemporary terminus position may have been either higher or up to 800 m lower (see Fig. 9.4.1 and section 9.4.5 below).

Naturally there have been attempts to show that these data conform to a pattern, but this can be simplistic, for example by plotting 'deglaciation age' against the altitude of the site where it was obtained. For a series of successively deglacierized sites at different altitudes on a single mountain or homogeneous region this may certainly yield a significant linear relationship. However, the ELA changes at only about half the rate of the terminus, so that a linear rise of a terminus with time is related to a linear rise of the ELA at half the gradient, the two lines intersecting near the summit altitude of the mountain. For different mountains with different summit heights these lines and their intersections will be displaced. Although some sites from different mountains, regions and continents, with various local ELAs, can be selected to show conformity, others will not.

In summary, almost every date of this kind is subject to unquantifiable additions for the period between ice melting at that site and organic matter formation, and the implications of ice melting at that site depend critically on its detailed topographic relationship to the ice extent. Any comparisons or unified conclusions about the dates of the previous glacial advances must therefore be very cautious. Most of the available dates, which span a range from 11 000 BP to 14 000 BP*, are not inconsistent with a common LGM at some time between 15 000 and 20 000 BP. However, the unrecorded gap before the formation of organic deposits is unlikely to put retreat back into a much earlier stage. Only in a few cases where there are datable soils or tephra above and below, or datable lake sediment cores affected by glacial outwash, is it possible to bracket the age more precisely.

At some sites there appear to have been earlier major advances at 25 000–30 000 BP, though these were usually followed by lesser ones before 12 000–15 000 BP. Hamilton & Taylor (1991), however, suggest that deglacierization dates of c. 11 000 BP on Elgon, which are closely matched by pollen changes and influxes of clay in swamps at lower altitudes elsewhere in Uganda, may reflect a definite warming just then, at a time when northern Europe and possibly Canada were experiencing the Younger Dryas cold period, which may also have occurred in the Andes (Francou *et al.*, 1995[4]), though the

evidence for this is slight and controversial (Heine & Heine, 1996). Beuning *et al.* (1998) report a distinct arid period coinciding with the Younger Dryas from a closely dated core in Lake Albert (Uganda), corroborating evidence from elsewhere in tropical Africa.

All these ages are now subject to confirmation or change by cosmogenic (^{36}Cl) dating of the associated moraines.

9.3 SUMMARY OF FIELD EVIDENCE OF FORMER GLACIATIONS IN THE TROPICS (EXCLUDING EAST AFRICA)

This condensed review is intended primarily to indicate the evidence for similarity of end-date and ELA change for the features and events which are here provisionally grouped under the head 'Last Glacial Maximum', but with brief indications of earlier or later advances. Secondly to provide leads to the most important source literature where further details and reviews are available (many references are concentrated in Table 9.3.1 and only inserted in the text where specially necessary). It attempts to include all the mountains between the Tropics of Cancer and Capricorn where evidence has been found (or might be expected) of glaciation in the upper Pleistocene and Holocene (except details of South America), and shows that there is considerable consistency within these limits. However, the examples used in the subsequent detailed discussion are mainly equatorial. Hollin & Schilling (1981) compiled an inventory of all LGM mountain glaciers and small ice-caps, including tropical ones, primarily to estimate the volumes of ice they contained, and Hastenrath (1984, 1985) and Mahaney (1990) have also provided concise summaries, but further information is now available. This general picture is supported by independent pollen evidence (Flenley, 1979; Walker & Flenley, 1979) from East Africa, tropical South America and Australasia which shows lowering of mountain vegetation zones to a minimum altitude in the period 15 000–20 000 BP. In Papua New Guinea this amounted to up to 1200 m difference from the present and from a warmer period at 30 000 BP. However, in East Africa and elsewhere this must be set against evidence of low lake levels at this time, indicating greater aridity, not greater rainfall or 'pluvial periods' as the early glaciologists there expected, and there is recent evidence of other factors affecting the vegetation. USGS (1999 etc.) gives many data about present tropical glaciers but limited data about Pleistocene ones.

These records are included in Table 9.3.1, the restricted aim

---

* All dates quoted are ^{14}C dates, unless specified as ^{36}Cl.

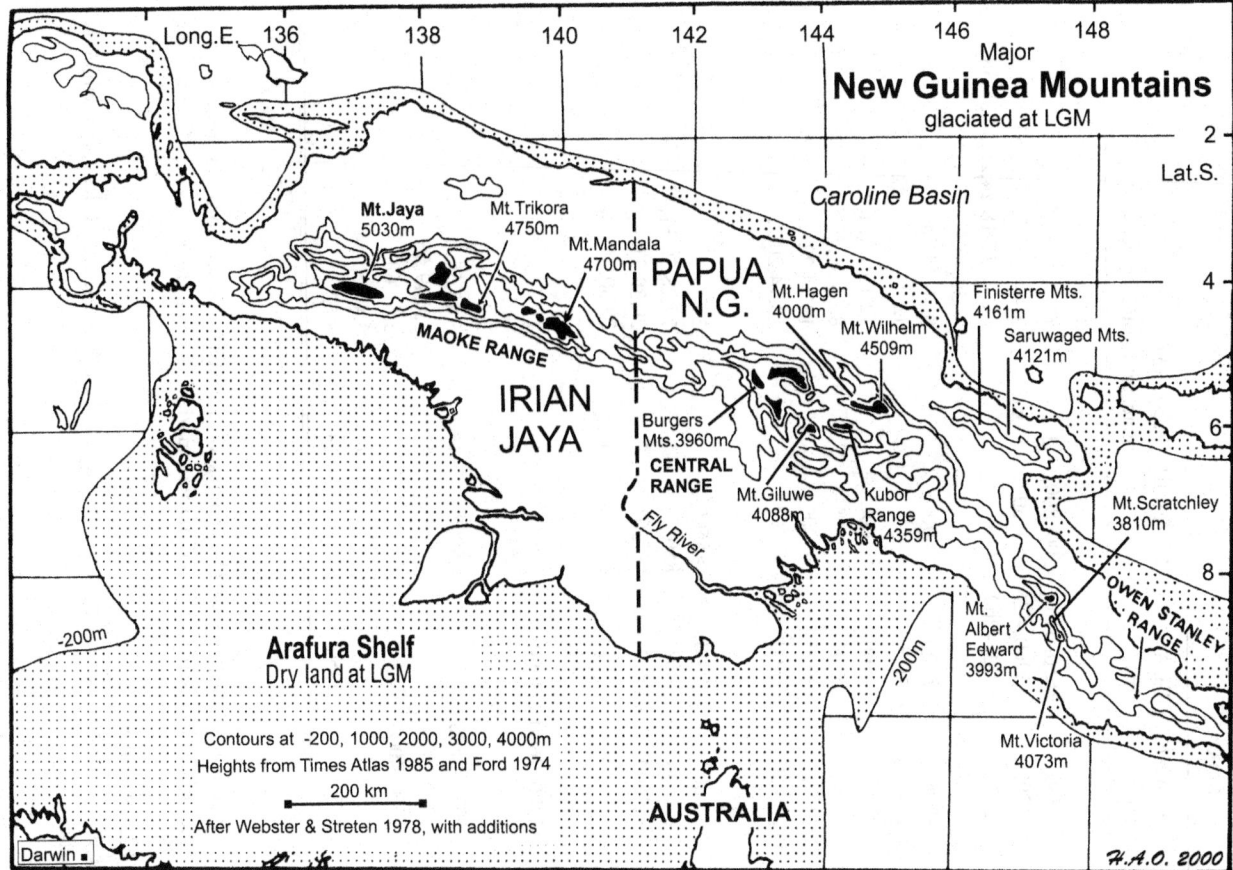

Fig. 9.3.1 Map of major New Guinea mountains glaciated at the LGM.

of which is to show that throughout the tropics most glaciers started to recede soon before 12000–15000 BP from their major advance positions in oxygen isotope stage 2. This temporal conformity, together with the similar form of most of the moraines of that advance, suggests that they were due, if not to a common cause, at least to a group of related environmental conditions. At a few sites there appear to have been earlier advances at 25000–30000 BP, though sometimes followed by lesser ones in the 12000–15000 BP bracket.

After summarizing the evidence for the tropics other than in East Africa, the latter will be discussed in considerable detail, particularly the Rwenzori, partly to complement the account of present and recent glaciers in chapter 6, partly because of the abundant data available.

9.3.1 New Guinea (Australasia)

All these mountains lie on the single island of New Guinea, comprising Irian Jaya and Papua New Guinea, and are in the moist equatorial zone, with high annual precipitations (Fig.

9.3.1). They have evidence of extensive glaciation at the LGM with ELAs below 4000 m. A few still have small glaciers with current ELAs of about 4500 m. At the LGM there were major regional changes in the climatic regime, arising from the global changes, due to the exposure of the whole Arafura Shelf between New Guinea and Australia by the lower sea-level; the southeast trades would have been substantially drier, affecting particularly western New Guinea and perhaps responsible for the higher ELA there at the LGM. Despite considerable Late Quaternary uplift of some areas, a few mountains have small exposures of old tills attributed to previous glaciations. Uplift rates such as 0.8–2.1 m ka^{-1} in the Finisterre Mountains (Abbott et al., 1997) and 1.9–3.3 m ka^{-1} in reefs on the Huon Peninsula (Ota & Chappell, 1999) could significantly affect the apparent ELAs of earlier glaciations. Reported peak altitudes vary; where that in the *Times Atlas* (1999) differs considerably from that in the literature, it is given in parenthesis.

IRIAN JAYA [PUNJAK JAYA] (INDONESIA)
Carstensz: Mount Jaya, 4884 m (5030 m T) and Idenberg Peak, 4717 m

The Carstensz is a limestone ridge 40 km long in the Merauke Mountains, part of the high mountain spine that runs the

Table 9.3.1 *Selected Quaternary tropical glaciers and glaciations*

Place	Max. alt. (m)	Exist. gl. area (km²) (date)[a]	Post-LGM glac. adv.	LGM End before[b] (years BP)	LGM Extent (km²)	LGM Lowest (m)	LGM ELA[c] (m)	LGM ELA fall[d] (m)	Pre-LGM glac. adv.	References
East Africa										
Kilimanjaro	5893	2 (2000)	2 or 3 stages	14–20000*	150	3230	3650–5180	500–800	3	8,26,27,34
Kenya	5198	0.7 (70)	2 or 3 stages	14–64000*	200	c.2900	c.3960–4260	450–600	1 or 2	6,14,35,37,53
Rwenzori	5108	2 (96)	2 or 3 stages	16000	260	2070	3500–4260	500[d]	2	1,4,5,14,32,33,42,53
										14,16,17,18,24,25,35,36,38, 39,40,47
Elgon	4318	0	1 or 2 stages	11000	75	3300	c.3950	>300[e]		9,10,14,34
Aberdares	3993	0	none	12200	c.10	c.3200	c.3600	>350[e]		14,41
Ethiopia, Simen	4543	0	none		13	3760	c.4100	>400[e]		12,13,14,21,22,34
Bale	4377	0	?		c.700	3200	c.3900	>400[e]		8,22,34,46,58
Bada	4170	0	?	11500	c.140	3650	c.3900	>400[e]		13,14,22,57,61
Kecha (Kaka)	4174	0	1 or 2 stages		c.50	c.3050	c.3600	>500[e]		13,14,20,22,46,61
SE Asia										
Kinabalu, Sabah	4094	0	?	9186	6	c.2800?	c.3500?	>550[e]		23,59
Carstensz, Irian Jaya	4884	7 (72)	1 stage?	13850	900	c.3400	3700?	850?	1	19,28,30,31,43
Wilhelm, Papua-NG	4509	0	some	12570	?	<3500	3700?	>750?	2	28,30,31
Giluwe, Papua-NG	4088	0		13050					1	29,30,31
C. America										
Citlatapetl, Mexico	5670	9 (75)		12100	c.100	3200–2600	c.4000	c.800?	1	11
										15
S. America, Andes										
Bolivar, Venezuela	5007	1 (97)		13000		3000–2600		c.800?	1+	2,3
Cocuy, Colombia	5493	39 (97)		27–12320		3000/4000			1+	3,45,55
Chimborazo, Ecuador	6310	20 (84)		34–15000		3600–4000			1	3,7,49,56
Cord.Ap.Real, Bolivia	6485	290 (97)		28–35000					1	3,48,50,54
Quelccaya, Peru	5645	55 (97)		12240		5050	c.5350		1+	3,52

Notes:

[a] date (19xx) refers to that at which the area was observed.

[b] The date refers to a radiometric date for basal lake/bog sediments formed soon after the retreat of the glaciers. Those marked* are ^{36}Cl dates on moraines (ref.53).

[c] Where a range of ELA altitudes is given, these are values for different aspects of the mountain. In addition, on Kilimanjaro and Mt Kenya, the ELAs on Mawenzi and Ithanguni, minor peaks to the east of each, were about 800 m and 600 m lower.

[d] Only on the Rwenzori has the ELA change been corrected for the slopes of the two ELA surfaces.

[e] On mountains now without glaciers, the minimum fall in ELA has been taken as the difference between the peak altitude and the ELA of the LGM, less 50 m which is the minimum height by which the glacierization limit is commonly lower than the peak.

References

1. Baker 1967
2. Clapperton 1991
3. Clapperton 1993
4. Coetzee 1964
5. Coetzee 1967
6. Downie 1964.
7. Gonzalez et al. 1965
8. Hamilton 1982
9. Hamilton & Perrott 1978
10. Hamilton & Perrott 1979
11. Hastenrath 1971
12. Hastenrath 1974[2]
13. Hastenrath 1977
14. Hastenrath 1984
15. Heine 1975
16. de Heinzelin 1952
17. de Heinzelin 1953
18. de Heinzelin 1962
19. Hope et al. 1976
20. Hövermann 1954[1,2]
21. Hurni 1982
22. Hurni 1989
23. Koopmans & Stauffer 1968
24. Livingstone 1962
25. Livingstone 1967
26. Livingstone 1975
27. Livingstone 1980
28. Löffler 1972
29. Löffler 1976
30. Löffler 1977
31. Löffler 1982
32. Mahaney 1982
33. Mahaney 1989
34. Messerli et al. 1980
35. Osmaston 1965
36. Osmaston 1975
37. Osmaston 1989[1]
38. Osmaston 1989[2]
39. Osmaston 1998
40. Osmaston & Pasteur 1972
41. Perrott 1982[1]
42. Perrott 1982[2]
43. Peterson et al. 1973
44. Potter 1976
45. Schubert 1975
46. Street 1979[1]
47. Whittow 1966
48. Lauer & Rafiqpoor 1986
49. Hoyos-Patiño 2000
50. Jordan 2000
51. Jordan & Hastenrath 2000
52. Morales-Arnao & Hastenrath 2000
53. Shanahan & Zreda 2000
54. Argollo 1982
55. Schubert & Clapperton 1990
56. Hammen et al. 1981
57. Street 1979[1,2]
58. Messerli et al. 1977
59. Flenley & Morley 1978
60. Hastenrath 1981
61. Gasse et al. 1980

length of New Guinea (Fig. 9.3.1). It was studied intensively by the Australian Universities Expedition (Hope *et al.*, 1976; see also Löffler, 1977, 1982). Only small glaciers still survive, mainly on the steep north face with a present ELA of 4580 m, but LGM moraines up to 150 m high extend over about 900 km², mostly descending to about 3400 m (moraines at exceptionally low altitude are reported at 2300 m in the south but this requires confirmation). This large extent is partly due to the local relief. The accumulation zones of the glaciers descended a steep rocky face but the ablation zones spread out on the Kemabu and adjoining plateaux as piedmont lobes not far below the ELA, which from cirque altitudes and the upper limits of moraines has been estimated at about 3700 m (Hope *et al.*, 1976), indicating a depression of 880 m (*c.* 5.7 °C). However, this would imply a very low THAR of 0.2, possibly due to the piedmont form, but suggesting that the LGM ELA has been underestimated and requires checking. The lack of contoured maps prevented application of an AAR or AABR method (see below). The present annual precipitation is about 3000 mm, changing little with altitude. On the glaciers incoming shortwave radiation is consequently much reduced by cloud, but even so the compensating longwave radiation causes most of the ablation (see also Allison & Kruss, 1977). A date of 13 850 BP was obtained from the base of a small former lake impounded by moraines of a minor retreat stage close to the LGM maximum moraines, supported by another of 13 260 BP from organic material on glaciated bedrock. Interpretation of deposits is complicated by strong karstic solution and collapse effects in the limestone resulting from the heavy rainfall. Spreads of till beyond the LGM moraines indicate probably one or more earlier glaciations and there is evidence of four post-LGM advances.

Other peaks in Irian Jaya
Mount Mandala (4680 m, 4700 m T) has evidence of extensive glaciation at the LGM and still had some ice cover recently (Verstappen, 1964) though that may now have disappeared (Peterson & Peterson, 1994). Mount Trikora (= Wilhelmina, 4730 m, 4750 m T) lost its remnants of ice a few decades ago and now only has traces of former glaciation. Other peaks over 4500 m are mapped but no information is available about them.

PAPUA NEW GUINEA
Löffler (1972) lists 20 mountains where there is evidence of Pleistocene glaciation totalling 600 km², mostly inferred from air photos, though none is still glacierized. Altitudes given by other sources (e.g. *Times Atlas*, 1999; Ford, 1974) often differ. Those with the largest glaciated areas (over 10 km²) are as in Table 9.3.2.

Table 9.3.2. *Papua New Guinea mountains, glaciated areas over 10 km²*

	Altitude (m)	Area (km²)
Mt. Giluwe	4368	188
Mt. Wilhelm	4509	107
Albert Edward	3990	90
Saruwaged Range	4121	80
Mt. Scratchley	3810	28
Kubor Range	4359	27
Mt. Hagen	4509	21
Mt. Victoria	4073	15
Finisterre Mts	4161	12
Burgers Mts	3960	10

Some had ice-caps on volcanic domes, while others are ridges with cirques and deep glaciated valleys. All have moraines of the LGM, recession from which is dated in some cases as just over 12 000–13 000 BP, and from which ELAs of 3500–3700 m have been inferred. These would imply minimum differences between present and LGM ELAs of 300–800 m. Mount Giluwe and Mount Wilhelm have been studied in detail and have traces of both previous (dated on Wilhelm to 290 000 BP and >380 000 BP by interpolated tephra) and later lesser advances (Löffler, 1972, 1976, 1977, 1982).

9.3.2 Mount Kinabalu (Sabah, East Malaysia), 4094 m

Mount Kinabalu is an upthrust pluton that rises steeply from lowland tropical rainforest with a spectacularly glaciated bare granite summit and a high rainfall. The sides are so steep that moraines are subject to rapid erosion. There are small LGM moraines but no traces of earlier or later stages (Koopmans & Stauffer, 1968). The basal sediments in a rock pool in the glaciated area have been dated at 9186 BP (Flenley & Morley, 1978).

9.3.3 Yu Shan (Taiwan), 3997 m

Yu Shan in south Taiwan is exactly on the tropic of Cancer. There is evidence of glaciation, including cirque lakes, polished surfaces and striations, moraines, etc., in three different periods dated at 44 and 18 ka BP and during the Late Glacial (Cui *et al.*, 2000).

9.3.4 Mauna Kea (Hawaii), 4206 m

Mauna Kea, the highest of five massive shield volcanoes that form the islands of Hawaii, is the only summit in the tropical mid-Pacific known to possess evidence of former glaciation.

A series of four tills are separated by dated tephra. The moraines of the last ice-cap are bracketed between tephra dated to 9000 and 29 000–41 000 BP. The ice-cap extended for about 5 km from the summit and covered an area of about 60 km² with an ELA at about 3700 m, not differing much on different aspects. Three older tills which were slightly more extensive are dated at between 41 000 and 69 000 BP, at about 174 000 BP, and between 122 000 and 278 000 BP (Stearns, 1945; Porter, 1979[1,2]; Gillespie & Molnar, 1995).

9.3.5 Central America

There are three still snowcapped volcanoes over 5000 m in Mexico, Citlaltepetl (5670 m), Popocatepetl (5452 m) and Ixtaccihuatl (5286 m) (see maps in Hollin & Schilling, 1981, and White, 1981). These and five others at 4000–4500 m show evidence of past glaciation, while others remain to be investigated. Several moraine stages have been identified and a 'standard Mexican glacial sequence' has been proposed by White (1962, 1981) and White & Velastro (1984) based on observations on Ixtaccihuatl and comprising five separate advances, the earliest being an ice-cap covering 100 km² with an ELA of about 4000 m. However, correlation between that and the sequence on Ajusco (3937 m) is poor and there is some conflict in the ¹⁴C dates. It is not clear which advance might correspond with the LGM, the last major advance appearing to date from before 25 000 BP. Heine (1975) reports several advances on Citlaltepetl and Malinche including a maximum dated at 33 000 BP and a slightly lesser one at 12 100 BP. White inferred a lowering of the ELA on Ixtaccihuatl by 900 m at the last major advance and a similar lowering on Ajusco but considered that all ELAs on the latter were 700–800 m lower than on Ixtaccihuatl; this large difference needs confirmation.

Hastenrath (1974[1], 1984) reported moraines on the Altos de Cuchumatanes (3993 m) in Guatemala, and on Chirripo (3820 m) in Costa Rica (1973, 1984), both indicating a lowering of the ELA by at least 300 m to about 3500 m.

9.3.6 Andes (South America)

The northern half of the Andean range extends for 5500 km across the tropics from Venezuela to Chile, with numerous parts still snow-capped and with extensive areas marked by the moraines and tills of Pleistocene glaciers which extended over more than 100 000 km², with snowlines at 3500–4500 m. Despite (or because of?) intensive studies by numerous workers, there are a number of contradictions in the reported observations and the conclusions drawn from them. Some deposits have initially been recorded as tills, subsequently reclassified as lahars, then once more considered to be tills.

There are also question of the effects of tectonic uplift of parts of the chain. Accordingly, recent critical reviews and syntheses (e.g. Schubert & Clapperton, 1990; Clapperton, 1991, 1993 (see particularly Table 14.1), 1998) should be consulted, and only a selection will be presented here. An extremely detailed technique for estimating ELAs at the LGM from cirque floors over the whole Andes was used by Nogami (1976) (see below). A more generalized approach to estimating present and former ELAs in the Central Andes (Klein & Isacks, 1998; Klein et al., 1999) used air photos and Landsat TM to identify present snowlines and Pleistocene glacier extents in the Central Andes (see below).

There is widespread evidence of a LGM somewhat older than 12 000–13 000 BP, but there is stronger support for a greater extension in 33 000–28 000 BP. Following these there were as many as four lesser retreat stages or readvances. More controversial are very extensive moraines and tills, many of the latter buried by volcanic or sedimentary deposits and hence invisible until exposed by excavation or erosion. Some of these are attributed to the penultimate glaciation c. 100 000 BP, but others are much older. In the Cordillera Real Apolobamba, (just north of Lake Titicaca, Bolivia) below outwash from the LGM are three major tills, interbedded with fluvioglacial sediments and two ignimbrites. The ignimbrite overlying the lowest till yielded a K–R age of 3.2 Ma, demonstrating that glaciers expanded even further than LGM ones in the late Pliocene and after (Clapperton, 1991).

The northernmost part of the Andes in Venezuela includes Pico Bolivar (5007 m) in the Mérida Cordillera (lat. 11° N) where the modern snowline is at 4700 m. Two main glacial stages have been distinguished here (Schubert, 1970, 1974, 1979; Schubert & Clapperton, 1990) with degraded moraines at 2600–2700 m and better preserved ones at 3000–3500 m respectively, but there is uncertainty whether both belong to the last glaciation or not. Peat interbedded with outwash of the latter was dated at 16 500–19 000 BP, indicating that they were LGM.

The Colombian cordilleras were extensively glaciated and have been the subject of intensive study, partly because of their proximity to exceptionally long cores from old lake basins, providing an opportunity for cross-checking glacial data with vegetation zone changes inferred from pollen records. During 21 000–14 000 BP the Andean treeline was 1200 m lower, apparently indicating a temperature lowering of at least 8 °C (Hammen et al., 1981; Helmens & Khury, 1995) (but see section 9.5.1). The most extensive glaciers reached down to 2700–3250 m, being lower on the eastern ranges and lower on the eastern slope of each range. The Sierra Nevada de Cocuy appears to have moraines of only two glaciations but there is still controversy as to their ages. The Ruiz Tolima massif now has only 35 km² of ice but at the LGM there were over 10 000 km², with an ELA over 1000 m lower (Thouret *et al.*, 1996).

In Ecuador (Hastenrath, 1981; Clapperton, 1987), the largest moraines lie at 3000–3600 m under peat dated at 38 000 BP, but others lie just short of them dating from before 12 000 BP and probably represent the LGM. A more extensive and highly weathered till appears to date from the penultimate glaciation.

In Peru, Mercer & Palacios (1977) and Wright (1983) have shown that the maximum extent of the last glaciation occurred before 28 000 BP, but that lesser advances date from before 12 000 BP. The LGM ELA was at about 3700 m in the northeast and 4000 m in the northwest, rising to 4300 m in central Peru and 4500 m in the south. The maximum Late Pleistocene ELA depression was 500 m in the western cordilleras but only 300 m in the east. There is also a much older till, probably of the penultimate glaciation. The Cordillera Blanca rising to Nevada Huascarán (6768 m) (see earlier chapters) now has the largest area of ice remaining in the tropics, with 722 glaciers covering 723 km² and an ELA of about 5100 m (Fig. 8.2.5), but at the LGM it was exceeded by several other cordilleras. At least four groups of moraines occur; although ¹⁴C dating indicates that the two youngest sets are late LGM, the ages of the others are not yet satisfactorily defined. The oldest are degraded, deeply weathered and lateritized, so probably date from the penultimate glaciation or before. Besides these there are Holocene moraines for which ¹⁴C dates (Röthlisberger, 1987) and ¹⁴C and lichenometric dates (Rodbell, 1992¹) indicated dates of 7000–6000 BP, 3350–1800 BP and 1250–4000 BP.

In southern Peru and northern Bolivia where the annual precipitation reaches 1000 mm, glaciers in the Apolobamba Real (6059 m) reached their maximum extent down to 3500 m elevation at 35 000–28 000 BP, inferred from dates on peat interbedded with moraine (Lauer & Rafiqpoor, 1986; Argollo, 1982, both quoted with some reserve in Clapperton, 1993). Dates from other similar material record a later advance at about 12 900 BP. In southern Bolivia where precipitation falls below 200 mm there are no present glaciers and only small traces of Pleistocene ones.

Clapperton (1993, p. 410) concludes that 'glaciers in the tropical Andes reached their greatest Last Glaciation extent between 35 000 BP and 28 000 BP, when temperature was low and precipitation was high, but shrank when severe global cooling and lower ocean temperatures at 18 000–14 000 BP caused greater aridity'.

9.3.7 Ethiopia (Northeast Africa)

The volcanic Ethiopian highlands include ten massifs above 4000 m with no current glaciers but high enough to have

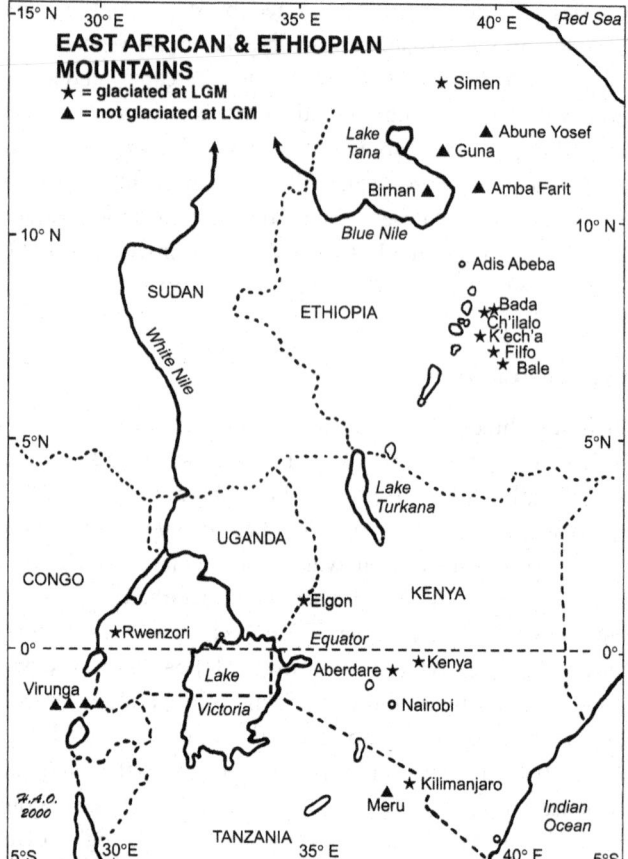

Fig. 9.3.2 Map of East African and Ethiopian mountains glaciated at the LGM.

possibly been glaciated at the last LGM (Fig. 9.3.2); indeed many of these have moraines but apparent evidence of only one glaciation. All these mountains are equatorial, with moderate to high annual precipitations (2000 mm), though on some this is sharply reduced at the highest levels. However, there is some ambiguity about the records. Various sources use up to three different names and five different altitudes (ranging over 300 m) for the same peak. Here H and T indicate altitudes derived respectively from Hurni (1989) and the *Times Atlas* (1999). The best studied is Simen (= Semyen) in the north (Hastenrath, 1974², 1977; Hurni, 1982 – clearly and fully, 1989) where Ras Dechen (4543 m H, 4620 m T) and several smaller peaks rise from the plateau basalts, with a combined glaciated area of 13 km², marked by small but conspicuous moraines down to 3500 m. These indicate an ELA of 4100 m on northerly aspects where most of the glaciers formed, but 4400 m on the south side. Dates from peats on these moraines indicated minimum ages of about 4100 years, but most workers correlate them with the LGM (Hurni, 1982, 1989). There are also smaller moraines at about 4250 m. There are four other mountains in the north which Hurni (1989) reports were just lower than the glacierization level there: Guliba Amba/Abuna Yosef (4284 m H, 4193 m T); Guna (4135 m H, 4231 m T); Birhan, Ch'ok'e Mountains (4052 m H, 4254 m T); Molle/Amba Farit (4247 m H, 3975 m T).

The mountains in the south of Ethiopia are less well studied but it is clear that the glacierization level there was lower. The Bale Mountains (summits Tullu Deemtu 4450 m H, Mount Batu 4377 m H, 4321 m T) have a large area at over 4000 m and preliminary estimates are that this had an ice-cap of over 700 km[2] at the LGM, with an ELA at about 3700 m and large glaciers extending down to 3200 m (Messerli *et al.*, 1977, 1980; Hurni, 1989). However, the evidence for these conclusions needs verification. A core from Danka Valley at 3830 m gave a basal date of only 7920 BP (Hamilton, 1982), but a basal age of 13 000 BP was obtained from a core taken at 3000 m in an intensely periglaciated area (Umer & Bonnefille, 1998).

A similar ice-cap occurred on Mount Bada, Arusi Mountains (4170 m H, 4136 m T; Potter, 1976; Hastenrath, 1977, 1984; Street, 1979[1,2]). Street (1979[1]) gives a detailed map of the moraines mainly from air photos, from which the glaciated area can be estimated at about 85 km[2] (not 140 km[2] as stated elsewhere, perhaps due to a scale error) with ELAs at 3700 m on the east and 3900 m on the west. The base of a bog in a glacial cirque at 4040 m was dated at 11 500 BP (Gasse & Descourtieux, 1979; Street, 1979[1,2]; Hamilton, 1982), from which it appears likely that the moraines are of LGM age, but there are a few slightly more extensive older moraines.

Reports of glaciation on other mountains in southern Ethiopia are controversial. Höverman (1954[1,2]) reported a 50 km[2] ice-cap with a low ELA on Mount K'ech'a (= Kaka/ Cacca, 4245 m, 4193 m T). Hastenrath (1977, 1984) also reported from air photos the presence of moraines there, as well as on Ch'ilalo (= Cilalo, 4005 m, 4139 m) and Filfo (= Enguolo, 3850 m, 4340 m); he mapped small ice-caps and cirque glaciers on all three, with areas estimated as 19 km[2], 18 km[2] and 2 km[2] respectively by Gasse *et al.* (1980). However, Hurni (1989) considered that none of these was glaciated; Street (1979[1]) considered that most of these features were lava flows or other non-glacial rock ridges; while Gèze (1974) considered the features on Ch'ilalo to be only nivational.

Reports of moraines at *c.* 2000 m on still lower mountains in Eritrea (Hövermann, 1954[2]) and in Godjam (=Gojjam, Mount Choke) (Kuls & Semmel, 1965) seem impossibly low.

Across the country there is a north–south downward slope to the glacierization limit, from about 4300 m to about 4100 m. No evidence of earlier glaciations has been reported other than the possibly older moraines on Bada. There are extensive periglacial features on all the high mountains and Hurni (1989) proposes a much drier climate, 7 °C colder at the LGM.

This is supported by the pollen evidence from Bada (Street, 1979[1]; Hamilton, 1982).

9.4 THE EAST AFRICAN MOUNTAINS

The three highest East African mountains, Kilimanjaro, Kenya and Rwenzori, all over 5000 m (Fig. 9.3.2), still carry small but rapidly diminishing glaciers, have conspicuous LGM moraines and have some evidence of several extensive former glaciations; Hastenrath (1984) gives useful maps. Two lower mountains over 4000 m, Elgon and Aberdare, have LGM moraines but no current glaciers, and evidence of only one glaciation. Hamilton (1982 and pers. comm.) reports that Mount Meru (Tanzania, 4565 m) appears to have moraines but this needs confirmation. However no evidence of glaciation has been found on the Virunga volcanoes (in Congo, Rwanda and Uganda, 4000–4500 m), (nor on Mount Cameroun, 4095 m, in West Africa), though on some of these it may have been obliterated by subsequent volcanic activity or erosion. A clear account has been given by Rosqvist (1990).

All these mountains are equatorial, with moderate to high annual precipitations, though on some this is sharply reduced at the highest levels. The five high East African mountains, Kilimanjaro, Kenya, Rwenzori, Elgon and Aberdare, which were glaciated at the LGM provide an exceptionally useful data set for studying the interactions of general and local climate and relief with glacier extent and ELA. Excellent air photos and 1:50 000 accurately contoured maps cover them. The moraines on each have been mapped, on some by several different workers. The glacial extents and estimated ELAs at different times show both remarkable differences and remarkable similarities between the mountains and between different aspects of each mountain. The areal extent of the LGM glaciers on the three highest mountains is in inverse order to their summit altitudes, reflecting partly their form and partly their humidity regime: the highest and driest, Kilimanjaro, has the smallest glaciated area.

Accounts of the vegetation zones of tropical mountains used to portray them as standard annular belts at successively higher altitudes. In 1951, Hedberg published his oft-reproduced diagram of the vegetation zones on East African mountains, showing that because of the differing local climates on different aspects of each mountain there are great differences in the vegetation belts. Some do not occur at all on certain mountains or only on certain aspects or at different altitudes. Naturally general appreciation of this variability has greatly affected tropical montane ecological and climatic studies.

Many glaciological accounts and diagrams still provide a single value for the glacial extent or ELA on a particular

Former glacier extents and ELAs on the East African Mountains

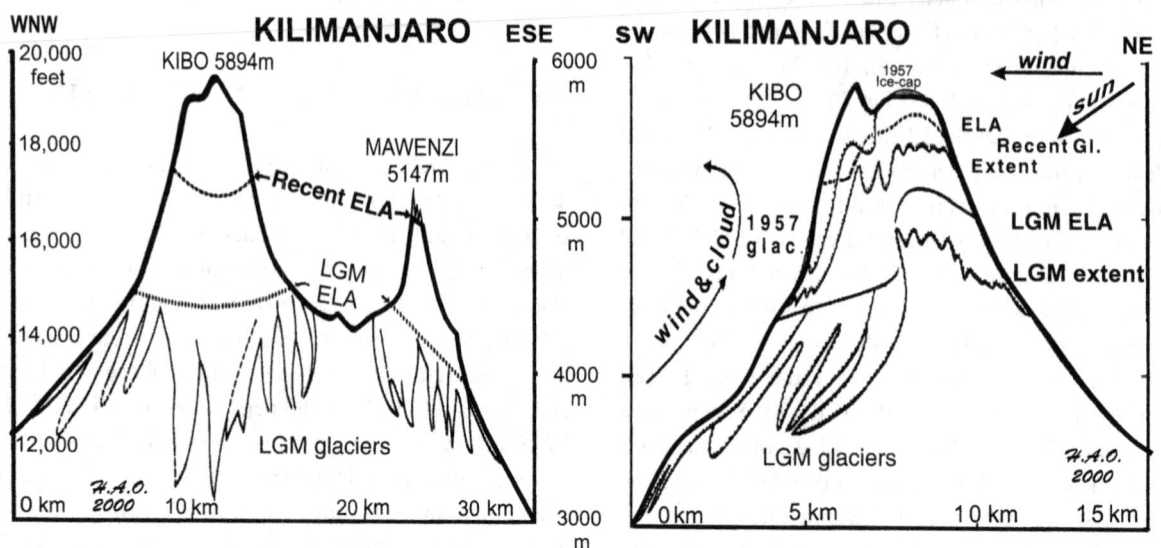

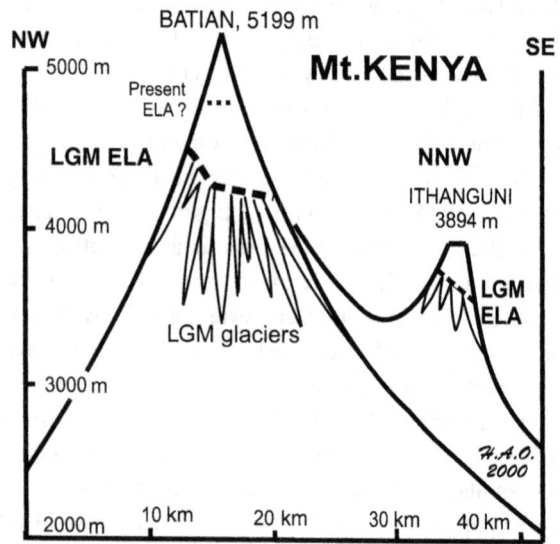

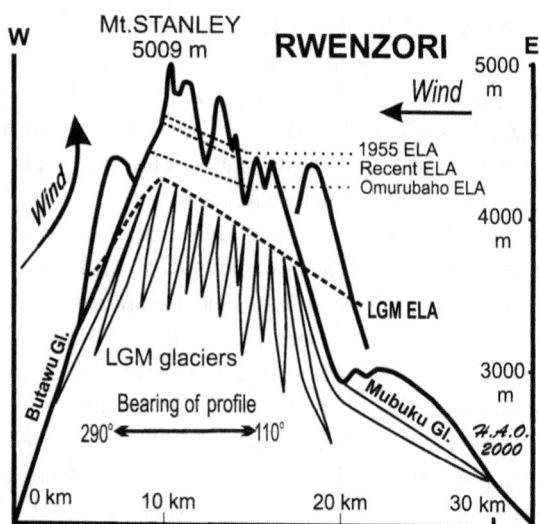

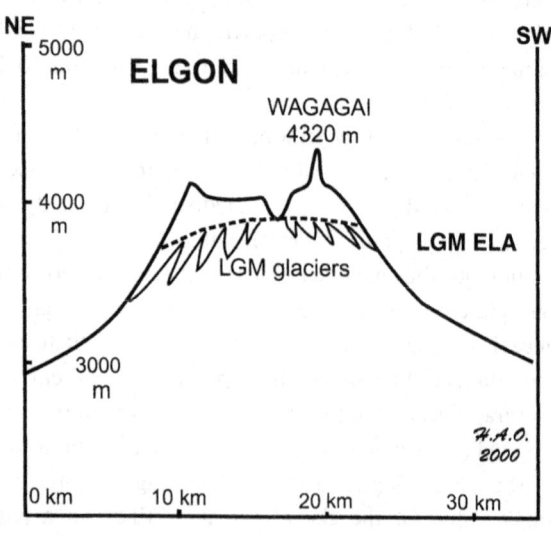

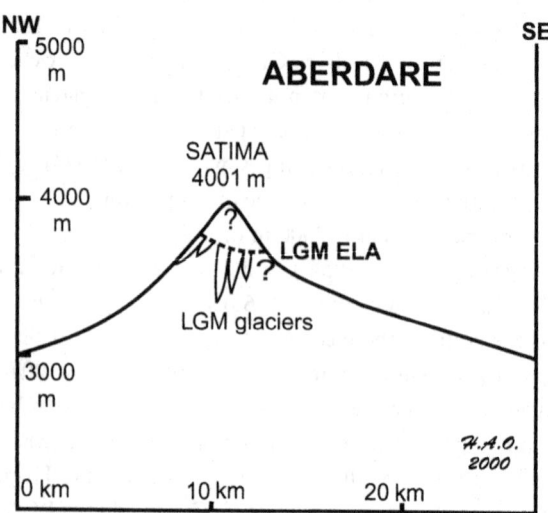

mountain or range when in fact a similar spatial variability applies (Hurni's figure for Simien (1982) is a notable exception). These effects can be seen with present and former glacier extents, glacierization levels and ELAs, and any discussion of them must take this into account. Fig. 9.4.1 shows in simplified form the LGM glacier extents and ELAs of glaciated mountains in East Africa (but see comments on Mount Kenya below). On both Kenya and Kilimanjaro there is also a minor outlying glaciated peak to the east upwind, Ithanguni (now without snow or glaciers) and Mawenzi (still with snow patches); on Rwenzori there were many such outliers. On each of these mountains it is clear that at the LGM the local ELA on the outliers was far below that on the main peaks and that the rise in ELA must either be assessed separately on the main and minor peaks, or preferably by taking into account the form of the ELA profile or trend surface.

9.4.1 Kilimanjaro, 5895 m

On Kilimanjaro the moraines of the 'Fourth or Main Glaciation', provisionally correlated with the LGM (Downie, 1964; Osmaston, 1965, 1989[1]), are conspicuous in the field and on air photos, covering 160 km^2 (Figs. 9.1.1, 9.4.1). A few relics of possibly three earlier glaciations were claimed by Downie, some preserved under subsequent dated lava flows: the First lies between lavas dated at 500 000 BP and 460 000 BP; the Second between lavas dated at 460 000 BP and 240 000 BP; the Third between the 240 000 BP lava and an undated one which preceded the LGM. Downie also recorded

a 'Little Glaciation' following and slightly less extensive than the Fourth, but Osmaston considered that these two could not be reliably distinguished on air photos. Some of the lavas show evidence of being erupted *either* into ice *or* on to ice-free ground, the latter indicating an interglacial. The crater was occupied by an ice-cap at the LGM but the outer sides are so steep that most of the outflows had typical valley glacier forms, though with expanded lobes where they flowed on to the Shira Plateau. Some decaying ice still survives in the crater and some rapidly retreating glaciers on the western slopes. Small moraines mark the limits of the Recent Glaciation a century or two old. Mawenzi, a lower outlying peak, was also glaciated in the Fourth Glaciation, but with a much lower ELA than on Kibo.

Important new evidence from the cosmogenic (^{36}Cl) dating of moraine boulders permits reassessment of these observations (Table 9.4.1; Shanahan & Zreda, 2000).

The estimated age of the landform is not simply the mean of the values for individual boulders and its standard deviation. It is derived from the individual values by a complex statistical simulation programme which uses the spread of values to make allowances for probable variations in surface erosion of the boulders and variations in probable exhumation date, thus yielding a 'best estimate of the age' with minimum variance. However, it is difficult to know what credence to give to the apparent precision of the landform ages in view of the wide spread of individual sample values. At this stage of the use of this technique, the most reliable conclusions that can be drawn from the results are that:

Fig. 9.4.1 Former glacier extents and ELAs of glaciated mountains in East Africa.

Notes:

1. Each mountain is viewed from a direction selected to show the maximum variation of the ELA profile, but on Mount Kenya this differs slightly from the direction of Ithanguni which has been superposed.

2. The vertical scale is the same for all, but the horizontal scales vary, so the vertical exaggeration ranges from 1:5 to 1:11.

3. Most of the mapping of moraines on all these mountains was done from aerial photos, with limited checking on the ground, but on volcanoes (all the mountains except Rwenzori) linear volcanic features can resemble moraines on air photos, especially under forest. Even on the ground it may be difficult to discriminate between moraines and lahars. On Kilimanjaro all three sources agree well except on the Saddle as noted in the text. On Mount Kenya there are some discrepancies, and I have disregarded a few of the lowest mapped moraines in the forest which appeared discordant with others. Similarly on Mount Elgon, where I have been unable to identify them on large scale air photos. On Rwenzori both sources agreed closely.

4. Except in the case of the Mubuku and Butawu Glaciers on the Rwenzori Mountains the drawn moraines are symbolic and do not represent individual glaciers, but their vertical extent is typical of those on that aspect of the mountain.

5. On Mount Kenya the moraines mainly represent Liki I stage which is now known to be much older than the LGM. However, the distributions of Liki II and III moraines have not everywhere been mapped and where they have been mapped their altitudes are often not very different from Liki I (see Fig. 9.4.2). Liki IIA and Liki IIIA are newly distinguished locally by Shanahan & Zreda (2000) by cosmogenic dating and are not generally mapped.

Map sources of moraine data etc. Kilimanjaro: Downie (1964), Osmaston (1965, 1989[1]), Hastenrath (1984). Mount Kenya: Baker (1967), Hastenrath (1984), Mahaney (1990). Rwenzori: Osmaston (1965, 1989[1,2]), Hastenrath (1984), Osmaston & Kaser (2001). Mount Elgon: Hamilton & Perrott (1979), Hamilton (1982), Hastenrath (1984). Aberdare: Hastenrath (1984).

Table 9.4.1. *Ages of moraines on Kilimanjaro estimated by* ^{36}Cl *analysis*

	Number of samples	Boulder age range (ka)	Estimated landform age (ka, ±SD)
Oldest (Second?) Glaciation moraines			
South Mawenzi	5	74–355	>360
Main = Fourth Glaciation moraines			
South Mawenzi	3	19–21	20±1
SW Mawenzi (= 'Saddle')	7	15–23	17.3±2.9
SE Kibo (probably a retreat stage)	4	12–17	13.8±2.3
Little Glaciation moraines			
SW Mawenzi (= 'Saddle')	12	11–19	15.8±2.5

Notes:

1. The ages of the younger landforms (<50 ka) are estimated as the mean and standard deviation of the ages of the individual boulders sampled on it, after excluding any boulders the age of which lies >2 SD outside the rest of the group.
2. The ages of the older landforms (>50 ka) are estimated from the individual boulder figures by statistical simulation models, allowing both for surface erosion of the boulder, and for surface erosion of the moraine which produces different exhumation dates. This indicates a range of landform ages over which the variance is minimized. This exceeds the apparent ages of the individual boulders.
3. The moraines recorded as 'Saddle' were deposited by a glacier that had flowed from Mawenzi across the southeast corner of the Saddle and was descending from there.
4. In addition two boulders recorded as 'on the Saddle' were sampled giving ages of 38 ka and 7 ka. The latter was discounted due to heavy spalling. Major element analyses of both boulders were consistent with an origin on Mawenzi, but both lie only just outside the moraines in Note 3 and are not necessarily representative of the central Saddle.

Source: Summarized from Shanahan & Zreda (2000).

1. As previously proposed, the Fourth or Main Glaciation was clearly LGM in the sense used in this chapter, with maximum extension at about 17–20 ka BP.
2. The Little Glaciation cannot be distinguished from it in age and probably represents retreat stages or minor readvances at 14–16 ka BP.
3. The single 38 ka BP boulder on the south margin of the Saddle gives limited support to my view that only the parts adjoining Mawenzi were glaciated in the Main Glaciation, and that the glaciers from Kibo and Mawenzi did not coalesce there then as supposed by others. The glaciers of earlier glaciations may or may not have covered it but have left no moraines visible on air photos.
4. There was at least one much older and slightly more extensive glaciation aged several 100 ka.

On Kibo peak the pattern of present and former glaciers, and their inferred ELAs, is very asymmetric, reflecting major differences in local climate. The annual precipitation is about 2000 mm on the southern slopes of Kilimanjaro up to 2000 m altitude, but half that on the drier northern slopes, and falls to less than 250 mm on the Saddle (4200 m) and above.

On Kibo, the main peak of Kilimanjaro, the present glaciers are nearly confined to the upper western slopes of the summit cone, because the eastern slopes receive the morning sun, and the northeast wind which prevails on the upper slopes during May–October is dry and cloudless whereas in the afternoon the western slopes are shaded by rising clouds (see Fig. 8.2.10). The ice masses in the crater are rapidly shrinking and the present hypothetical ELA there is above the level of the summit (5894 m) though on the south and west it may be lower. Hastenrath & Greischar (1997) show that the ice cover diminished from 12 km² (1912) to 6.6 km² (1953) and to 3.3 km² (1989) in almost exactly the same proportions as that on the Rwenzori over the same periods. In the Recent Glaciation the ELA lay at 5700 m on the east slopes of the peak and as low as 5200 m on the west. The extents of the LGM glaciers and their estimated ELAs were also very asymmetric, stretching at a nearly uniform 4500 m across the southern and western faces of the peak, but being much higher on the east and northeast than elsewhere for probably similar reasons to now. This eastern side is also in a precipitation shadow from the subsidiary peak of Mawenzi. (This interpretation is based on Osmaston (1965, 1989[1]) with ELAs estimated by the AABR method. It differs from that of Downie (1964) and Hastenrath (1984) who claimed that large glaciers had descended from Kibo and Mawenzi on to the Saddle and coalesced there at the LGM. However, neither author mapped moraines there, and such an extent would greatly differ from the forms of the recent and present glacial extents which conform closely elsewhere. Recent cosmogenic dating partially confirms this.)

On Mawenzi peak, further to the east, the ELA of the LGM sloped steeply eastwards, with a gradient of 90 m km⁻¹ (see Table 9.4.4 below) reaching a lowest altitude of 4250 m. In the precipitation shadow between Mawenzi and Kibo there is no clear slope of the ELA but on Kibo itself there is an ELA discontinuity of 500 m between the northeast and the southeast aspects.

Table 9.4.2. *Ages of moraines on Mount Kenya estimated by* [36]*Cl analysis*

	Number of samples	Boulder age range (ka)	Estimated landform age (ka, ±SD)	Previous estimated age (ka) **not** obtained by [36]Cl analysis (Mahaney, 1990)
Liki I moraines				
Gorges valley	3	329–432	355–420	c. 50
Teleki moraines				
Teleki valley	7	134–531	255–285	>100
Naro Moru till				
Gorges valley	8	36–101	55±23	>Teleki
Liki II moraines				
Teleki Valley	10	19–135	64±40	>15
Gorges Valley	6	24–32	28±3	>15
Liki IIA moraine				
Gorges Valley	2	14–15	14.6±1.2	Not distinguished from L II
Liki III moraines				
Gorges Valley	3	11–14	14.1±0.6	>12.5
Teleki valley	6	9.5–10.9	10.2±0.5	
Liki IIIA moraine				
Teleki Valley	7	6–12	8.6±0.2	Not distinguished from L III

See Notes 1 and 2 in Table 9.4.1.

Source: Summarized from Shanahan & Zreda (2000).

These differences emphasize the dependence of ELAs on differing macro- and meso-scale factors such as the general aspects of the various sides of the mountain, while the Mawenzi moraines of adjoining glaciers shown in Fig. 9.1.1, with almost identical form and extent, show how uniformly the meso-scale factors can affect glaciers with the same aspect and similar headwalls and substrate.

9.4.2 Mount Kenya, 5198 m

The Quaternary history of Mount Kenya has been the subject of prolonged and intensive studies by Mahaney (assembled in 1990, with some subsequent additions). Maps of the complete glaciated area were prepared by Baker (1967) and Hastenrath (1984) with numerous partial maps by Mahaney (Fig. 9.4.1). There is a nearly complete ring of large lateral moraines in the radial valleys and on their interfluves, apparently similar to those of the Main Glaciation on Kilimanjaro and enclosing 230 km². Mahaney (1990) distinguished a few smaller but lower moraines as the Teleki Stage of unknown age, probably more than 100 ka. He subdivided the remainder into Liki Stages I–III and inferred minimum ages (from

[14]C dated bog cores) of 15 ka BP and 12.5 ka BP for the last two, provisionally proposing an age of 50 ka BP for Liki I.

There are still older diamicts, some of moraine-like form in valleys, and others that occur mainly on interfluves, which may not be of glacial origin. There are also small moraines not far below the remaining glaciers attributable to two brief advances in the last millennium. Small glaciers still survive and the rapidly diminishing Lewis Glacier has been the subject of intensive, prolonged and important glaciological studies (Hastenrath, assembled in 1984, with subsequent additions).

Important new evidence from the cosmogenic ([36]Cl) dating of moraine boulders necessitates radical reassessment of these observations (Table 9.4.2) (Shanahan & Zreda, 2000). The most remarkable result of these figures is the reversal of the relative ages of the main moraine systems. Liki I appears to be the oldest by far, while Liki IIA, III and IIIA are left to represent the LGM (IIA and IIIA being new subdivisions introduced to conform with the new dates). The measurement of very few dates on the Liki I moraines was unfortunate, due to an assumption in the field that they were not significantly older than Liki II, though in retrospect their much greater degree of weathering contradicted this (T. M. Shanahan, pers. comm.).

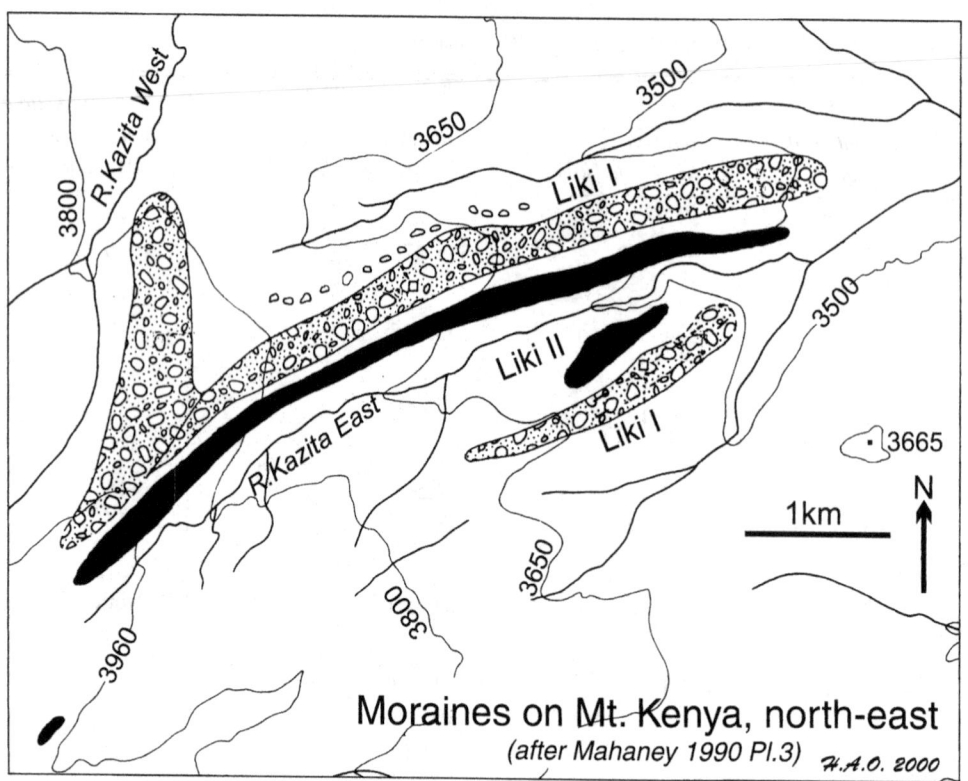

Fig. 9.4.2 Liki I and II moraines in the River Kazita catchment on the east side of Mount Kenya. This shows how the Liki I moraines lie on interfluves, though it is possible that they were formerly more extensive on a less dissected surface (as were the oldest Rwenzori tills and moraines), before the erosion of the valleys in which the Liki II and later moraines were deposited. The extent of the two stages appears to have been similar in many valleys, especially that illustrated here, so that Fig. 9.4.1 and some of the conclusions drawn in the text from Liki I moraines may well be representative also of Liki II and III.

Their positions well up on the valley sides or on interfluves between them also invite comparison with the oldest glaciation on the Rwenzori (Katabarua and Rwamya–Rwanoli stages) which occur solely on interfluves. As there too, these positions permitted the valley glaciers of later glaciations to flow past them and sometimes still further down the more deeply incised valleys to lower altitudes (Fig. 9.4.2).

It is also unfortunate that because of previous assumptions about relative ages (and implied importance), the Liki II, IIA, III and IIIA moraines have not been well distinguished and mapped throughout the mountain, though in general they are not very different from the altitudinal distribution of Liki I which has been repeatedly mapped. The moraines drawn in Fig. 9.4.2 before publication of these results represent mainly Liki I, not Liki II and III moraines, but until the new Liki I age is confirmed and more detailed and discriminating mapping of

the younger moraines is available this must stand as a substitute. From the information available it seems likely that it represents approximately the asymmetric distribution, form and ELA of the LGM (Liki II–IIIA) glaciers.

Careful reassessment in the field of all the moraines currently designated as Liki and Teleki is urgently needed, accompanied by the dating of more samples, especially from Liki I moraines and from a range of localities including the outlying peak Ithanguni. The large size and extensive preservation of Liki I compared with the relatively modest-sized Liki II–IIIA moraines differ markedly from the inverse characteristics of similar aged moraines on Kilimanjaro and possibly on Rwenzori. It also throws uncertainty on the general conclusion hitherto (based on minimum ages of last glaciation from ^{14}C dates in bogs) that on Elgon, Aberdare and the Ethiopian mountains, where only one glaciation can generally be traced, this was indeed the LGM; and if so renews the question as to why there are no traces there of an earlier one which appears to have been strong and extensive on the three highest East African mountains, with an ELA similar to or even lower than later ones.

This confirmation of previous relative age estimates on Kilimanjaro and their radical revision on Mount Kenya has been discussed in detail to emphasize two points. First, the importance of very careful observation and consideration of all the field evidence when attempting to assess relative ages without cosmogenic dating. Second, the probability that

Table 9.4.3. *Equilibrium line altitudes at different aspects on Mount Kenya (5199 m) at the LGM*

Aspect	N	NW	W	SW	S	SE	E	NE
Altitude of moraine toes (m)	3800	3800	3400	3400	3600?[a]	3200	3300	3700
ELA (m) THAR 0.5	4500	4500	4300	4300	4100?[a]	4200	4250	4450

Notes:

Estimated by THAR = 0.5 from map of moraines by Hastenrath (1984). For THAR = 0.4, ELAs are 150–200 m lower.
[a] The glaciers on the south did not rise on the main peaks but on a lower ridge at 4600 m, hence their smaller extent. The identification of their moraines is also uncertain. Hastenrath's map does not distinguish the various Liki stages.

further tentative and even apparently well-established glacial sequences and correlations will be changed when cosmogenic dating is available.

As on Kibo, the glaciers are now restricted to the steep western side of the central former volcanic plug, and the present and recent ELAs are lower on the west than on the east, due to the shelter obtained there from the morning sun; clouds often rise in the afternoons resulting in less ablation on the western slopes (see Fig. 8.2.10). The older moraines were mapped by Baker (1967), but with a different nomenclature; by Hastenrath (1984); but different stages were not distinguished and some low ones in the south are dubious; and by Mahaney (1990) but only in parts. In contrast to the present, at each of Liki I, II and III the ELA was lower on the southeast than the northwest, presumably because this was (as now) the direction from which the main snow-bearing winds came and shade was less important on a large ice-cap. Mahaney (1990, fig. 11.8) emphasizes the variation of ELA with both aspect and time for these stades: both the relative differences between the ELAs of glaciers of the same stade on different aspects of the mountain, and the different orientations of the highest and lowest ELAs of different advances. Unfortunately the absolute values of the ELAs are not reliable as Mahaney estimated the ELAs from 'the mean altitudes of lateral moraines', but his map of the northwestern moraines (Mahaney, 1977, fig. 2) shows that the upper limit of the Liki I/II moraines (a minimum estimate of the ELA) in that sector averaged 4200 m, i.e. 200 m higher than that shown on his fig. 11.8. At the LGM there were valley-bottom cirque-headed glaciers rather than a continuous ice-cap and my own THAR estimates with probably appropriate factors of 0.4 and 0.5 give estimates about 300 m and 500 m higher than Mahaney's on all aspects (Table 9.4.3). The other lesser advances may also need smaller corrections. It is likely that in the Liki I glaciation the valleys were less incised and there was more of an ice-cap with broad lobes descending from it so an appropriate THAR of 0.3 would put that ELA slightly lower. For these reasons the altitude differences between Mahaney's curves are not accurate estimates of the changes in ELA.

Because of the reversal of ELA slope, the lateral displacements of the equilibrium lines and the lack of present glaciers and hence an ELA on the east, any single figure of ELA change for the mountain as a whole is problematic, but taking the present ELA as ranging from 4900 m to 4700 m on aspects from north through west to southwest (Mahaney, 1990) the rise there since the LGM has been about 400 m, rather than the 1000 m indicated by Mahaney's fig. 11.8.

As on Kilimanjaro, there is an outlying glaciated peak (7 km²) to the east, Ithanguni, which similarly had a much lower ELA than the main peak. Comparative data for each are presented in Table 9.4.4. The conspicuous ring of moraines round Ithanguni was mapped by Mahaney (1989, Pl.4) as Liki II, with Teleki moraines outside them. These stages and the lack of Liki I need confirmation. Since Ithanguni is so much smaller than Mawenzi it did not exert a similar precipitation shadow effect, and there is approximately a single gradient of the ELA across the main and minor peak. The outline of the main peak of Mount Kenya is drawn NW–SE the direction of maximum difference of ELA; however, Ithanguni actually lies to the ENE so possibly the very low ELA there suggests a strengthening of the northwest monsoon.

The analysis and dating of pollen cores from lakes well below the moraines indicate a great lowering (equivalent to 5.1–8.8 °C, probably accompanied by drier conditions) during the period preceding this retreat and a subsequent rise of the vegetation zones (Coetzee, 1964, 1967), but this evidence is now open to question (see section 9.5.1). The present annual precipitation rises to a maximum of about 1500–1750 mm at 3000 m, then falls off to 800–1000 mm above 4000 m (Thompson, 1966; Hastenrath, 1984); the northeast slopes of the mountain are much the driest.

9.4.3 Elgon (Uganda and Kenya), 4321 m

This large Miocene collapsed caldera (Fig. 9.4.1), 8 km in diameter, has a conspicuous ring of apparently LGM moraines descending from the outer walls of the crater.

Table 9.4.4. *Comparison of ELAs on main and minor peaks of Mount Kenya and Kilimanjaro*

Main peak	Batian, Kenya	Kibo, Kilimanjaro
Summit altitude	5199 m	5894 m
Lowest LGM moraines	3400 m (W & SW)	3200 m (SSW)
Lowest estimated LGM ELA	4200 m (W & SW)	4600 m (SSW)
Lowest estimated recent ELA	4700 m (W & SW)	5300 m (SSW)
Estimated rise in ELA since LGM	500 m	700 m
Estimated LGM ELA on nearest side to minor peak	4000 m (NE)	4600 m (SE)
Minor peak	**Ithanguni**	**Mawenzi**
Summit altitude	3894 m	5147 m
Lowest LGM moraines	3200 m (E)	3350 m (SE)
Lowest estimated LGM ELA	3550 m (E)	4250 m (SE)
Lowest estimated recent ELA	>3900 m	c. 5000 m
Estimated rise in ELA since LGM	>450 m	c. 850 m
Estimated LGM ELA on nearest side to major peak	3700 m (W)	4600 m (SW)
Distance between estimated positions of ELAs on major and minor peak	13 km	12 km
ELA difference, major to minor peak	300 m	0 m
ELA slope, main to minor peak	**23 m km^{-1}**	**0 m km^{-1}**
Distance between estimated position of ELAs on E and W aspects of minor peak	2 km	4 km
ELA difference between E and W of minor peak	150 m	350 m
ELA slope, E to W across minor peak	**75 m km^{-1}**	**90 m km^{-1}**

Notes:

On the E face of Kibo there is a discontinuity of 500 m in the estimated ELA between SE and NE – see Fig. 9.4.1.

ELAs were estimated by AABR method on Kilimanjaro, by THAR (0.5) on Mount Kenya.

Particularly on the northern slopes, the moraines are not so large as those of major glaciers on Kilimanjaro or Mount Kenya, partly because the glaciers here were relatively small, mostly spanning an altitude range of only about 400 m. The extent of glaciation (70 km²) is also indicated by a marked change in valley morphology from smooth U-shapes to V-shapes with incised streams. Within the crater, moraines are small and few, though Hamilton & Perrott (1979; see also Hamilton, 1982, 1987) attribute an extensive scoured shelf at the base of the crater walls to glacial erosion. They have travelled the mountain frequently and their map of the moraines is more accurate than Hastenrath's (1984), as confirmed by my new check against large scale air photos. They closely analysed the extent of the glaciers which descend further on the north (3400 m) and northeast than on the south (3700 m) and estimated their ELAs from the altitudes of the upper ends of their moraines, concluding that they were at 3560 m in the north and 3900 m in the south. Comparative estimates using the more reliable THAR method with a factor of 0.5 indicate an ELA of 3700 m in the north, 150 m higher (a THAR of 0.4 would only reduce this by 50 m) but show precise agreement in

the south. This discrepancy is probably due to the small size of the moraines on the north, so that their upper ends are below the ELA. This does not invalidate their conclusion that the LGM ELA was tilted towards the north (though by a lesser amount), but it is curious that the larger glaciers there are less incised and had smaller moraines than on the south.

The present annual rainfall on the southern upper slopes is about 2000 mm but the rainfall pattern indicated on the *Atlas of Uganda* (1962) is not reliable as there are very few records; the vegetation, however, clearly indicates that the north and northeast is now the driest aspect of the mountain (the bamboo belt is lacking and there is dry cedar forest; see Dale, 1940; Langdale Brown *et al.*, 1964; Hamilton, 1987). Hamilton & Perrott (1979) suggest that the apparent reversal at the LGM may have been due to reduced humidity in the southerly winds due to desiccation of Lake Victoria. At the LGM the mountain had a mainly grassy vegetation, and the trees were species characteristic of drier conditions but there was little evidence of temperature change (Hamilton, 1987).

The difference between the revised ELAs and the altitude (mostly 4000–4100 m) of the crater rim at the relevant valley heads (rather than the highest peaks on the rim), averages about 200 m. Applying a lapse rate of 6.5 °C/1000 m, this provides a crude estimate of minimum temperature depression of only 1.3 °C at the LGM. Even with an addition of a 2 °C allowance for increased dryness, this is well below the likely total from comparisons with the higher mountains.

On the south side there are uncertain indications of a post-LGM period of cirque formation, and an age of 11 000 BP for the LGM glaciers has been obtained from the base of a bog in an ice-eroded basin at 4150 m (Hamilton & Perrott, 1978, 1979; Hamilton, 1982). This provides a minimum date for deglacierization following the LGM, but may also represent the end of a subsequent period of cirque occupation.

Although there is no positive evidence of earlier glaciation, Hamilton & Perrott (1979) hypothesized a series of previous glaciations, each more intense on the north than the south, which resulted in the different forms of the crater rim, valleys and moraines in the north and south, but the underlying glaciological assumptions of this are uncertain.

9.4.4 Aberdare (= Mount Nyandarua) (Kenya), 4001/3993 m

The broad summit of this high volcanic ridge running along the flank of the Eastern Rift (Fig. 9.4.1) was only just high enough to be glaciated in the LGM. Hastenrath (1984) maps three groups of small moraines, extending to 3200 m on the NE, 3600 m on the NW, 3400 m on the SW, but none on the SE. It seems that they represent outflows from three small ice-caps (c. 22 km²) but may have formed a single larger one of possibly 60 km². A date from a cirque bog at 3800 m near the summit of Mount Satima gives 12 200 BP as the minimum age of the LGM (Perrott, 1982[1]). There are no traces of previous or subsequent glaciation.

9.4.5 Rwenzori (Uganda and Dem. Rep. of Congo), 5009 m

See Figs. 9.4.1, 9.4.3. This range lies between the equator and 1 °N with a high annual rainfall of 2500 mm up to the snowline (Osmaston, 1989[2]; see Fig. 6.2.2 and on back of map included with this book), so is truly wet equatorial and had the most extensive glaciations of all the East African mountains (with the possible exception of the Bale Mountains, Ethiopia). These covered up to 500 km² and the LGM is recorded by large moraines, retreat from the maximum extent of which is estimated at approximately 16 000 BP (see below). Apart from Mount Kenya where the new dates for Liki I need wider confirmation and revised field mapping, there is clearer and more substantial depositional evidence on the Rwenzori than has

been found on any other African or Southeast Asian tropical mountain of a much earlier glacial extension which clearly had the form of an ice-cap, besides erosional evidence of another major extension of valley glaciers between these two stages. The present glaciers are small and have lost three-quarters of their area in the last century.

Unlike most of the other high mountains in East Africa, the Rwenzori Mountains are not volcanoes. The surface on which the glaciers developed was an uptilted block of old basement rock at the east side of the Western African Rift Valley which, curiously, seems not to have been much dissected before glaciation. The field evidence for these glaciations has previously been described in detail (de Heinzelin, 1953, 1962; Osmaston, 1965, 1967, 1989[2]; Hastenrath, 1984) so it will only be summarized here. The sequence of glaciations or glacial advances which have been recognized are summarized in Table 9.4.5 and shown in detail in the map included with this book, which also has more detailed notes on the back.

THE KATABARUA GLACIATION
Extensive landforms and deposits of two types occur in several separate areas, showing similarities to each other but marked differences from later glaciations. Hence they are tentatively correlated with each other as representing the same episode. They are both characterized by their sites on broad interfluves, separated from each other by the deeply incised valleys of the Rwimi Basin and Mahoma Glaciations. The Katabarua Stage is marked by moderately sized frontal moraines on the east of the range; the Rwamya–Rwanoli tills consist of varved silty clays and gravels on distinctively smooth topography on the north and west (Figs. 9.1.2, 9.4.3, 9.4.4). Both were apparently deposited at or near the margin of a broad ice-cap, which in area was the most extensive ice-cover of all, covering some 500 km², before the development of the valleys in which subsequent glaciers were channelled. Thus the character of this glaciation differed fundamentally from subsequent ones which mainly occupied new valleys, accompanied by some major changes to the drainage system. Supraglacial moraine would have been less, but subglacial more, resulting in the till spreads of Rwamya–Rwanoli. From fragmentary remains on interfluves there are similar indications from Mount Kenya and Kilimanjaro and elsewhere that there too the pre-LGM glaciers were more of this nature than valley glaciers, which is only to be expected if they developed on previously unglaciated uplands. Possible correlation with Liki I moraines on Mount Kenya indicates an age >300 000 BP.

THE RWIMI BASIN GLACIATION
The evidence for the existence of this glaciation is primarily erosional and has been inferred from the situation of the

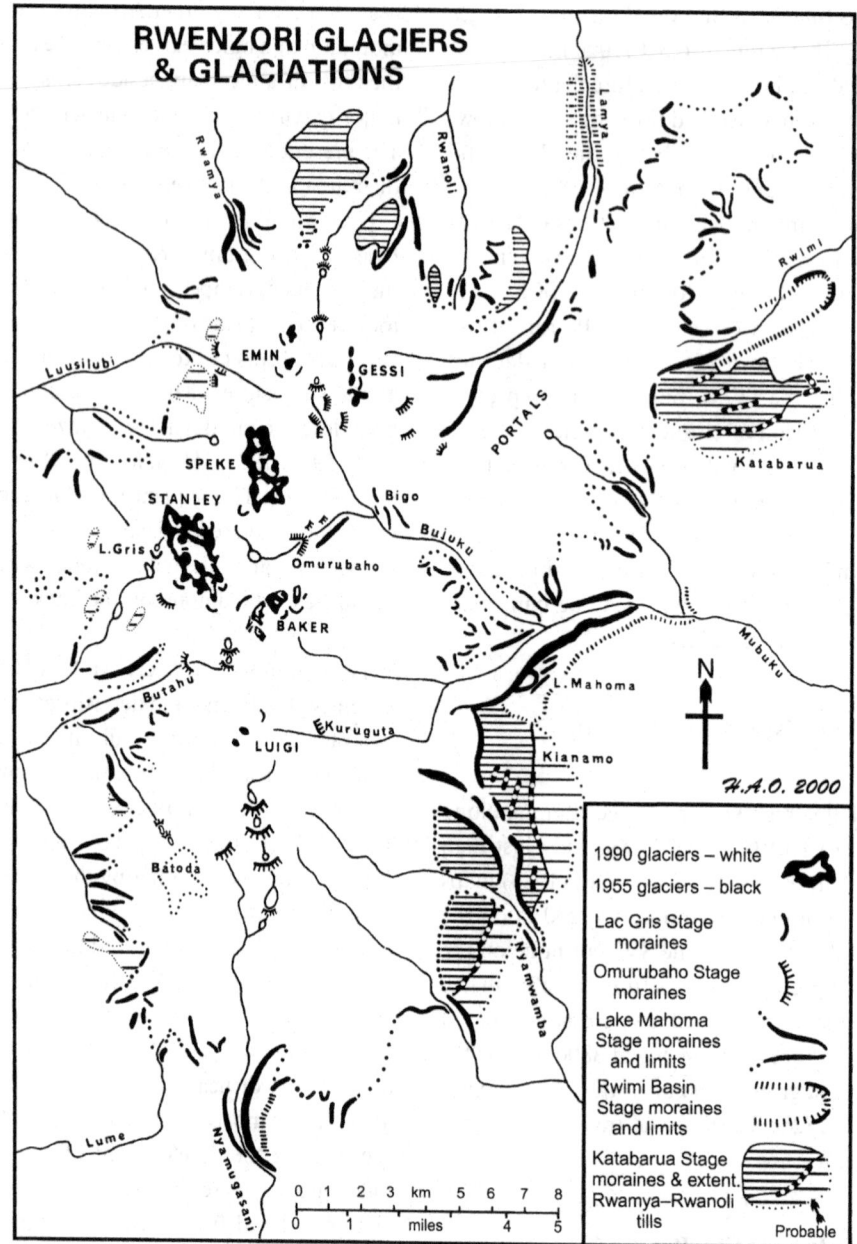

Fig. 9.4.3 Map of Rwenzori glaciers and glaciations.

Mahoma stage moraines, which lie in broad, fairly straight, but deeply cut valleys which apparently existed before they were occupied by the smaller Mahoma glaciers and their moraines (see Figs. 9.4.4, 9.4.5). On the other hand these valleys apparently did not exist at the time of the Katabarua glaciation. In the valley of the River Rwimi in particular, the Katabarua moraines of the earliest glaciation appear to have been deposited by an ice-sheet which flowed across the line of the later Mahoma Stage valley and modern valley before this was excavated. It is unlikely that a long period of fluvial erosion could have produced this effect, as the valley forms are

quite broad and straight, considering the steep relief. On the air photos there appears to be a moraine of this stage in the Rwimi valley beyond the limit of the Lake Mahoma Stage, but this needs confirmation on the ground.

THE LAKE MAHOMA GLACIATION = LGM
See Figs. 9.1.2, 9.4.1, 9.4.3, 9.4.4, 9.4.5. This is the glaciation which has left the most abundant evidence on the mountain. It has left very large moraines in all the main valleys which descend from the main peaks down to a lowest altitude of 2000 m, and smaller moraines higher in tributary valleys (Fig. 9.4.3). These contrast with the present small extent of the glaciers of which the lowest tongue only reaches 4400 m. These

Table 9.4.5. *Glaciers and glaciations on the Rwenzori: their chronology and extent*

Glacial Stage	Age	Area	Lowest	ELA
1991	—	1.7 km²	c. 4400 m	>4800 m
1955	41 yrs	4.1 km²	4200 m	>4600 m
1906	90 yrs	7.5 km²	c. 4100 m	?
Lac Gris	c. 300 BP	10 km²	4000 m	c. 4500 m
Omurubaho	c. 10 000 BP	70 km²	3630 m	c. 4300 m
Mahoma	end c. 15 000 BP	260 km²	2070 m	c. 4000 m
Rwimi Basin	>100 000 BP?	c. 300 km²	1900 m	c. 3800 m
Katabarua	>300 000 BP?	c. 500 km²	2900 m	c. 3900 m

Source: After Osmaston (1989[2]) and Kaser (1996) with additions by H.A. Osmaston.

moraines occur on all sides of the mountain, but descend to different altitudes, depending partly on the height and extent of the accumulation zone and partly on the aspect (those on the east having larger catchments situated above lower ELAs, descended lower than in similar sized valleys on the west), this effect being accentuated by the tilt of the ELA surface towards the east. Sediment from near the base of the small intermorainal Lake Mahoma (2960 m, Fig. 9.4.5) yielded a frequently quoted date of 14750 BP (Livingstone, 1962). However, it is almost universally overlooked that the core extended a further 60 cm with an estimated basal age of

16000 BP (Livingstone, 1967), and that Lake Mahoma is 800 m higher than the lowest extension of the main glacier in the Mubuku valley. It lies within a side-loop or kettle-hole of the same terminal-lateral moraine so is roughly contemporaneous, but it certainly represents a time near the start of retreat from the lowest position, rather than complete or even substantial deglacierization of the mountain. The pollen near the base has a high proportion of *Dendro-Senecio* (tree groundsel) pollen which matches present surface deposits 1000 m higher, suggesting a temperature at least 4 °C and probably about 6 °C lower. This conforms well with the estimate from ELA change, and is not subject to the uncertainties involved in assessing the implications of changes between forest and grassy vegetation. However, this could have been influenced by cold katabatic winds, which still occur in this valley on cold nights and are likely to have an effect on the local climate at the LGM.

The main glaciers which left these moraines were primarily a system of valley glaciers in deep glaciated valleys, draining a group of ice-caps which crowned the six main central mountains, Stanley, Speke, Baker, Emin, Gessi and Luigi. On subsidiary peaks and ridges there were many minor glaciers, commonly marked by a small source valley which was only moderately eroded and a pair of converging lateral moraines. Most glacial lakes in the Rwenzori are in the main valleys, a few being rock-basins but mostly dammed by the moraine of a later minor advance.

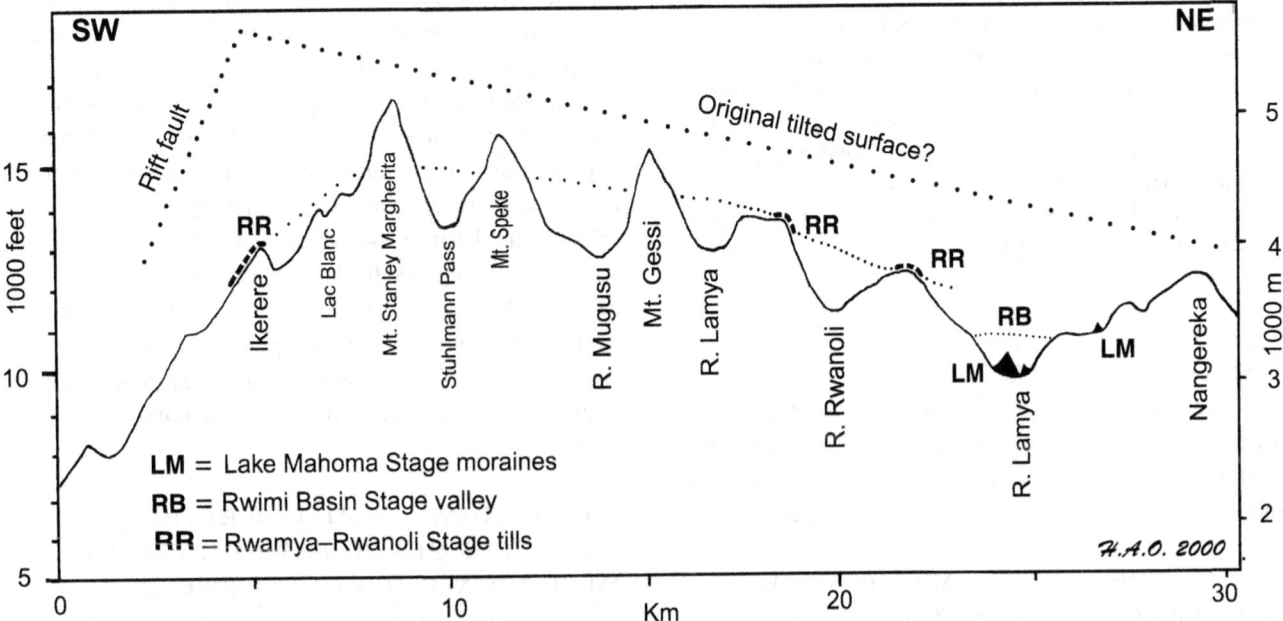

Fig. 9.4.4 Rwenzori Mountains, Uganda and Dem. Rep. of Congo. Section SW–NE showing the topographic relationship between deposits of three former major advance stages.

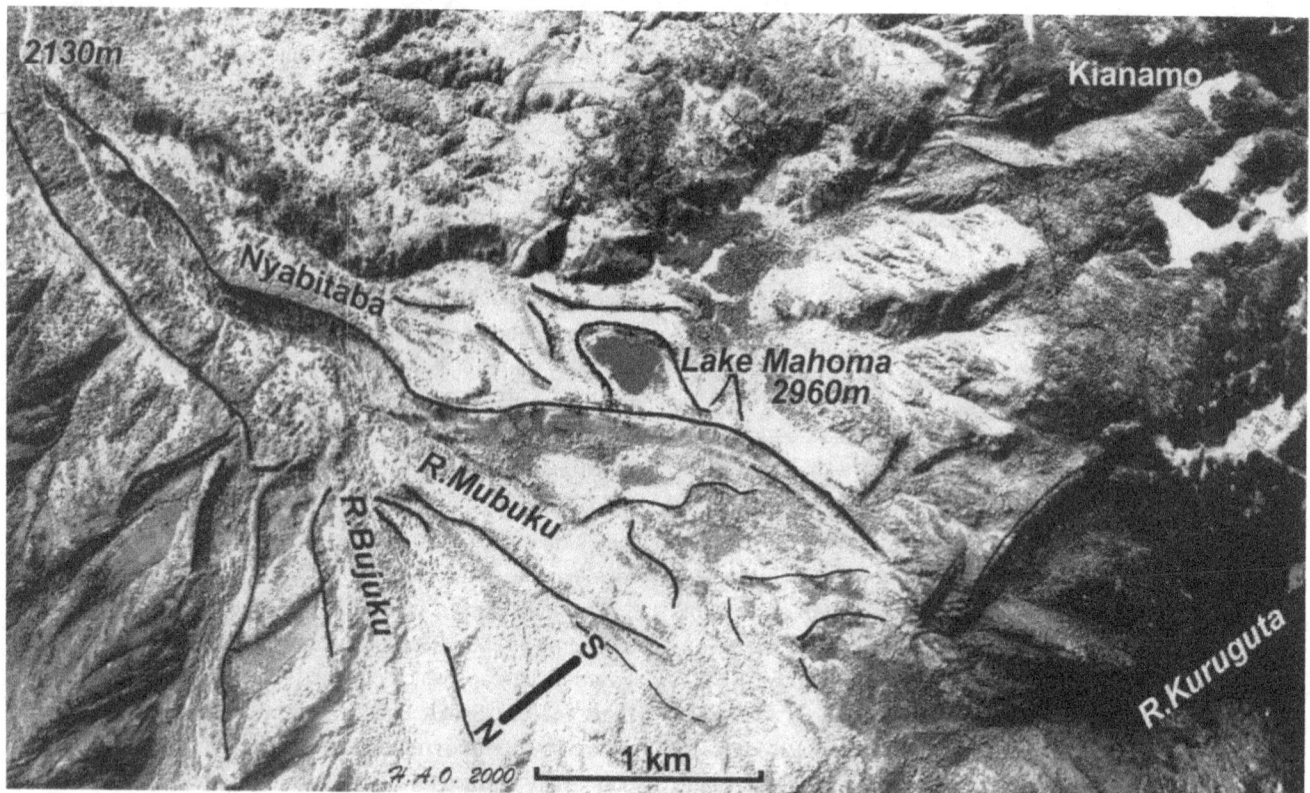

Fig. 9.4.5 Lake Mahoma Stage moraines. Air photo of Lake Mahoma, Rwenzori, a 6 m core from which yielded a date of 14750 BP from organic sediment 60 cm above bottom of basal silt, thus postdating the formation of the basin (>16000 BP ?) (Livingstone, 1967). The lake lies at the side of the main Mubuku valley and its moraines, in a side loop of moraine, 800 m above the lowest terminus of the glacier, and about 700 m below the local contemporary ELA, so the age and altitude do not have a precise glaciological significance. The scarp on the far side of the Nyabitaba moraine may represent the valley side of the previous Rwimi Basin Stage. At Kianamo are some moraines of the Katabarua Stage. The Kuruguta valley contributed the largest ice discharge to the combined Mubuku Glacier. See section 9.5.1 in text. (Adapted from DOS/Huntings air photo 15.UG.31 No.023, f = 6 in, alt. 28000 ft, October 1955.)

POST-LGM ADVANCES

There are two distinct sets of moraines formed by later advances, well correlatable in different valleys: the Omurubaho Stage probably a few thousand years old, and the Lac Gris Stage dating from the nineteenth century or just before.

ANALYSIS OF THE LAKE MAHOMA STAGE GLACIERS ON THE RWENZORI

The information about the LGM (= Lake Mahoma Stage) comprises the moraines of 75 glaciers of a wide range of sizes, shown on good quality air photos and accurately mapped on 1:50000 contoured maps. Then and at other stages including the present, the glaciers were well spread out on several different mountains, thus permitting the construction of ELA trend surfaces. Thus they comprise an exceptional data set for methodological testing, as representative of the equatorial LGM. It has already been shown above that in general morphology the moraines conformed more closely to those of some modern non-tropical glaciers than modern tropical ones. Examination can next be extended to testing what parameters appear to be most appropriate for estimating ELAs from their more detailed morphology, and comparing these with modern values. It has been stressed above that these statistical parameters should be applied to glacier populations that are statistically valid in terms of consistency and group size. In conformity with this, the most appropriate means of testing the results that they provide is also statistical, since the means of testing them by direct observations is impossible for past glaciers.

TOE–HEADWALL ALTITUDE RATIO (THAR) AND ACCUMULATION AREA RATIO (AAR) METHODS APPLIED TO THE RWENZORI

The accounts of the previous two mountains have emphasized the differences between groups of glaciers on the various aspects of a single mountain. With the complex topography of this range it is even more necessary to divide them into

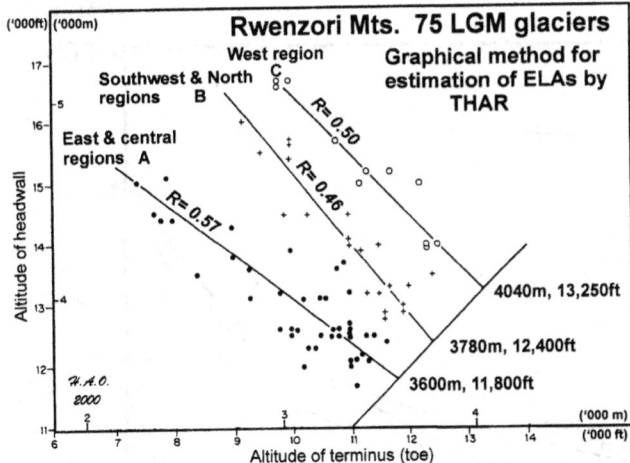

Fig. 9.4.6 Graphical estimation of ELA by THAR method for 75 LGM glaciers in three groups on the Rwenzori.

homogeneous groups, but it is also necessary to compromise between a small group for maximum homogeneity and a large group for better statistical significance. It can then be assumed that the macroclimatic factors will be fairly uniform over the whole group and in the ideal case all should have the same ELA. However, besides variations in macroclimate at the margins of the group, the microclimatic influences on each glacier will differ slightly and result in a slightly different morphology and indicated ELA. The application of the same particular value of THAR or AAR to each of these glaciers will yield a population of individual ELA estimates centred around a group mean. The differences of these from the mean represent the differing effects on individual glaciers of local relief factors such as inclination, shading, snow-drift, avalanches, etc., while the mean represents the mean effect of these local factors superimposed on the more uniform effects of larger scale relief and climatic factors. It is not possible to separate these two components, so the best estimate of their combined value will be one which has the minimum variance.

Therefore the THAR or AAR which minimizes the variance from this mean is the most appropriate one (and similarly for the AABR discussed below). A simple graphical method of estimating the most appropriate THAR is shown in Fig. 9.4.6. Each glacier is plotted as a point given by its toe and headwall altitudes, and these fall into well-defined locality groups, through which a best-fit line can be drawn to meet the diagonal through the origin. The slope of this line indicates the best-fit THAR, a 45° slope indicating a THAR of 0.5. This figure indicates that best-fit THARs of three groups of LGM glaciers on the Rwenzori are 0.55, 0.46 and 0.50. This is more than the value of 0.40 found by Meierding (1982) for modern non-tropical glaciers, and than that implied by Kaser in chapter 5.4

for modern tropical glaciers. This method also gives a good graphical estimate of the ELA, the altitude at which the line through the glacier points intersects the diagonal, and yielded almost identical ELA results to the more rigorous methods described below.

It has been proposed in chapter 5.4 that because of increased ablation on tropical glacier tongues, an AAR of 0.82 is more appropriate than the value of 0.67 used for mid-latitude ones. This was tested, using a simple spreadsheet program, which accepts the hypsometric data for several groups of glaciers, calculates their individual and group mean ELAs, and the group standard deviations, for a succession of selected AARs. This was used for the three main groups of the Rwenzori LGM glaciers, to calculate their individual and mean ELAs for a series of values of AAR between 0.01 and 1.0 (which represent cases when the ELA was at the top or bottom of the glacier). The standard (RMS) deviations have been plotted in Fig. 9.4.7 and show clearly that for all three groups this criterion indicates that an AAR of 0.65–0.70 is the most appropriate. This conforms with the value found by Meierding but not with Kaser's conclusions for modern tropical glaciers.

Contrary perhaps to intuitive ideas, it is not the largest glaciers which give the most precise estimates of former mountain-scale ELAs, since they have larger uncertainties such as the thickness of the ice at the equilibrium line, and small differences in ratio will move the indicated ELA by large amounts. It is the medium-sized and smaller glaciers which give the most precise estimate of their own ELA, as it must be narrowly confined between their top and bottom, and is not greatly affected by choice of ratio. However, the very smallest glaciers, and especially those confined to cirques, are liable to be affected by individual microclimatic variation in aspect, shading, etc., which may increase their variance from the group mean.

AREA–ALTITUDE BALANCE RATIO (AABR = AHA) METHODS APPLIED TO THE RWENZORI

As with the ratio methods, the appropriate tests for the AABR method are statistical. First the simple method, assuming a linear mass balance gradient, was applied to each of the 75 glaciers which were divided into homogeneous regional groups based on their approximate ELAs. The standard deviations showed an improvement of about 30% on the THAR method. This rather small improvement suggests that use of the AABR method was not required for most of the former Rwenzori LGM glaciers because they did not seriously depart from a 'typical' Alpine valley glacier morphology with a linear mass balance gradient, not one that was much steeper below the ELA. A notable exception is the Rukenga glacier which

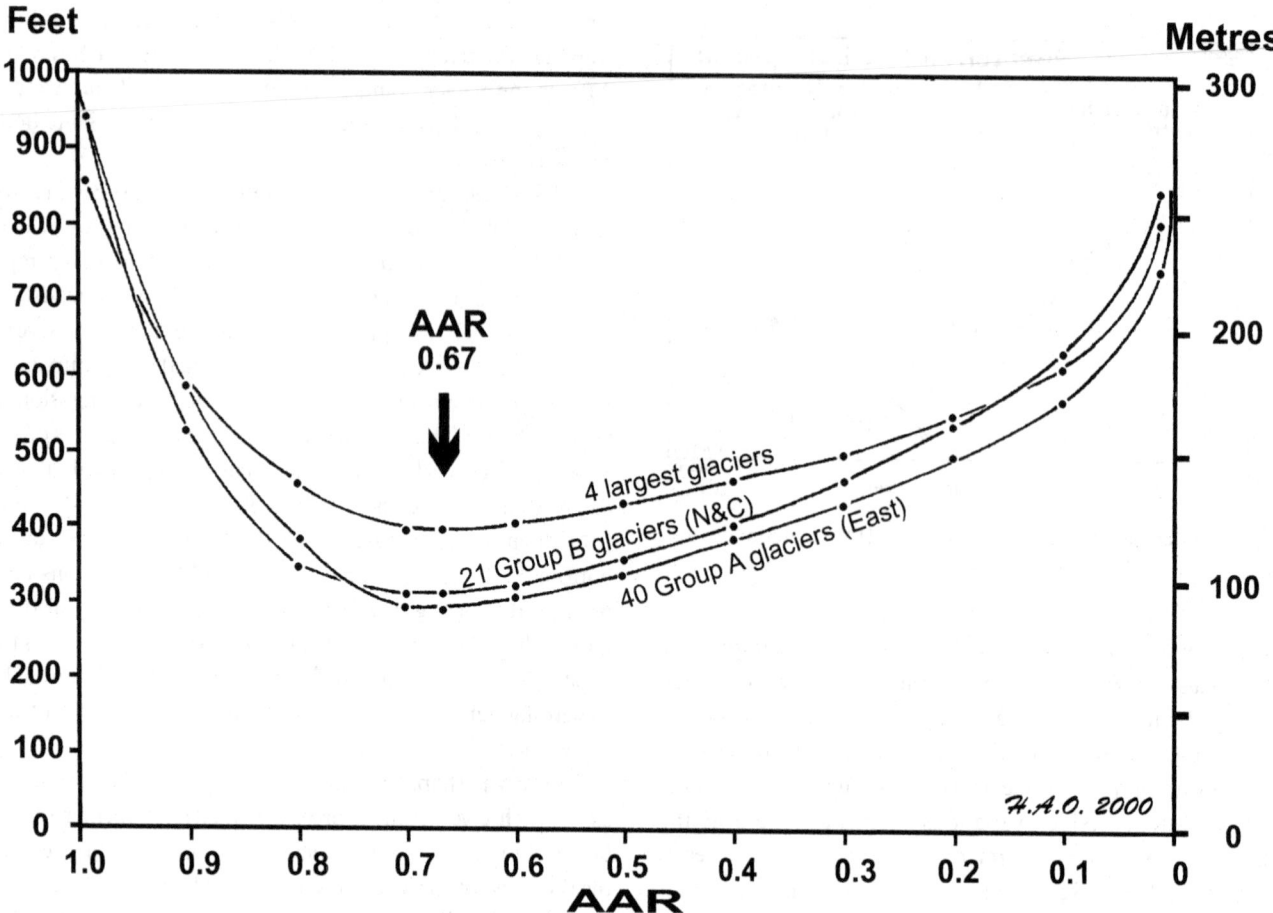

Fig. 9.4.7 Graphical test for the AAR for three groups of LGM glaciers on the Rwenzori, with the criterion of minimum standard deviations from group means. Areas are for present valley floors, not reconstructed glacier surfaces.

had a large piedmont lobe spreading over the shelf-like Kianamo area. The ELA indicated by the AABR method was 3650 m, which agrees exactly with the upper limit of the conspicuous moraines; at this altitude, however, only 0.45 of the glacier area is above the ELA, an exceptionally low AAR.

Application of the advanced AABR method using a mass balance gradient below the ELA twice that above it ($F = 2$) gave group mean ELA estimates about 70 m lower than with a linear gradient but with almost identical standard deviations of about 100 m. The resulting standard errors of the means did not justify discriminating between these two gradients, and confirmed lack of support for the hypothesis of a much steeper gradient below the ELA.

THE RESULTS ON THE RWENZORI
The finally preferred estimates of the former ELAs of the LGM on the Rwenzori were obtained by a statistical treatment only made possible by the large number and wide distribution

of the former glaciers. Instead of being placed in discrete regional groups, their individual ELAs estimated by the AABR method and located on the map (after adjustments for probable ice thickness) were treated as a single population to which a trend surface could be fitted (Osmaston, 1965, 1975, 1989[2]). The outcome of the fitting method is given in Fig. 9.4.8 which shows that it forms an elongated dome following roughly the shape of the mountain. The computed quadratic surface shows only a slight improvement over grouping, but the standard deviation of individual ELAs from the manually fitted surface is only 35 m, showing strong support for the AABR method of defining a quite simple and climatically plausible surface. The vertical distance between this surface and a similar surface representing the ELA at the time of the recent Lac Gris Stage is about 500 m, although the difference between the means of the two sets of ELAs is over 1000 m, showing the importance of correcting for slope.

Table 9.4.6 shows the great improvement in precision that can be obtained by simply grouping the glaciers into homogeneous regions, which is superior to that provided by the best computed quadratic trend surface and the remarkable precision of 97% explanation of the total variance, which can be

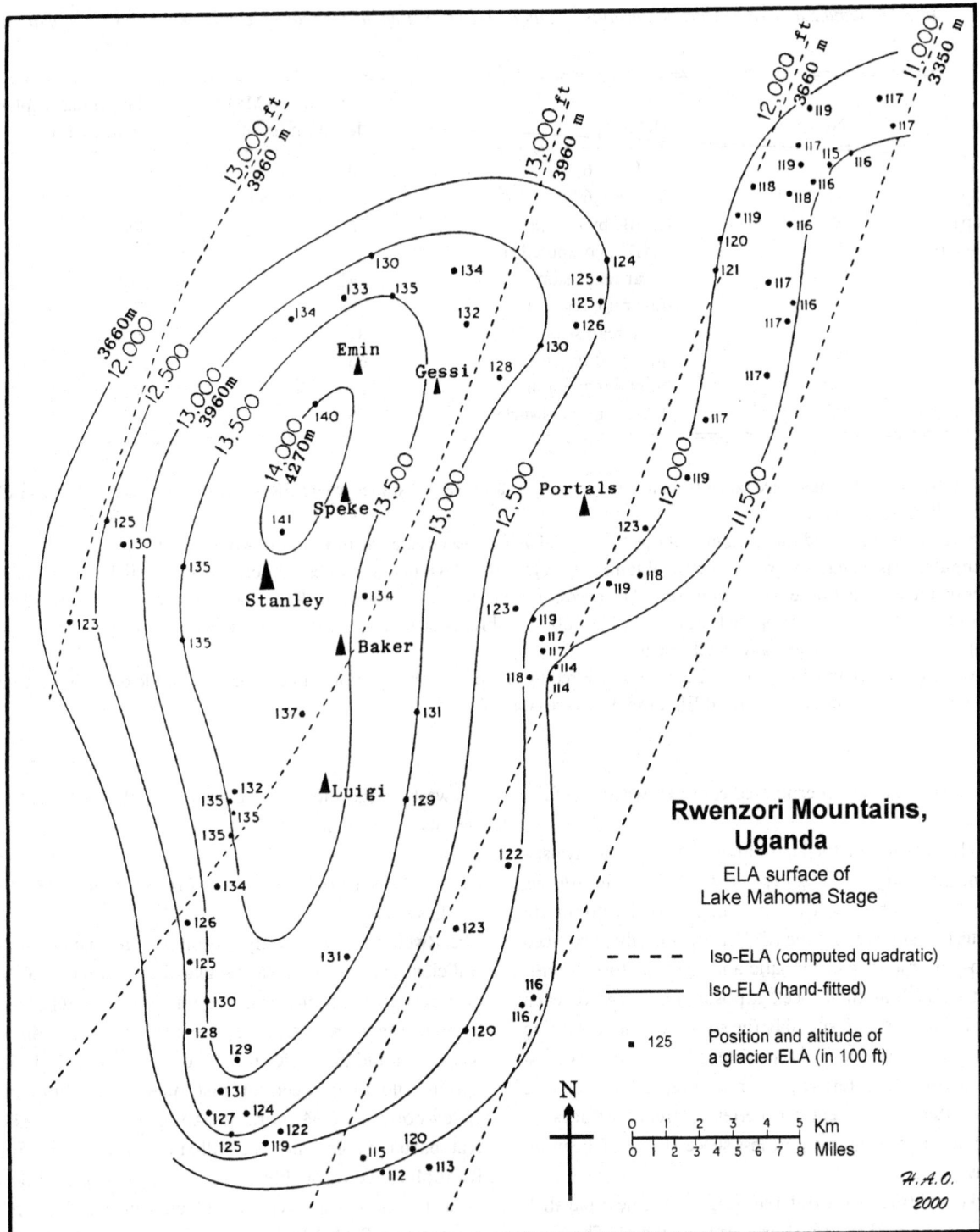

Fig. 9.4.8 Trend surfaces of ELAs of the Lake Mahoma Stage on
the Rwenzori Mountains (computed by the AABR method,
adjusted for ice thickness).

Table 9.4.6. *Lake Mahoma Stage, Rwenzori Mountains, Uganda and Dem. Rep. of Congo: precision of various ELA estimation methods*

Population	Number	Method	Standard (RMS) deviation (σ, m)	Percentage explanation Adjusted R^2
A. All glaciers	73	AAR = 0.67	200	0
B. 4 groups	40, 22, 9, 2	AAR = 0.67	87, 94, 115, 84	86
C. 4 groups	39, 25, 9, 2	AABR by groups	76, 99, 68, 116	88
D. All glaciers	75	AABR, computed surface		
		linear regression	147	66
		quadratic regression	108	76
		cubic regression	81	86
		quartic regression	66	90
		quintic regression	58	91
E. All glaciers	75	AABR, manual surface	35	97

Notes:

1. In the AABR method values were taken from a mean of BR = 1 and BR = 2 values where BR = ratio of the mass balance gradient below the ELA to that above it.

2. The same glaciers were used throughout, but one compound glacier was treated as three in the AABR analysis.

3. The computed least squares regressions were obtained through MINITAB (using a special procedure for the higher level ones). Apparently much higher order regressions would be necessary to obtain a precision and R^2 similar to the manually fitted surface shown in Fig. 9.4.8. Computer plotted contours of the cubic and higher order regressions are simplified versions of the manually fitted surface, but with increasing resemblance to it.

4. Deviations from the manually fitted surface were measured by comparing the position and calculated altitude of each ELA point with that indicated by the superimposed ELA trend surface contours.

obtained from manually interpolated contours when carefully fitted.

The high precision of the latter, and its smooth and consistent form, an elongated, slightly tilted and distorted dome, reflecting the general form of the range and the dominant wind patterns, indicate that the ELAs are primarily influenced by macro- and meso-scale climatic and relief factors. Micro-scale factors such as the aspect, exposure and slope of each individual glacier are presumably the source of the remaining 3% of the total variance. This consistency also provides reassurance that all these terminal moraines represent the same advance. Attempts to generate better fit trend surfaces by various computer programs were less successful than the manually drawn one.

A marked feature of the different stages is the near parallelism between their ELA trend surfaces (Fig. 9.4.1). These are probably determined by regional wind patterns, and indicate a marked consistency in these which does not appear to accord well with other evidence for a change from a moist to a drier climate. The present precipitation shown on the *Atlas of Uganda* (1962) is seriously in error; more accurate figures are given in Osmaston (1989[2]) and Osmaston *et al.* (1998), which conform with the vegetation belts, having a dry zone high on the west slopes, and with the pattern of the ELA at the LGM (see also back of included map).

ELAS OF OLDER GLACIATIONS ON THE RWENZORI

The Rwenzori are the only mountains in East Africa where sufficient remains of an earlier glaciation than the LGM have been recognized to make any estimate of its extent and ELA (though the estimates made above for the LGM on Mount Kenya, should now refer rather to the old Liki I glaciation). On interfluves between or just outside the limits of the valley-bottom LGM moraines, the frontal moraines of the Katabarua Stage and the till-spreads of the Rwamya–Rwanoli Stage (probably different facies of the same stade) indicate an ice-cap type of glacierization, for which an appropriate THAR factor would have been about 0.3 or 0.4; this would place the ELA below that of the Lake Mahoma Stage.

The more limited and debatable erosional evidence of the Rwimi Basin Stage would indicate an ELA similar to that of the Lake Mahoma Stage.

Although the morphology of the faults surrounding this horst and the continuing frequency of earthquakes indicate that it is still rising, the rate is unknown so its effect on the apparent past ELAs cannot be estimated.

9.5 TROPICAL CLIMATES AT THE LGM

9.5.1 Types of evidence

The primary purpose in making estimates of past ELAs is to draw from them inferences about past climates. This cannot be done without considering other lines of evidence which affect, support or conflict with the indications from ELAs. For several decades the principal sources from which quantitative information could be drawn about tropical climates in the last 100 000 years fell into two categories. One, including changes of glacier ELAs, levels of lakes in closed basins and oxygen isotope ratios in foraminifera in ocean cores, appeared to have direct physical links with climatic variables. The other used pollen analysis of lake sediment cores to trace the vertical shifts of vegetation zones on tropical mountains, and inferred that this was primarily attributable to changes in mean temperature. The problem has been to reconcile the differing results of these approaches. Inference from both ELAs and vegetation zones depends primarily on the value selected for the mean atmospheric temperature lapse rate in the tropics, so this needs to be considered first.

TROPICAL ATMOSPHERIC TEMPERATURE LAPSE RATES (ATLRS)

The ATLR is a key variable in estimating past changes of atmospheric temperature from proxy data which themselves are strongly temperature dependent such as ELAs and the altitudes of montane vegetation belts. However, the ATLR in the troposphere is not a constant. Except in temporary inversions, it is constrained by strict physical laws to be between a theoretical maximum of the 'dry adiabatic lapse rate' of 9.8 °C/1000 m altitude and a minimum of the 'saturated adiabatic lapse rate' which ranges from about 3 °C/1000 m at 40 °C (i.e. sea level in the tropics) to 7 °C/1000m at 0 °C (i.e. at or just above the ELA of a moist tropical glacier). The commonly used global 'mean lapse rate' of about 6.5 °C/1000 m is therefore a 'black-box' combination of various regional, seasonal and altitudinal effects from which there are considerable deviations.

Although ATLRs can in theory lie anywhere between the dry and saturated adiabatic rates, observations show that in wet climates it often lies very near the latter, but even in dry climates it does not approach the high values of the dry adiabatic rate. Webster & Streten (1978) showed that in moist equatorial Papua New Guinea the free air ATLR lay close to the saturated rate, without changing even in the dry season, and that

Table 9.5.1. *Free air ATLRs along a 70°/80° W meridian (°C)*

Altitude (km)	20° N	10° N	Equator	10° S	20° S
5–10	6.8	6.8	6.8	6.0	6.9
0–5	5.1	5.0	4.8	4.3	3.6

Source: Rex (1969).

a similar ATLR prevails at very dry Darwin (north Australia).

The actual mean annual ATLRs along a 70°/80° W meridian, calculated from tables of mean monthly atmospheric free air temperatures 1957–1960 at heights up to 30 km given by Rex (1969), are as shown in Table 9.5.1 for the dry tropics and equatorial zone. The summits of some tropical mountains rise into the 5–10 km altitude zone, but most ELA and relevant vegetation zone changes are in the range of 3.5–5 km altitude so included in the lower line of Table 9.5.1.

Some of these values are depressed by the regional wind and ocean currents along the South American coast, but they give no support to hypotheses of high equatorial ATLRs in drier periods of the Pleistocene.

Similar estimates are given by general circulation models. For the central Andes GEOS-1 (Goddard Earth Observation System) indicates an ATLR just below 6° C in the humid east and just above 6.5° C in the dry west (Klein *et al.*, 1999). Rind & Peteet (1985) concluded that at the LGM the ATLR did not differ significantly from the present.

Ground surface lapse rates are available for some equatorial mountains. On the wet Carstensz Mountains the ATLR rose from 4.5 °C between sea-level and 1.5 km, to 7.5 °C between 3.6 and 4.2 km (Hope *et al.*, 1976). On dry Kilimanjaro and the wet Rwenzori Mountains, Osmaston (1989[1,2]) estimated ATLRs of 7.0 °C and 6.5 °C respectively. According to Graf (1981) the ATLR in the Ecuadorean Andes is 6.5 °C. The ground-level 0 °C isotherm is often lower than the free-air one, so that the ground-level and free-air ATLRs are not identical.

It is clear that evidence that tropical climates generally were dry at the LGM does not provide grounds for hypothesizing a lapse rate then of over 7.5 °C; that 6.5 °C remains the 'best guess'; and that it could have been as low as 5 °C.

ELA CHANGES

The ELA is determined primarily by mean temperature and snowfall, but cloudiness is also important, both in reducing incoming shortwave radiation and through its link with precipitation. By applying a selected ATLR, a change in ELA can be interpreted as a corresponding temperature change, under the assumption that all other factors remain the same. Although that is unlikely to be true, such a temperature estimate is valuable in providing a reliable reference point from which departures due to other factors can be estimated. Kaser (1995[2]) has concluded from climatic modelling that in the humid tropics, where the ELA lies close to the 0 °C isotherm, its altitude is mainly affected by temperature changes, whereas in the dry tropics where ELA lies well above the 0 °C isotherm it is more affected by precipitation changes. Thus mountains in the moist equatorial zone are the most likely to give good estimates of atmospheric cooling. However, the situation is further complicated when independent evidence suggests a simultaneous shift from moist to dry conditions. Kruss (1981, 1983, and in Hastenrath, 1984) has used numerical models of the Lewis Glacier on Mount Kenya to show that the changes in that glacier over the twentieth century (a reduction in area from 63 ha to 26 ha) could have been caused by changes of (a) 1 °C in mean temperature OR (b) 1 m water equivalent in snowfall OR (c) one tenth in cloudiness OR (d) 10% in relative humidity OR (e) 3% in albedo. In practice, changes of climate involve changes in all these variables simultaneously. Kaser (previous chapters; 1999, in press) considers that recent changes in air humidity have been more important than temperature rises in the widespread retreat of tropical glaciers. Either a decreased air humidity or an increased air humidity can cause a retreat of the glaciers. Decreased humidity causes a decrease in cloudiness and precipitation and an increase in incoming shortwave radiation. Increased humidity and cloudiness, combined with a moderate increase in air temperature such that more precipitation falls as rain, decreases snowfall on the ablation zone and raises the incoming longwave radiation markedly. Moreover, energy-consuming sublimation will be decreased. When considering moist equatorial glaciers at the LGM, when palaeolake, pollen and sea-surface temperature evidence indicate substantially drier and colder conditions, these last two factors would have been acting in opposition, so making a greater temperature reduction necessary to achieve the advances. This confirms the importance of supplementing crude estimates of past temperature decreases from ELAs and the ATLR by additions for increases in dryness. However, the estimation of the latter remains problematical; estimates of the past water balances of closed lakes can provide indications, though again it is difficult to distinguish the effects of temperature and humidity changes.

GLACIER HYPSOMETRY, MASS BALANCE GRADIENT RATIOS AND MORAINE MORPHOLOGY

These have been discussed above.

LAKE LEVEL CHANGES

In many parts of the tropics, especially East Africa, there is evidence that lake levels in closed basins were lower at the LGM due to greater aridity, and attempts have been made to quantify the climatic effects concerned, e.g. Street (1979[1]); even Lake Victoria, now a major contributor to the Nile, was completely dry at the LGM (Johnson et al., 1996). Greater aridity would affect both ELAs and vegetation zones. In the Central Andes, however, there is some evidence of the converse (Clayton & Clapperton, 1995) but discrimination between the similar effects of cooling and increased rainfall remains a problem. This issue is discussed by Klein et al. (1999) who conclude that over the moist eastern Andes the ELA depression of 1200 m was primarily mediated by temperature, whereas in the dry western Andes the depression of 800–1000 m was partly caused by an increase in precipitation. The latter does not mean that the temperature change was less, merely that it was less effective in causing ELA lowering and required supplementary snowfall to achieve the observed effect.

OXYGEN ISOTOPE RATIOS

In the 1970s and 1980s those studying past climates on tropical mountains were constrained by estimates of tropical sea-surface temperatures at the LGM only 1–2 °C cooler than at present, derived from isotope analysis of foraminifera from ocean cores (CLIMAP Project, 1976; Rind & Peteet, 1985). These were seriously discordant with estimates from palynology and ELAs of much greater temperature lowering on the mountains amounting to 4–8 °C or more. Some were driven to propose unjustifiably increased ATLRs to accommodate this (e.g. Walker & Flenley, 1979; Fox & Bloom, 1994); the latter hypothesized ATLRs as high as 10 °C km⁻¹ to accommodate inferred ELA lowering of 1400 m in the Andes. This problem has now been overcome by larger sea-surface temperature lowering estimates of about 5 °C, derived from corals (Flenley, 1997; Guilderson et al., 1994; Guilderson & Schrag, 1999); it appears that foraminifera partition oxygen isotopes misleadingly.

MONTANE VEGETATION BELTS

Pollen analysis of lake sediments and bogs throughout the tropics (summarized in Flenley, 1979) indicates downward shifts of up to 1200 m in mountain vegetation zones at the LGM, which have generally been interpreted in terms of temperature, though there is good evidence of greater aridity as a contributory cause and this is supported by pollen analysis at lower altitudes especially in East Africa, confirming the lake evidence there (Coetzee, 1964; Livingstone, 1975, 1980; Messerli et al., 1980; Hamilton, 1982; Hamilton & Taylor, 1991). Hamilton (1987) from his work on Elgon asserted 'the general unsuitability of pollen analysis as a technique for establishing past changes in temperature through altitudinal movements of vegetation in regions which have been subject to massive changes in rainfall' and Street-Perrott *et al.* (1997) warned against using the forest/grassland changes as indicating temperature or humidity changes (see below). However, Hamilton & Taylor (1991), in their review of the palynological and other evidence for climate change in East Africa, showed that in many cases the evidence of increased aridity rested not just on the balance between grass and forest but on the better grounds of the specific composition of the vegetation. The same may be said of inferred temperature changes where these rest on, for instance, inferred movements of the montane heather zone.

NEW EVIDENCE: GROUND-WATER, ICE-CORES, $\delta^{18}O_{Si}$ AND CARBON ISOTOPES

Further evidence has now been obtained from independent sources. Analyses of noble gases in ^{14}C dated 'fossil' ground-water in the Amazon basin indicate LGM temperatures 5 °C lower than now at the LGM, supported by an almost identical result from New Mexico (Stute *et al.*, 1995[1,2]).

Following the recovery of a 1500 year ice-core from the Quelccaya ice-cap (Thompson *et al.*, 1986), two cores over 160 m long were taken from the Huascarán ice-cap (Thompson *et al.*, 1995). Although the lowest layers of these were not positively dated, counts to lesser depths and correlations with other cores worldwide indicate that the basal ice was formed in the Late Glacial and also shows a short cooling episode, probably the Younger Dryas. These layers have low levels of $\delta^{18}O$ indicating that temperatures were 5–6 °C lower in the Late Glacial, possibly 8–12 °C at the LGM, and that this was accompanied by high dust levels that could have contributed to the cooling. The high dust levels accord with similar observations on Greenland ice-cores.

Sediment cores from Lake Hausberg high on Mount Kenya which show low levels of rock-flour input and high $\delta^{18}O_{Si}$ values in diatom silica at 2800–2500 BP have been interpreted as indicating cold-based glaciers then (Karlen *et al.*, 1999; Rietti-Shati *et al.*, 1998). In view of the possible major cooling mentioned in the last paragraph, a similar scenario must be considered for LGM glaciers, or at least their accumulation zones.

Other cores from lakes high on Mount Kenya and Mount Elgon have demonstrated that the vegetation changes there, previously inferred from pollen analysis, also represent shifts from dominance of C4 vegetation (typically forest) to C3 vegetation (typically grasses). C4 plants are favoured by high atmospheric pCO_2, and C3 plants by lower ones, so this evidence corroborates other evidence (e.g. from air bubbles in ice-cores) of low pCO_2 at the LGM (Olago *et al.*, 1999;

Street-Perrott *et al.*, 1997). It has two important implications: first that low pCO_2 and hence reduced 'greenhouse effect' contributed to cooling at the LGM; second that if the vegetation changes were partly due to pCO_2 changes, the temperature changes may have been less than previously inferred and that treeline elevation should be used with caution as an indicator of palaeo-temperatures. This factor may also have affected the changes from forest to savanna at lower altitudes at the time of the LGM that have been inferred from widespread pollen evidence.

9.5.2 Climates in East Africa

There is clear evidence on the Rwenzori that the high THARs, moderate AARs and low mass balance gradient ratios of the LGM glaciers resembled more those of present non-tropical glaciers than those of present tropical ones which are discussed in chapter 5. If the 0 °C isotherm is lowered by global cooling, but due to increased aridity the ELA is at a colder isotherm and higher altitude, this would have these effects. All three differences would be compatible with a development of debris-covered glaciers. The common occurrence of large and long lateral moraines may also indicate debris-covered glaciers.

One reason for the lack of supraglacial moraine today is the lack of very low temperatures in the accumulation zone. Records show that at points below the Rwenzori glaciers at about 4000 m, the mean minimum temperature is only -1 °C (Osmaston, 1989[2]), so that in the accumulation zone it is probably about -3 °C, hardly sufficient to cause much rock-shattering. A climate which would give rise to severe frost-shattering and debris-covered glacier tongues must have been colder, and hence probably less cloudy and drier. There is abundant evidence from past lake levels and pollen analysis in East Africa that the regional climate was drier at 20 000–15 000 BP, i.e. during the latter part of the last glacial (Livingstone, 1975, 1980; Messerli *et al.*, 1980; Hamilton, 1982; Hamilton & Taylor, 1991).

As an experiment, G. Kaser (pers. comm.) has taken his Balance Profile Model for the mid-latitudes and modified it to represent the equatorial zone, but cooler and drier than now, by making the following changes:

(a) extending the duration of the ablation period to 365 days per year;

(b) making the gradient of albedo zero, i.e. allowing for debris cover;

(c) making the ratio which splits the available energy into sublimation and melting 0.5: this means that half of it is consumed by sublimation, making up about 6% of the

Table 9.5.2. *Estimated temperature reductions on East African mountains at the LGM*

	Rwenzori	Mt Kenya	Kilimanjaro
Recent ELA	4500 m	4700 m	5300 m
0 °C ground level isotherm now[a]	4600 m	4600 m	4600 m
Temp. at Recent ELA[b]	1 °C	− 1 °C	− 4 °C
Est. rise in ELA: Recent–LGM	500 m	500 m	700 m
Crude temp. depression at LGM[b]	3 °C	3 °C	4.5 °C
Climate change now/at LGM	wet/moist	moist/drier	drier/dry
Addition for ELA/0 °C change	2 °C	2 °C	2 °C
Adjusted temp. depression at LGM	5 °C	5 °C	6.5 °C

Notes:

[a] Trapnell & Griffiths (1960) and Osmaston (1989[1,2]). The free air 0 °C isotherm is much higher.

[b] ATLR of 6.5 °C/1000 m used.

ablation. This represents the dryness of the climate and is comparable to northern Himalayan conditions.

These reasonable assumptions result in the same balance profile as for modern Alpine glaciers, with the AAR consequently being about 0.66. The very different hypsometry of former long valley glaciers and modern relic glaciers on upper slopes could add to the differences between now and the LGM. This provides theoretical support for the characteristics of LGM tropical glaciers inferred from field evidence.

If a lapse rate of 6.5°C is applied to the ELA difference of 500 m between the Lac Gris (the last definitely stationary position) and Lake Mahoma Stages, the indicated temperature depression is about 3°C. This must then be adjusted by a further depression of the 0°C ground level isotherm below the ELA due to the drier climate. However, the great extent of the LGM glaciers suggests that there was still considerable snowfall then, so that the climate was only moderately drier (a postulated drying of the climate must not be carried too far: in the south of Bolivia it is so dry and snowfall so deficient that even peaks of 6000 m are not glacierized now and had only small glaciers at the LGM). At present there is a difference of 5°C between the isotherms of the wet Rwenzori and the dry Kilimanjaro ELAs (Table 9.5.2). If this is spread evenly over an arbitrary qualitative scale from 'wet' (the climate of Rwenzori now), through 'moist', 'drier', to 'dry' (the climate of upper Kilimanjaro at the LGM) it results in steps of about 2°C for each difference in climate. If it is assumed that the climate of each mountain changed by one such step at the LGM, and the corresponding temperature steps of 2°C are added to the crude estimates of temperature change made simply from the ELA change, the result should be more accurate.

Table 9.5.2 shows that this method yields results for the three East African mountains which accord better than some

previously proposed, with a spread of only from 5°C to 6.5°C and a mean of 5.5°C. This would be sufficient to promote more frost-shattering than at present.

This figure also accords well with the new estimates of about 5°C cooling of ocean surface temperatures, and for the Brazilian basin, the sources of most of the snow for the Andes. Such cooling may have been an important factor in reducing moisture uptake from the Indian Ocean and so reducing that deliverable by the monsoon to the African mainland, resulting in aridity there. At present some of the precipitation on the Rwenzori is derived from Lake Victoria besides from the Indian Ocean, so the well-established drying out of the lake (Johnson *et. al.*, 1996) would have accentuated this effect.

On the Rwenzori, the similarities between the ELA profiles of different stages shown in Fig. 9.4.1, and the like similarity between their trend surfaces at the LGM and at other times (Osmaston, 1965, 1989[2]), indicate that there has been no great change in the main moisture-bearing winds, these being dominated by the easterly monsoons at both times, though westerlies from the Congo also affect this range.

On Kilimanjaro the resemblance of the ELA patterns at successive stages indicate that then as now the main accumulation was on the south and west, major factors being afternoon cloud and precipitation on the west of Kibo coupled with morning sun on the east. However, this dependence on local meteorological effects makes inferences about major wind patterns uncertain, but the much lower ELA on Mawenzi and its steep gradient suggest that the easterly monsoon was still an important source of snow at medium altitudes.

On Mount Kenya the reversal of the ELA gradient between the LGM and now does not reflect a change in the dominant wind, but rather that the influence of snow-bearing winds on large glaciers on the slopes were a more important factor then, in contrast to insolation on the vestigial glaciers clinging to the peaks now. The NW–SE tilt of the LGM ELA on the main peak suggests that the southeast monsoon was still dominant as it is now.

On Elgon, however, there was a shift from a dry/wet: north/south axis to one with the lowest ELA on the northeast, and Hamilton & Perrott (1978) proposed that this indicated a weakening in effectiveness of the southeast monsoon relative to the northeast one. On Aberdare the moraine pattern is not very clear but suggests that the lowest ELA was on the northeast as on Elgon.

These inferences drawn from five rather different individual mountains spread over a wide region show satisfactory mutual support in the indications they give of the regional temperature at the LGM; the majority indicate that major winds have remained fairly similar, though with perhaps a shift in importance from the southeast trades to the northeast ones. However, they also show the importance of local relief in modifying local meteorology and ELAs, often over-ruling regional climatic patterns. Similar detailed studies are needed on other mountains in other tropical regions. Even in the nearby mountains of Ethiopia many of the reports are incomplete and inconsistent.

9.5.3 Climates elsewhere in the tropics

Nowhere else in the tropics have LGM ELAs been estimated by fully rigorous and critical methods. On the wet equatorial mountains of Southeast Asia and Australasia, estimated ELAs indicate a depression of up to 800 m (see above) and thus a crude cooling estimate of 5 °C. Adjustment for less wet conditions due to the drying out of the Arafura shelf might raise this to 7 °C or 8 °C, but the ELA estimates need careful checking. In Central America similar estimates have been made and similar comments apply.

In the Andes many estimates have been made of ELAs mostly on a regional basis, and ELA changes of up to 1200 m have been inferred (e.g. Rodbell, 1992[1]; Klein & Isacks, 1998; Klein et al., 1999) which would imply crude temperature changes of 8 °C or more. However, they suggest that, in such western areas which are dry at present, ELA lowering was primarily mediated by increased moisture due to wind circulation changes. In the wetter areas on the east where the ELA lies close to the 0 °C isotherm and the depression was only 500 m, it was primarily temperature mediated, Seltzer (1993) estimating a temperature depression of only 3.5 °C. Hastenrath & Kutzbach (1985) show that palaeolakes on the Altiplano contemporaneous with glacial extensions indicate increases of precipitation by 50% or more; this is supported by Seltzer (1994). Even though Klein & Isacks' method enabled them to estimate the spatial variability of the ELA in more detail than

many other workers, it still has the general problems of regional ELAs which have been discussed above and which tend to result in overestimates of changes of regional ELAs and temperature. One reason for the acceptance hitherto of large figures for ELA changes has been similarly large estimates from palynological studies. These provide less support in the face of further evidence of the changes and effects of atmospheric pCO$_2$ concentrations. The new evidence of reductions of about 5 °C in ocean surface temperatures and from lowland Brazil (see above) also suggest that ELA depressions in the Andes should conform with these.

9.5.4 Is there an LGM equatorial glacier type?

Possible differences between the LGM climates and glaciers of the moist equatorial belt and those of the peripheral dry tropics have only been touched on above and the concentration has been on the former which are a better defined entity. Superficially it might appear that a simple lowering of temperature by 5 °C and a drier climate could account for the shift in the nature of equatorial glaciers towards that now seen in the subtropics and mid-latitudes. Important though this difference was, it would be simplistic to regard LGM equatorial glaciers as just copies of modern subtropical ones, since other aspects of the equatorial climate still prevailed, particularly the lack of seasonality. I therefore propose the recognition of an 'LGM equatorial valley glacier type' which builds on Tricart's proposal (1971) that the large Quaternary moraines, left by the lack of a melt season and consequent spates to remove them, should be regarded as the key characters of such a type. To this, however, should be added the probability that many of the glaciers were debris covered, the characteristic morphological (THAR c. 0.5, AAR c. 0.67) and mass balance gradient (ratio c. 1:1 to 1:1.5 for below ELA:above ELA) parameters of the glaciers as described above, and the persistence in a modified form of the other effects of the equatorial climate described by Kaser (see section 9.5.2 above). The moderate values of these parameters, less extreme than those found on many present debris-covered glaciers (see section 9.2.2), may represent compromises between the opposing effects of debris cover and the equatorial climate. However, as with most 'types' it must be accepted that some glaciers within the region may not show these characteristics but that others outside it may do so. The reason for restricting the field to LGM valley glaciers is that ice-caps and piedmont glaciers and their moraines have other forms and parameters, and there is evidence that on many equatorial mountains earlier glaciations occurred on less dissected mountain-tops.

9.6 SUMMARY

Numbers in parenthesis refer to sections in this chapter.

9.6.1 The LGM

1. On many tropical mountains in Australasia, Taiwan, Hawaii, Central and South America, and Africa, there are clear traces of a pantropical 'LGM', which ended about 15000 BP. This is usually marked by conspicuous moraines extending far below the present glaciers (if any). (9.1, 9.3, 9.4)

2. In some areas there are also limited traces of earlier major advance stages, and often of smaller readvances subsequent to the LGM. (9.1, 9.3, 9.4)

3. On four East African mountains the evidence of past glaciations has been repeatedly studied in detail, and these provide important evidence about the nature of past equatorial glaciers. (9.4)

4. The recent cosmogenic (^{36}Cl) dating of moraines on Kilimanjaro and Mount Kenya has demonstrated new possibilities of confirming (on the former) or radically changing (on the latter) accepted glacial sequences and estimated relative or absolute ages. (9.4.1, 9.4.2)

9.6.2 ELA estimation

5. The commonest indirect methods of ELA estimation (THAR, AAR) using morphological ratios of height and area depend on ratios determined from sources which are often not demonstrated to be appropriate. Moreover, if the glaciers did not conform to typical Alpine valley glacier morphology special ratios must be used. (9.2.3, 9.2.4)

6. The area–altitude balance ratio (AABR) method, using the area and height data (hypsometry) for the glacier and an assumed or statistically inferred mass balance ratio between the gradients above and below the ELA, gives the most accurate estimates of ELAs of former glaciers. All three methods can also be used for present glaciers if other information is lacking. (9.2.4, 9.2.5)

7. Direct observations on present glaciers often show steeper gradients of the mass balance gradient below the ELA than above it. Simple calculations of the ELAs of former glaciers using an assumed balance ratio of 1 can often provide estimates with a precision little exceeding the likely observational errors, but the AABR method also provides for the testing of other values of the balance ratio which may give a better precision and indicate a somewhat higher or lower ELA. (9.2.4, 9.2.5)

8. Such statistical tests on the ELA estimates obtained by whatever method, THAR, AAR or AABR, can show whether an appropriate parameter has been used, and indicate the parameter of choice – that which gives estimates with the lowest variance. These may be tests within localized groups, or within a larger population used for a trend line or surface. (9.2.4)

9. There are numerous constraints on the use of these indirect methods of ELA estimation which are often ignored, though the AABR method avoids most of them. The commonest, which affects all three, is the assumption that the glacier is stationary: this is seldom so with present glaciers but may be justifiable in the case of an apparently contemporaneous past maximum advance of several glaciers in the past. (9.2.5)

10. Examination of the data from 75 former LGM glaciers on the Rwenzori Mountains indicates that both in the geomorphology of their moraines and in the statistics of their morphology in relation to ELA estimation, they resemble present subtropical and mid-latitude glaciers more closely than they do present tropical ones. (9.4.5)

11. In particular there is statistical evidence that they may have had THARs of 0.4–0.5 and AARs of about 0.67 like present mid-latitude glaciers, and mass balance gradients not greatly steeper below the ELA than above it. (9.4.5)

9.6.3 Debris cover

12. These parameters and the large size of the LGM moraines suggest that the LGM glaciers had debris-covered tongues with lowered ablation, the extension of which left large moraines and modified the glacier morphology and mass balance gradient. This in turn suggests that the climate was one that produced more rock-shattering and that global temperature lowering was generally accentuated on a local scale by a drier climate with clearer skies and increased radiation cooling. This is supported by regional evidence from lake levels and pollen analysis. (9.2.1, 9.2.2, 9.5)

13. This may also be the consequence of the convergence whereby, both on the currently moister and on the driest mountains, changes in temperature and precipitation were potentially such as to increase rock-shattering and result in debris-covered glaciers. (9.5)

9.6.4 Local relief

14. Comparisons of the estimated ELA patterns on the East African mountains, where there are exceptionally good data, show that local relief and the consequent local

meteorology has a major effect on wind patterns, snow-fall, insolation and ELAs. This often greatly modifies the effects of the regional climate. (9.4)

9.6.5 ELA surface gradients and ELA differences

15. Many estimates of the lowering of the ELA during the LGM and its predecessors are in the range of 450–800 m (some up to 1200 m), but such estimates have often taken inadequate account of the numerous uncertainties and constraints involved in such estimates for former glaciers, some of which result in systematic overestimates of differences. Serious errors result from ignoring the slopes of the ELA surfaces, and by taking different sample populations at different times. (9.2.6)

16. Numerous examples show that, where sufficient observations permit the construction of local ELA profiles or trend surfaces, these show gradients of up to 100 m km^{-1}, and even where (as often) these are sub-parallel in different glacial stages it is essential to take these gradients into account when estimating ELA changes. (9.2.6, 9.4)

17. Theoretical modelling supported by practical examples shows that where peaks of various sizes have been glacierized in one glacial stage but only some of them in another stage, in a region with a significant ELA slope, simple differences of regional ELAs yield overestimates of regional ELA changes. Instead these must be estimated by means of detailed local studies. (9.2.6, 9.4)

18. Observations and analysis, after allowing for ELA surface slopes and for differences on different aspects, indicate that at the LGM the ELA was lower than at the last stationary position (about a century ago) by mean amounts of 500 m on Mount Kenya and the Rwenzori, and 700 m on Kilimanjaro. (Table 9.5.2)

9.6.6 Climatic change

19. Having estimated changes in ELA, inferences about climatic change depend on information or assumptions about the ATLR, about changes in atmospheric humidity and about the consequent relationship between the altitudes of the ELA and 0 °C level. These inferences must then be critically compared with estimates from other sources of evidence. (9.5)

9.6.7 Atmospheric temperature lapse rates

20. Inference about climatic change from ELA changes depends on assumptions about the tropical atmospheric temperature lapse rate (ATLR) at the LGM. Extensive data from the tropics indicate that the most likely figure is indeed the generally used figure of 6.5 °C/1000 m. Raised estimates of LGM tropical ocean temperatures have removed previous incentives to hypothesize higher lapse rates. (9.5.1)

9.6.8 Estimates of temperature change

21. Crude estimates of temperature changes made from ELAs and the ATLR require adjustments for changes in precipitation, cloudiness, etc., inferred from other sources. Widespread indications of drier conditions, and hence a rise of ELAs above the 0 °C isotherm, necessitate increases in the estimates of temperature depression on currently wet to moderately dry mountains. However, the driest mountains are thought to have become less dry, so that the crude estimate of temperature depression from ELAs must be reduced. On these an increase in humidity could have increased rock-shattering. It is important to use measurements or estimates of the ground level 0 °C isotherm rather than the free air one which is higher. (9.5.1)

22. Estimates of temperature change from glacial evidence should desirably conform to the best estimates from other sources. Some now indicate a depression of about 5 °C at the LGM, though the Huascarán ice-core suggests more. Previous estimates from the lowering of vegetation zones now appear to be too large owing to the effects of lower atmospheric pCO$_2$. (9.5.1)

23. Careful estimation of the ELA changes on the three highest East African mountains, making allowance for moderate decreases in humidity indicated by other evidence, suggest temperature depressions at the LGM of 5 °C to 6.5 °C. (9.5.2). This conforms with the vegetational evidence from fossil pollen on the Rwenzori. (9.4.5)

24. Theoretical modelling using adaptations of a model for present mid-latitude glaciers confirms that the inferred former differences in equatorial climate would probably have had the glaciological effects described above. In particular it indicates that the presence of a debris cover, which on modern glaciers outside the tropics commonly results in BRs <1, would not necessarily have had this effect on tropical ones at the LGM, thus removing potential conflict in the evidence. (9.5.2)

9.6.9 Future studies

25. Similar detailed studies of moraines and ELAs to those described on the East African mountains should be carried out on other tropical mountains.

26. Further cosmogenic dating of moraines is needed throughout the tropics, and especially on the third well-studied East African mountain, Rwenzori, to correlate this with Kilimanjaro and Mount Kenya.

9.6.10 A new glacier type

27. An *LGM equatorial valley glacier type* is proposed, subjected to a cooler, drier climate, and characterized by large moraines, the probability that many of the glaciers were debris covered, and moderate morphological (THAR 0.4–0.5, AAR *c.* 0.67) and mass balance gradient ratio (*c.* 1 to 1.5) parameters. (9.5.4)

Prospect

The interpretation of tropical glaciers as climatic indicators is the core purpose of this book. The nature of the tropical climate requires special examination. In the Part I, assessment was attempted through a theory-oriented analysis of key variables of the glacier–climate relationship. In Part II, the processes of modern glacier fluctuations in tropical high mountain regions were examined with the help of various information and observations. The first forms an abstract model analysis which requires further thorough verification. The results of the second are incomplete. However, the two approaches lead the way to a better understanding of the ice and water balances of tropical glaciers in future investigations that may concentrate on field measurements.

Sensitivity analyses which are based on measurements of the budget variables and parameters in areas of differing degrees of humidity or dryness can produce additional knowledge. They help to recognize and interpret spatial differences of glacier fluctuations, not only between windward and lee sides on mountains, but also along a north–south profile from the tropics over the outer tropics and the subtropics to the mid- and high latitudes. Above all, the American Cordilleras, which span through all climate zones and obstruct the global circulation of the atmosphere as an almost continuous barrier, present themselves for this type of investigation.

In addition to the ability to interpret these 'climate meters', the spatial and temporal reconstruction of former glacial fluctuations, both modern and Quaternary, and their careful evaluation and synthesis can contribute to a detailed and advanced understanding of the global climate and its fluctuations.

Part III provided a critical evaluation and development of the techniques for assessing the evidence left by extensive Quaternary tropical glaciers. On mountains where detailed data are now available, these indicate that, though these former tropical glaciers differed in some respects from modern ones, they provide means of estimating how their contemporary climates differed from present ones. If the necessary data are collected, these methods can be applied to many other glaciated mountains in the tropics to provide a firmer basis for our knowledge of climatic change in the Quaternary.

References

Abbott, L.D., E.A. Silver, R.S. Anderson, R. Smith, J.C. Ingle, S.A. Kling, D. Haig, E. Small, J. Galewsky & W. Sliter (1997). Measurement of tectonic surface uplift rate in a young collisional mountain belt. *Nature*, 385, 501–507.

Alean, J. & A. Ames (1994). Ein Gletscherinventar für Peru. *Die Alpen* (70/3), 122–137.

Allison, I. (1976). Glacier regimes and dynamics. In: Hope, G.S., J.A. Peterson, U. Radok & I. Allison (eds.): *The Equatorial Glaciers of New Guinea. Results of the 1971–1973 Australian Universities' Expeditions to Irian Jaya: survey, glaciology, meteorology, biology and paleoenvironments*. A.A. Balkema, Rotterdam, 39–58.

Allison, I. & J. Bennett (1976). Climate and microclimate. In: Hope, G.S., J.A. Peterson, U. Radok & I. Allison (eds.): *The Equatorial Glaciers of New Guinea. Results of the 1971–1973 Australian Universities' Expeditions to Irian Jaya: survey, glaciology, meteorology, biology and paleoenvironments*. A.A. Balkema, Rotterdam, 61–80.

Allison, I. & J.A. Peterson (1976). Ice areas on Mt. Jaya. Their extent and recent history. In: Hope, G.S., J.A. Peterson, U. Radok & I. Allison (eds.): *The Equatorial Glaciers of New Guinea. Results of the 1971–1973 Australian Universities' Expeditions to Irian Jaya: survey, glaciology, meteorology, biology and paleoenvironments*. A.A. Balkema, Rotterdam. 27–38.

Allison, I. & P. Kruss (1977). Estimation of recent climate change in Irian Jaya by numerical modeling of its tropical glaciers. *Arctic and Alpine Research*, 9(1), 49–60.

Alpenverein (year unknown). *Cordillera Blanca 1: 200.000*.

Ambach, W. & H. Eisner (1966). Analysis of a 20 m firn pit on the Kesselwandferner (Ötztal Alps). *J. Glaciol.*, 6(44), 223–231.

Ames, A. (1985). *Estudio de mediciones glaciológicas efectuadas en la Cordillera Blanca por ELECTROPERÚ S.A. – Variación y balance de masas de los glaciares y su contribución en el caudal de las cuencas*. Laboratoire de Glaciologie & Géophysique de l'Environnement, C.N.R.S., Grenoble, publication No. 457.

Ames, A. (1994). Die Gletscher der Cordillera Blanca in Perú – Glaciar Kinzl und Glaciar Schneider. *Mitteilungen des Österreichischen Alpenvereins*, 2/94, 49(119), 10–11.

Ames, A. (1998). A documentation of glacier tongue variations and lake development in the Cordillera Blanca, Perú. *Z. Gletscherk. Glazialgeol.*, 34(1), 1–36.

Ames, A. & B. Francou (1995). Cordillera Blanca, Glaciares en la Historia. *Bull. Inst. fr. études andines*, 24(1), 37–64.

Ames, A. & S. Hastenrath (1996). Mass balance and iceflow of the Uruashraju Glacier, Cordillera Blanca, Perú. *Z. Gletscherk. Glazialgeol.*, 32(2), 83–89.

Ames, A., G. Muñoz, J. Verástegui, R. Vigil, M. Zamora & M. Zapata, (1989). *Glacier Inventory of Peru*. Part I. Hidrandina S. A. Unit of Glaciology and Hydrology. Huaraz – Perú.

Ames, A., J. Alean & G. Kaser (1994). Gletscherschwankungen in der Cordillera Blanca, Perú. *Die Alpen*, 3(70), 138–152.

Argollo, J. (1982). *Evolution du Piedmont Ouest de la Cordillère Royale (Bolivie) au Quaternaire*. Thèse 3c, Fac. Sc. Luminy, Univ. d'Aix-Marseille II, France.

Asnani, C.G. (1993). *Tropical Meteorology, Vol. 1 & 2*. Indian Institute of Tropical Meteorology, Pune.

Atlas of Uganda (1962). Department of Lands and Surveys, Uganda.

Baker, B.H. (1967). *Geology of the Mt. Kenya area*. Geological Survey of Kenya, Report No.79, Nairobi.

Barry, R.G. (1992). *Mountain Weather and Climate*. Routledge, London, N.Y.

Benn, D.L. & A.M.D. Gemmell (1997). Calculating equilibrium line altitudes of former glaciers by the balance ratio method: a new computer spreadsheet. *Glacial Geology and Geomorphology*. http://ggg.qub.ac.uk/ggg/

Benson, C. S. (1961). Stratigraphic studies in the snow and firn of the Greenland Ice Sheet. *Folia Geographica Danica*, 9, 13–37.

Berger, P. (1989). Rainfall and agroclimatology of the Laikipia Plateau, Kenya. *Geographica Bernensia*, African Studies Series, A7, University of Berne.

Bergström, E. (1955). The British Ruwenzori Expedition, 1952. *J. Glaciol.*, 2(17), 468–476.

Beuning, K.R.M., K. Kelts & J.C. Stager (1998). Abrupt climatic changes associated with the arid Younger Dryas interval in Africa. In J.T. Lehman (ed.) *Environmental Change and Response in East African Lakes*, Kluwer, Netherlands. 147–156.

Bhatt, N., S. Hastenrath & P. Kruss (1981). Ice thickness determination at Lewis Glacier, Mount Kenya: seismology, gravimetry, dynamics. *Z. Gletscherk. Glazialgeol.*, 16(2), 213–228.

Bonner, W.D. (1968). Climatology of low level jet. *Monthly Weather Review*, 96, 83–85.

Borchers, B. (1935¹). *Die Weiße Kordillere*. Berlin.

Borchers, B. (ed.) (1935²). *Cordillera Blanca und mittleres Santa-Tal 1:100.000*. DÖAV.

Bozhinsky, A.N., M.S. Krass & V.P. Popovnin (1986). Role of debris cover in the thermal physics of glaciers. *J. Glaciol.*, 32, 255–266.

Broggi, J.A. (1943). La desglaciación actual de los Andes del Perú. *Boletín de la Sociedad Geológica del Perú*, XIV/XV, 59–90.

Brückner, E. (1886). Die Vergletscherung des Salzachgebietes. *Geogr. Abh.*1, 1–183.

Budd, W.F. & D. Jensen (1975). Numerical modelling of glacier systems. Proceeding of the Moscow Symposium on Snow and Ice in Mountainous Regions, 1971, *IAHS publ*. no. 104, 257–291.

Burbank, D.W. (1991). Late Quaternary snowline reconstructions for the southern and central Sierra Nevada, California and a reassessment of the 'Recess Peak Glaciation'. *Quaternary Research*, 36, 294–306.

Busk, D.L. (1954). The southern glaciers of the Stanley Group of the Ruwenzori. *The Geographical Journal*, CXX(2), 137–145.

Busk, D.L. (1957). *The Fountains of the Sun: Unfinished journeys in Ethiopia and the Ruwenzori*. Parrish, London.

Callendar, G.S. (1961). Temperature fluctuations and trends over the Earth. *Q. J. Roy. Meteorol. Soc.*, 87(371), 1–12.

Clapperton, C.M. (1972). The Pleistocene moraine stages of West–Central Perú. *J. Glaciol.*, 2(62), 255–263.

Clapperton, C.M. (1987). Maximal extent of the Wisconsin glaciation in the Ecuadorian Andes. *Quaternary of South America & The Antarctic Peninsula*, 5, 165–179.

Clapperton, C.M. (1991). Influence of tectonics on the extent of Quaternary glaciation in the Andes. *Bol. IG-USP, Publ.Esp.*, 8, 89–108.

Clapperton, C.M. (1993). *Quaternary Geology and Geomorphology of South America*, Elsevier, Amsterdam.

Clapperton, C.M. (1998). Late Quaternary glacier fluctuations in the Andes: testing the synchrony of global change. *Quaternary Proc.*, 6, 65–73.

Clark, D.H., M.M. Clark & A.R. Gillespie (1994). Debris-covered glaciers in the Sierra Nevada, California, and their implications for snowline reconstructions. *Quat. Res.*, 41, 139–153.

Clayton, J.D. & C.M. Clapperton (1995). The last glacial cycle and palae-olake synchrony in the southern Bolivian Altiplano: Certo Azanques case study. *Bull. Inst. fr. Etudes andines*, 24, 563–571.

CLIMAP Project Members (1976). The surface of the Ice-Age Earth. *Science*, 191, 1131–1137.

Coetzee, J.A. (1964). Evidence for a considerable depression of the vegetation belts during the Upper Pleistocene on the East African mountains. *Nature*, 204, 564–566.

Coetzee, J.A. (1967). Pollen analytical studies in Eastern and Southern Africa. *Palaeoecology of Africa*, 3.

Crutcher, H.L. (1969). Temperature and humidity in the troposphere. In: D.F. Rex (ed.): *Climate of the Free Atmosphere*. vol. 4 of Landsberg, H.E. (ed.) *World Survey of Climatology*, Elsevier Scientific, Amsterdam, 45–83.

Cui, Z.J., J.F. Yang, G.G. Liu, X. Wang & G.C. Song (2000). Discovery of Quaternary glacial evidence of Snow Mountain in Taiwan, China. *Chinese Science Bulletin*, 45, 566–571.

D.L.S.U. (1970). *Central Ruwenzori 1:25.000*. Department of Lands and Surveys, Uganda, Entebbe.

D.O.S. (1962). *Central Ruwenzori 1:25.000*. Directorate of Overseas Surveys.

Dale, I.R. (1940). The forest types of Mount Elgon. *J. East African Nat. Hist. Soc.*, 15, 574–82.

David, J.J. (1904). Forschungen über das Okapi und am Runssoro. *Globus*, 68(4), 61–63.

David, J.J. (1909). Mondgebirge (Runssoro, Ruwenzori). *Jahrb. Schweiz. Alp. Club*, 45, 153–181.

De Filippi, F. (1908). *Ruwenzori*. Constable, London.

De Filippi, F. (ed.) (1909). *Der Ruwenzori. Erforschung und erste Ersteigung seiner höchsten Gipfel*. Von L.A. v. Savoyen, Herzog der Abruzzen, F. A. Brockhaus. In Beilage: *Topographische und geologische Karte der Ruwenzori-Kette 1:40.000*.

de Grunne, V. (1933). Ruwenzori from the West. *Alpine J.*, 45, 275–289.

de Grunne, V., L. Hauman, L. Burgeon & P. Michot (1937). *Vers les glaciers de l'Equateur, le Ruwenzori*. Dupriez, Brussels.

de Heinzelin, J. (1951). Le retrait des glaciers du flanc ouest du massif Stanley (Ruwenzori). *Union Géod. Géoph. Inst. Ass. Gén. Bruxelles*, 1, 203–205.

de Heinzelin, J. (1952). Glacier recession and periglacial phenomena in the Ruwenzori Range. *J. Glaciol.*, 2(12) 137–140.

de Heinzelin, J. (1953). Les stades de recession du Glacier Stanley occidental. *Inst. Parcs Nat. Congo Belge, Expl. P.N.A.*, 2.

de Heinzelin, J. (1962). Carte des extensions glaciaires du Ruwenzori (versant Congolais). *Biul. Peryglacjalny*, 11, 133–139.

Deutscher Alpenverein (ed.) (1939). *Cordillera de Huayhuash 1:50.000*. Innsbruck.

Deutscher Alpenverein (ed.) (1945). *Cordillera Blanca Südteil 1:100.000*. Innsbruck.

Dirección Generál de Reforma Agraria y Asentamiento Rurál (1972). *Arbeitskarte 1:25.000*.

Downie, C. (1964). Glaciations of Mt. Kilimanjaro, Northeast Tanganyika. *Geol. Soc. Amer. Bull.*, 75, 1–16.

Drake, N. & A. Jones (1987). *Report of the 1987 Reading University Expedition to the Ruwenzori*. Mimeo.

Drygalski, E. von & F. Machatschek (1942). Gletscherkunde. In *Enzyklopädie der Erdkunde*. Deutich.

Eisenmann, E. (1939). Die Ruwenzori–Kundfahrt 1937/38 des Zweiges Stuttgart. *Z. d. Deutschen Alpenvereins*, 70, 40–49.

Escher-Vetter, H. (1980). *Der Strahlungshaushalt des Vernagtferners als Basis der Energiehaushaltsberechnung zur Bestimmung der Schmelzwasserproduktion eines Alpengletschers*. Meteorologisches Institut der Universität München, Wissenschaftliche Mitteilungen, 39.

Evans, I.S. (1990). Climatic effects on glacier distribution across the S. Coast Mountains, BC, Canada. *Annals of Glaciol.*, 14, 58–64.

Evans, I.S. & N.J. Cox (1995). The form of glacial cirques in the English Lake District, Cumbria. *Zeitschrift für Geomorphologie*, N.F., 39, 175–202.

Fantin, M. (1968). *Sui Ghiacciai dell' Africa*. Cappelli, Bologna.

Fernández, J. (1957). El problema de las lagunas de la Cordillera Blanca. *Boletín de la Sociedad Geológica del Perú*, 32: 87–95.

Finsterwalder, S. (1897). Der Vernagtferner. Seine Geschichte und seine Vermessung in den Jahren 1888 und 1889. *Wissensch. Ergänzungshefte zur Z. d. Deutschen u. Österr. Alpenvereins*. I (I).

Finsterwalder, R. (1987). Der Gletscherrückgang in der Cordillera Real (Bolivien) seit 1928 im Vergleich zu dem in den Ostalpen. *Z. Gletscherk. Glazialgeol.*, 23(2), 143–153.

Fiory-Ceccopieri, M.R. (1981). Dal Caucaso al Himalaya 1889–1909. Vittorio Sella fotografo alpinista esploratore. *Touring Club Italiano*, Milan.

Firmin, K. (1945). Ruwenzori: The Mountains of the Moon. *East Afr. Ann.* 1944/45, 25–35.

Flenley, J.R. (1979). *The Equatorial Rainforest: a Geological History*. Butterworth, London.

Flenley, J.R. (1997). The Quaternary in the tropics: an introduction. *J. Quat. Sc.*, 12, 345–346.

Flenley, J.R. & R.J. Morley (1978). A minimum age for the deglaciation of Mt. Kinabalu, East Malaysia. *Modern Quaternary Research in S.E. Asia*, 4, 57–61.

Fletcher, R.D. (1945). The general circulation of the tropical and equatorial atmosphere. *J. Meteorol.*, 2, 167–174.

Fliri, F. (1968). Beiträge zur Hydrologie und Glaziologie der Cordillera Blanca (Peru). *Alpenkundliche Studien*. Veröffentl. Univ. Innsbruck, 1, 26–52.

Ford, E. (ed.) (1974). *Papua New Guinea Resource Atlas*. Jacaranda Press, Milton, Queensland.

Fosberg, F.R., B.J. Garnier & A.W. Küchler (1961). Delimitation of the humid tropics. *Geographical Review*, LI(3), 333–347.

Fountain, A.G., K.J. Lewis & P.T. Doran (1999). Spatial climatic variation and its control on glacier equilibrium line altitude in Taylor Valley, Antarctica. *Global and Planetary Change*, 22, 1–10.

Fox, A.N. & A.L. Bloom (1994). Snowline altitude and climate in the Central Andes (5°-28°S) at present and during the Late Pleistocene glacial maximum. *J. Geogr. (Japan)*, 103, 867–885.

Francou, B. (1983). Les regimes termiques et pluviometriques de Pachachaca. *Bull. Inst. fr. études andines*, 12(1–2), 17–53.

Francou, B., P. Ribstein, E. Tiriau & R. Saravia (1995[1]). Monthly balance and water discharge on an intertropical glacier: the Zongo Glacier, Cordillera Real, Bolivia, 16°S. *J. Glaciol.*, 41(137), 61–67.

Francou, B., P. Ribstein, H. Semiond, C. Portocarrero & A. Rodríguez (1995[2]). Balances de glaciares y clima en Bolivia y Perú: Impacto de los eventos ENSO. *Bull. Inst. fr. études andines*, 24(3), 661–670.

Francou, B., B. Pouyaud & P. Wagnon (1995[3]). *Rapport de mission en Cordillere Blanche, Andes de Perou*. ORSTOM – La Paz.

Francou, B., P. Mourgiart & M. Fournier (1995[4]). Phase d'avancée des glaciers au Dryas récent dans les Andes de Pérou. *C.R. Acad.Sc., Paris*, 320 IIa, 593–599.

Funk, M., (1985). Räumliche Verteilung der Massenbilanz auf dem Rhonegletscher und ihre Beziehung zu Klimaelementen. *Zürcher Geographische Schriften*, 24. Geogr. Inst. der ETH Zürich.

Furbish, D.J. & J.T. Andrews (1984). The use of hypsometry to indicate long term stability and response of valley glaciers to changes in mass transfer. *J. Glaciol.*, 30, 199–211.

Gasse, F. & C. Descourtieux (1979). Diatomées et evolution de trois milieux Ethiopiens d'altitude differente, au cours du quaternaire superieur. *Palaeoecology of Africa*, 11, 117–134.

Gasse, F., P. Rognon & F.A. Street (1980). Quaternary history of the Afar and Ethiopian Rift lakes. In: M.A.J. Williams & H. Faure (eds.): *The Sahara and the Nile*. Balkema, Rotterdam. 361–400.

Georges, Ch. (1996). *Untersuchungen zu den rezenten Gletscherschwankungen in der nördlichen Cordillera Blanca, Perú*. Diplomarbeit, Institut für Geographie, Innsbruck.

Gèze, F. (1974). *La région centrale du Rift Ethiopien*. Thèse, Paris, Sorbonne.

Gillespie, A. & P. Molnar (1995). Asynchronous maximum advances of mountain and continental glaciers. *Reviews of Geophysics*, 33 (3), 311–364.

Gonzalez, E., Th. van der Hammen & R.F. Flint (1965). Late Quaternary glacial and vegetational sequence in Valle de Lagunillas, Sierra Nevada de Cocuy, Colombia. *Leidse Geologische Mededelingen*, 32, 157–182.

Good, W. (1982). Structural investigations of snow and ice on Core III from the drilling on Vernagtferner, Austria, in 1979. *Z. Gletscherk. Glazialgeol.*, 18(1), 53–64.

Graf, K. (1981). On the altitude pattern of the sub-nival step in the tropical Andes, especially in Bolivia and Ecuador. *Zeitschr. Geomorphology* Supp.Band, 37, 1–24.

Graf, K. (1986). *Klima und Vegetationsgeographie der Anden. Grundzüge Südamerikas und pollenanalytische Spezialuntersuchungen Boliviens*. Universität Zürich.

Gross, G. (1987). Die Flächenverluste der Gletscher in Österreich 1850–1920–1969. *Z. Gletscherk. Glazialgeol.*, 23(2), 131–141.

Gross, G., H. Kerschner & G. Patzelt (1977). Methodische Untersuchungen über die Schneegrenze in alpinen Gletschergebieten. *Z. Gletscherk. Glazialgeol.*, 12, 223–251.

Grove, J.M. (1988). *The Little Ice Age*. Methuen, London.

Gruber, A. (1972). Fluctuations in the position of the ITCZ in the Atlantic and Pacific Oceans. *J. Atm. Sc.*, 29, 193–197.

Guilderson, T.P., R.G. Fairbanks & J.L. Rubenstone (1994). Tropical temperature variations since 20,000 years ago: modulating inter-hemispheric climatic change. *Science*, 263, 663–665.

Guilderson, T.P. & D.P. Schrag (1999). Reliability of coral isotope records from the western Pacific warm pool: A comparison using age-optimized records. *J. Paleoceanography*, 14, 457–464.

Gutmann, J. (1948). *Beobachtungs- und Meßmethoden des Wetterdienstes*. Zentralanstalt für Meteorologie und Geodynamik, Publ. Nr. 158. Wien.

Haeberli, W. (1992). Zur Stabilität von Moränenseen in hochalpinen Gletschergebieten. *wasser, energie, luft – eau, énergie, air*, 84(11/12). 361–364.

Haenke, T. (1901). *Descripción del Perú*. Biblioteca del Perú, Lima.

Hagen, V.W. (ed.) (1959). *The Incas of Pedro de Cieza de León*. Transl.: H. de Onis. Univ. of Oklahoma Press.

Hamilton, A.C. (1982). *Environmental History of East Africa*. Acad. Pr., London.

Hamilton, A.C. (1987). Vegetation and climate of Mt. Elgon during the late Pleistocene and Holocene. *Palaeoecology of Africa*, 18, 283–304.

Hamilton, A.C. & R.A. Perrott (1978). Date of deglacierisation of Mt. Elgon. *Nature*, 273, 49.

Hamilton, A.C. & R.A. Perrott (1979). Aspects of the glaciation of Mt. Elgon, East Africa. *Palaeoecology of Africa*, 11, 153–161.

Hamilton, A.C. & D. Taylor (1991). History of climate and forests in tropical Africa during the last 8 million years. *Climatic Change*, 19, 65–78.

Hammen, T. van der, J. Barelds, H. de Jong & A.A. de Veer (1981). Glacial sequence and environmental history in the Sierra Nevada del Cocuy (Colombia). *Palaeogeography, Palaeoclimatology, Palaeoecology*, 32, 247–340.

Hansen, J. & S. Lebedeff (1987). Global trends of measured surface air temperature. *J. Geophys. Res.*, 92, 13354–13372.

Hastenrath, S. (1967). Observations on the snowline in the Peruvian Andes. *J. Glaciol.*, 6, 541–550.

Hastenrath, S. (1971). On snowline depression and atmospheric circulation in the tropical Americas during the Pleistocene. *S. African Geogr. J.*, 53, 53–69.

Hastenrath, S. (1973). On the Pleistocene glaciation of the Cordillera de Talamanca, Costa Rica. *Zeitschr. für Gletscherkunde und Glazialgeologie*, 9, 105–121.

Hastenrath, S. (1974[1]). Spuren pleistozäner Vereisung in den Altos de Cuchumatanes, Guatemala. *Eiszeitalter und Gegenwart*, 25, 25–34.

Hastenrath, S. (1974[2]). Glaziale und periglaziale Formbildung in Hoch-Semyen, Nord Aethiopien. *Erdkunde*, 28, 176–186.

Hastenrath, S. (1977). Pleistocene mountain glaciation in Ethiopia. *J. Glaciol.*, 18, 309–313.

Hastenrath, S. (1978). Heat budget measurements on the Quelccaya Ice Cap, Peruvian Andes. *J. Glaciol.*, 20, 85–97.

Hastenrath, S. (1981). *The Glaciation of the Ecuadorian Andes*. Balkema, Rotterdam.

Hastenrath, S. (1984). *The Glaciers of Equatorial East Africa*. D. Reidel Publishing Company, Dordrecht.

Hastenrath, S. (1985). A review of Pleistocene to Holocene glacier variation in the tropics. *Zeitschr. für Gletscherkunde*, 21, 183–194.

Hastenrath, S. (1989). Ice flow and mass changes of Lewis Glacier, Mount Kenya, East Africa: observations 1974–86, modelling, and predictions to the year 2000 A.D. *J. Glaciol.*, 35(121), 325–332.

Hastenrath, S. (1991[1]). *Climate Dynamics of the Tropics*. Updated edition from Climate and Circulation of the Tropics. Atmospheric Sciences Library. Kluwer Academic Publishers, Dordrecht.

Hastenrath, S. (1991[2]). *Glaciological Studies on Mount Kenya 1971–83–91*. University of Wisconsin, Madison.

Hastenrath, S. (1995). Glacier recession on Mount Kenya in the context of the global tropics. *Bull. Inst. fr. études andines*, 24(3), 633–638.

Hastenrath, S. & Ames A. (1995[1]). Recession of Yanamarey Glacier in Cordillera Blanca, Perú, during the 20th century. *J. Glaciol.*, 41(137), 191–196.

Hastenrath, S. & Ames A. (1995[2]). Diagnosing the imbalance of Yanamarey Glacier in the Cordillera Blanca of Perú. *J. Geoph. Research*, 100(D3), 5105–5112.

Hastenrath, S. & L. Greischar (1997). Glacier recession on Kilimanjaro, East Africa, 1912–1989. *J. Glaciol.*, 43, 455–459.

Hastenrath, S. & P. Kruss (1982). On the secular variation of ice flow velocity at Lewis Glacier, Mount Kenya, Kenya. *J. Glaciol.*, 28(99), 333–339.

Hastenrath, S. & P. Kruss (1992). The dramatic retreat of Mount Kenya's glaciers 1963–87: greenhouse forcing. *Ann. Glaciol.*, 16, 127–133.

Hastenrath, S. & J. Kutzbach (1985). Late Pleistocene climate and water budget of the South American Altiplano. *Quaternary Research*, 24, 249–256.

Hastenrath, S., R. Rostom & R.A. Caukwell (1989). Variations of Mount Kenya's glaciers 1963–87. *Erdkunde*, 43, 202–210.

Hauman, L. (1933). Esquisse de la végétation des haute altitudes sur le Ruwenzori. *Bull. Ac. Roy. Belg.*, Cl. Sc. 5 ser, 19, 602–616, 702–717, 900–917.

Hedberg, O. (1951). Vegetation belts of East African Mountains. *Svensk Botanisk Tidschrift*, 45, 140–195.

Heine, K. (1975). *Studien zur jungquartären Glazialmorphologie mexikanischer Vulkane*. Mexiko Projekt DFG, 7, Steiner Verlag, Wiesbaden.

Heine, K. & J.T. Heine (1996). Late glacial climatic fluctuations in Ecuador: glacier retreat during Younger Dryas time. *Arctic and Alpine Research*, 28, 496–501.

Helmens, K.F. & P. Khury (1995). Glacier fluctuations and vegetation change associated with Late Quaternary climatic oscillations in the Andes near Bogota, Colombia. *Quaternary of South America & the Antarctic Peninsula*, 9, 117–140.

Höfer, H. (1879). Gletscher und Eiszeit Studien. *Sitz.–Ber. d. kais. Akad. Wiss. Wien, Mathem.-naturw. Classe*, 79.

Hofmann, G. (1963). Zum Abbau der Schneedecke. *Arch. Met. Geoph. Biokl.*, B, 13(1), 1–20.

Hofmann, G. (1965). Zur Rolle des Wärmehaushalts bei der selektiven Ablation. *Carinthia*, II(24), Sonderheft: Bericht über die 8. Tagung für alpine Meteorologie, 9.

Hoinkes, H. (1970). Methoden und Möglichkeiten von Massenhaushaltsstudien auf Gletschern. *Z. Gletscherk. Glazialgeol.*, VI(1–2), 37–90.

Hollin, J.T. & D.H. Schilling (1981). Late Wisconsin-Weichselian mountain glaciers and small ice caps. Chap. 3 in G.H. Denton & T.J. Hughes (eds.): *The Last Great Ice Sheets*. Wiley, New York.

Hope, G.S. (1976). Mt. Jaya: The area and first exploration. 1–14. In: Hope, G.S., J.A. Peterson, U. Radok & I. Allison (eds.): *The Equatorial Glaciers of New Guinea. Results of the 1971–1973 Australian Universities' Expeditions to Irian Jaya: survey, glaciology, meteorology, biology and paleoenvironments*. A.A. Balkema, Rotterdam.

Hope, G.S., J.A. Peterson, U. Radok & I. Allison (eds.) (1976). *The Equatorial Glaciers of New Guinea. Results of the 1971–1973 Australian Universities' Expeditions to Irian Jaya: survey, glaciology, meteorology, biology and paleoenvironments*. A.A. Balkema, Rotterdam.

Hövermann, J. (1954[1]). Uber die Höhenlage der Schneegrenze in Aethiopien und ihre Schwankungen in historischer Zeit. *Nachtrichten Akad. Wiss. in Gottingen*, 6.

Hövermann, J. (1954[2]). Uber glaziale und 'periglaziale' Erscheinungen in Erithrea und Nordabessinien. *Veroffentlichungen der Akademie für Raumforschung und Landesplanung, Abhandlung*, 28, 87–111.

Hoyos-Patiño, F. (2000). Glaciers of Colombia. In *Satellite image atlas of glaciers of the world: glaciers of S.America*. USGS P 1386-I.

Huamán, A. (1985). *La laguna en formación en el glaciar 513 A (Hualcán)*. Internal report of Hidrandina S.A., Huaraz.

Humphreys, G.N. (1927). New routes on Ruwenzori. *Geogr. J.*, 69(6), 516–531.

Humphreys, G.N. (1933). Ruwenzori flights and further exploration. *Geogr. J.*, 82, 481–514.

Hunting Aerosurveys (1955). Aerial photographs; June and October 1955. No. 15UG 13, 14, 29, 31, 33.

Hurni, H. (1982). Climate and the dynamics of altitudinal belts from the last cold period to the present day, Simien Mountains, Ethiopia (in German, English summaries). Pt. II co-authored with Peter Stähli, Geographica Bernensia, G13.

Hurni, H. (1989). Late Quaternary of Simien and other mountains in Ethiopia. In: W.C. Mahaney (ed.): *Quaternary and Environmental Research on East African Mountains*. Balkema, Rotterdam, 105–120, 7–30.

Hydrographisches Zentralbüro im BM f. Land- und Forstwirtschaft (ed.) (various years). *Hydrographisches Jahrbuch von Österreich*. Wien.

IGM (Instituto Geográfico Militar Lima) (1970). *Carta Nacional del Perú* 1:100.000.

IGN (Instituto Geográfico Nacional) (1986). *Carta Nacional del Perú* 1:100.000.

Inoue, J. & M. Yoshida (1980). Ablation and heat exchange over the Khumbu Glacier. *Seppyo, J. Jap. Soc. Snow & Ice*, Sp. Issue 41, 26–33.

Iwanow, N.N. (1959). Kontinentalitätsgürtel auf der Erde. *Izwest. Wsesoj. Geogr. Obschtsch.*, 91, 410–423 (Russian; cited after Paffen, 1967).

Jóhannesson, T., C. Raymond & E. Waddington (1989[1]). Time-scale for adjustment of glaciers to changes in mass balance. *J. Glaciol.*, 35(121), 355–369.

Jóhannesson, T., C. Raymond & E. Waddington (1989[2]). A simple method for determining the response time of glaciers. In: J. Oerlemans (ed.): *Glacier Fluctuations and Climatic Change*. Kluwer Academic Publishers, 343–352.

Johnson, A. M. (1976). The climate of Peru, Bolivia and Ecuador. In: Schwerdtfeger: *World Survey of Climatology*, 12. Elsevier, Amsterdam, 147–218.

Johnson, T.C., C.A. Scholz, M.R. Talbot, K. Kelts, R.D. Ricketts, G. Ngobi, K. Beuning, I. Ssemmanda & J.W. McGill (1996). Late Pleistocene desiccation of Lake Victoria and rapid evolution of Cichlid fishes. *Science*, 273, 1091–1093.

Jordan, E. (1979). Grundsätzliches zum Unterschied zwischen tropischem und außertropischem Gletscherhaushalt unter besonderer Berücksichtigung der Gletscher Boliviens. *Erdkunde*, 33, 297–309.

Jordan, E. (1983). Die Vergletscherung des Cotopaxi – Ecuador. *Z. Gletscherk. Glazialgeol.*, 19(1), 73–102.

Jordan, E. (1991). *Die Gletscher der Bolivianischen Anden. Eine photogrammetrisch – kartographische Bestandsaufnahme der Gletscher Boliviens als Grundlage für klimatische Deutungen und Potential für die wirtschaftliche Nutzung*. Franz Steiner Verlag, Stuttgart.

Jordan, E. (2000). Glaciers of Bolivia. In *Satellite Image Atlas of Glaciers of the World: Glaciers of S. America*. USGS P 1386-I-5.

Jordan, E. & S.L. Hastenrath (2000). Glaciers of Ecuador. In *Satellite Image Atlas of Glaciers of the World: Glaciers of S. America*. USGS P 1386-I.

Karlen, W., J.L. Fastook, K. Holmgren, M. Malmström, J.A. Matthews, E. Odada, J. Risberg, G. Rosquist, P. Sandgren, A. Shemesh & L-O. Westerberg (1999). Glacier fluctuations on Mt. Kenya since ~6000 cal. years BP: implications for Holocene climatic change in Africa. *Ambio*, 28, 409–418.

Kaser, G. (1983[1]). *Verdunstung von Schnee und Eis*. Dissertation, Universität Innsbruck.

Kaser, G. (1983[2]). Über die Verdunstung auf dem Hintereisferner. *Z. Gletscherk. Glazialgeol.*, 19(2), 149–162.

Kaser, G. (1988). Report on field work in the Cordillera Blanca (Peru). Preparatory investigations and proposals for a glaciological, climatological and hydrological project. *Unidad de Glaciologia e Hidrologia – Hidrandina S.A. Huaraz*. Unpublished report.

Kaser, G. (1989). Vom drohenden Ausbruch eines Gletschersees in der Cordillera Blanca, Perú. *GW – Unterricht*, 33, 35–42.

Kaser, G. (1993). Le fluttuazioni dei ghiacciai del Ruwenzori (Africa Orientale) dalla spedizione del Duca degli Abruzzi (1906) sino agli anni '90. *Geogr. Fis. Dinam. Quat.*, 15(1/2), 121–125.

Kaser, G. (1994). 513A, oder ein See gegen eine Stadt. Die Geschichte eines bedrohlichen Gletschersees in der Cordillera Blanca. *Mitteilungen des Österreichischen Alpenvereins*, 2/94, 49(119), 6–7.

Kaser, G. (1995[1]). How do tropical glaciers behave? Some comparisons between tropical and midlatitude glaciers. In: Ribstein, P. & B. Francou (eds.) *Aguas glaciares y cambios climaticos en los andes tropicales. Conferencias y posters del seminario internacional, La Paz, 13.–16.6.1995*. 207–218.

Kaser, G. (1995[2]). Some notes on the behavior of tropical glaciers. *Bull. Inst. fr. études andines*, 24(3), 671–681.

Kaser, G. (1996). *Glaciological Field Work in the Cordillera Blanca 1995*. Report to the Unidad de Glaciologia e Hidrologia, Electroperu S.A., Huaraz. Institut für Geographie d. Universität Innsbruck.

Kaser, G. (1999). A review of the modern fluctuations of tropical glaciers. *Global and Planetary Change*, 22, 93–103.

Kaser, G. (in press). Glacier–climate interaction at low latitudes. *J. Glaciol.*

Kaser, G. & Ch. Georges (1997). Changes of the equilibrium line altitude in the tropical Cordillera Blanca (Perú) between 1930 and 1950 and their spatial variations. *Ann. Glac.*, 24. Papers from the International Symposium on Changing Glaciers held in Fjærland, Norway, 24–26 June 1996.

Kaser, G. & B. Noggler (1991). Observations on Speke Glacier, Ruwenzori Range, Uganda. *J. Glaciol.*, 37(127), 313–318.

Kaser, G. & B. Noggler (1996). Glacier fluctuations in the Ruwenzori Range (East Africa) during the 20th century. A preliminary report. *Z. Gletscherk. Glazialgeol.*, 32, 109–117.

Kaser, G., A. Ames & M. Zamora (1990). Glacier fluctuations and climate in the Cordillera Blanca, Peru. *Ann. Glaciol.*, 14, 136–140.

Kaser, G., Ch. Georges & A. Ames (1996[1]). Modern glacier fluctuations in the Huascarán–Chopicalqui Massif of the Cordillera Blanca, Perú. *Z. Gletscherk. Glazialgeol.*, 32, 91–99.

Kaser, G., S. Hastenrath & A. Ames (1996[2]). Mass balance profiles on tropical glaciers. *Z. Gletscherk. Glazialgeol.*, 32, 75–81.

Kerschner, H. (1990). Methoden der Schneegrenzbestimmung. In: H. Liedtke (ed.): *Eiszeitforschung*. Wissenschaftliche Buchgesellschaft Darmstadt, 299–311.

Kinzl, H. (1935[1]). Gegenwärtige und eiszeitliche Vergletscherung in der Cordillera Blanca (Peru). *Verhandlungen und wissensch. Abh. des 25. Deutschen Geographentages Bad Nauheim, 1934*, 41–56.

Kinzl, H. (1935[2]). Aufgaben und Reisen des Geographen. In: B. Borchers: *Die Weiße Kordillere*. Berlin. 180–203.

Kinzl, H. (1935[3]). Die Landschaft der Cordillera Blanca. In: B. Borchers: *Die Weiße Kordillere*. Berlin. 213–239.

Kinzl, H. (1935[4]). Altindianische Siedlungsspuren im Umkreis der Cordillera Blanca. In: B. Borchers: *Die Weiße Kordillere*. Berlin. 324–343.

Kinzl, H. (1937[1]). Die Kordillere von Huayhuash (Peru). *Z. d. Deutschen und Österr. Alpenvereins*, 68, 1–20.

Kinzl, H. (1937[2]). Der Nevado Acrotambo. In: K. Körner : Marine Trias am Nevado de Acrotambo (Nordperu). *Palaeontographica*, 86(A), 145–149.

Kinzl, H. (1940[1]). La ruptura del lago glacial en la Quebrada Ulta en el año 1938. *Boletín del Museo de Historia Natural 'Javier Prado'*, 4, 153–167.

Kinzl, H. (1940[2]). Las tres expediciones del Deutscher Alpenverein a las cordilleras peruanas. *Boletín del Museo de Historia Natural 'Javier Prado'*, 4, 3–24.

Kinzl, H. (1940[3]). Los glaciares de la Cordillera Blanca. *Revista de la Ciencia Natural*, 43, 417–440.

Kinzl, H. (1941). Die Anden Kundfahrt des Deutschen Alpenvereins im Jahre 1939. *Z. d. Deutschen Alpenvereins*, 72, 1–17.

Kinzl, H. (1942). Gletscherkundliche Begleitworte zur Karte der Cordillera Blanca (Peru). *Z. Gletscherk.*, 28 (1/2), 1–19.

Kinzl, H. (1943). Die anthropogeographische Bedeutung der Gletscher und die künstliche Flurbewässerung in den peruanischen Anden. *Sitzungsberichte europ. Geographen, Würzburg 1942, Leipzig 1943*, 353–380.

Kinzl, H. (1944). Die künstliche Bewässerung in Peru. *Zeitschrift für Erdkunde*, 12, 98–110.

Kinzl, H. (1949). Die Vergletscherung der Südhälfte der Cordillera Blanca (Perú). *Z. Gletscherk. Glazialgeol.*, 1, 1–28.

Kinzl, H. (1950). Die Cordillera Blanca (Peru). Das Arbeitsfeld dreier Alpenvereins – Expeditionen. *Jahrbuch des Österr. Alpenvereins*, 75, 37–48.

Kinzl, H. (1954). Ein Jahr geographischer Forschung in Peru. *Mitt. Geogr. Ges. Wien*, 96, 321–329.

Kinzl, H. (1955[1]). Neues von der Huayhuash Kordillere (Peru). *Jahrbuch Österr. Alpenverein*, 80, 123–131.

Kinzl, H. (ed.) (1955[2]). *Cordillera Huayhuash*. Tiroler Graphik, Innsbruck.

Kinzl, H. (1965). Die altindianischen Bewässerungsanlagen in Perú nach der Chronik des Pedro de Cieza de León (1553). *Mitt. Österr. Geogr. Ges.*, 105(III), 331–339.

Kinzl, H. (1970[1]). Gründung eines glaziologischen Institutes in Perú. *Z. Gletscherk. Glazialgeol.*, VI(1–2), 245–246.

Kinzl, H. (1970[2]). Bedrohte Natur in den peruanischen Anden. *Colloquium Geographicum*, 12, 253–270.

Kinzl, H. & A. Wagner (1938). Pilotaufstiege in den peruanischen Anden. *Gerlands Beiträge zur Geophysik*, 54(1), 29–55.

Kinzl, H., E. Schneider & F. Ebster (1942). Die Karte der Kordillere von Huayhuash (Peru). *Z. Ges. f. Erdkunde*, 1–9, 18–35.

Kinzl, H., F. Ebster, E. Gotthardt, K. Heckler & E. Schneider (1964). Begleitworte zur Karte 1:100.000 der Cordillera Blanca (Perú). Südteil. *Wissensch. Alpenvereinshefte*, 17.

Kirkbride, M.P. (1989). *The role of sediment budget on the geomorphic activity of the Tasman Glacier, Mt. Cook National Park, New Zealand*. Ph.D. Thesis, Univ. Canterbury, New Zealand.

Kirkbride, M.P. & V. Brazier (1998). A critical evaluation of the use of glacier chronologies in climatic reconstruction, with reference to New Zealand. *Quaternary Proc.*, 6, 55–64.

Kirkbride, P.M. & C.R. Warren (1999). Tasman Glacier, New Zealand: 20th-century thinning and predicted calving retreat. *Global and Planetary Change*, 22, 11–28.

Klein, A.G. & B.L. Isacks (1998). Alpine glacial geomorphological studies in the central Andes using Landsat thematic mapper images. *Glacial Geology & Geomorphology*, http://ggg.qub.ac.uk/ggg/.

Klein, A.G., G.O. Seltzer & B.L. Isacks (1999). Modern and last local glacial maximum snowlines in the central Andes of Peru, Bolivia and Northern Chile. *Quaternary Science Reviews*, 18, 63–84.

Klute, F. (1915). Beobachtungen über Zackenfirn (Büßerschnee) und dessen Entstehung am Kilimandscharo. *Z. Gletscherk.*, IX, 289–305.

Koopmans, B.N. & P.H. Stauffer (1968). Glacial phenomena on Mt. Kinabalu, Sabah. *Malaysian Geol. Survey Bull.*, 8, 25–35.

Körner, H.J. (1983). Zur Mechanik der Bergsturzströme vom Huascarán, Perú. In Patzelt, G. (ed.): Die Berg und Gletscherstürze vom Huascarán, Cordillera Blanca, Perú. *Hochgebirgsforschung–High Mountain Research*. Arbeitsgemeinschaft für vergleichende Hochgebirgsforschung, 6.

Kotlyakov, V.M. & I.M. Lebeveda (1974). Nieve and ice penitentes. Their way of formation and indicative significance. *Z. Gletscherk. Glazialgeol.*, 10, 111–127.

Krapf, L. (1849). Von der Afrikanischen Ostküste. *Zeitschrift der Deutschen Morgenländischen Gesellschaft*, 3, 310–321.

Krapf, L. (1858). *Reisen in Ostafrika, ausgeführt in den Jahren 1837–1855*. 2 Bände; Kronthal: im Selbstverlage des Verfassers, Stuttgart: in Commission bei W. Stroh.

Krapf, L. (1860). *Travels, Researches, and Missionary Labors, During the Eighteen Years' Residence in Eastern Africa*. Trübner, London.

Kraus, E.B. (1955). Secular changes of tropical rainfall regimes. *Q. J. Roy. Meteorol. Soc.*, 81(348), 198–210.

Kraus, H. (1966). Freie und bedeckte Ablation. *Ergeb. Forschungsuntern. Nepal Himalaya*, Liefg. 3, Springer, 203–236.

Kraus, H. (1972). Energy Exchange at Air–Ice Interface. *IAHS Publ. no.* 107(1), 128–164.

Kruss, P. (1981). *Numerical modelling of climatic change from the terminus record of Lewis Glacier, Mt. Kenya*. Ph.D.diss., Dept. Meteor. Univ. Wisconsin, Madison, USA.

Kruss, P. (1983). Climate change in East Africa: a numerical simulation from the 100 years of terminus record at Lewis Glacier, Mount Kenya. *Z. Gletscherk. Glazialgeol.*, 19(1), 43–60.

Kruss, P. (1984). Terminus response of Lewis Glacier, Mount Kenya, Kenya, to sinusoidal net-balance forcing. *J. Glaciol.*, 30(105), 212–217.

Kruss, P.D. & S. Hastenrath (1987). The role of radiation geometry in the climate response of Mt. Kenya's glaciers, Part I: Horizontal reference surfaces. *J. Climatology*, 7, 493–505.

Kruss, P.D. & S. Hastenrath (1990). The role of radiation geometry in the climate response of Mt. Kenya's glaciers, Part 3: The latitude effect. *Intern. J. of Climate*, 10, 321–328.

Kuhn, M. (1979). On the computation of heat transfer coefficients from energy balance gradients on a glacier. *J. Glaciol.*, 22(87), 263–272.

Kuhn, M. (1980[1]). Die Reaktion der Schneegrenze auf Klimaschwankungen. *Z. Gletscherk. Glazialgeol.*, 16(2), 241–254.

Kuhn, M. (1980[2]). Climate and glaciers. Sea level, ice and climatic change (Proceedings of the Canberra Symposium, Dec. 1979) *IAHS Publ.* No. 131, 3–20.

Kuhn, M. (1987). Micro-meteorological conditions for snow melt. *J. Glaciol.*, 33(113), 24–26.

Kuhn, M. (1989). The response of the equilibrium line altitude to climatic fluctuations: theory and observations. In: J. Oerlemans (ed.): *Glacier Fluctuations and Climatic Change*. Kluwer Academic Publishers, 407–417.

Kuhn, M., G. Kaser, M. Markl, H.P. Wagner & H. Schneider (1979). *25 Jahre Massenhaushaltsuntersuchungen am Hintereisferner*. Inst. f. Meteorologie u. Geophysik, Universität Innsbruck.

Kuhn, M., G. Markl, G. Kaser, U. Nickus, F. Obleitner & H. Schneider (1985). Fluctuations of climate and mass balance: different responses of two adjacent glaciers. *Z. Gletscherk. Glazialgeol.*, 21, 409–416.

Kuhn, M. & A. Herrmann (1990). Schnee & Eis. In: Baumgartner, A. & H.J. Liebscher (eds.): *Lehrbuch der Hydrologie, Bd. 1–Allgemeine Hydrologie – Quantitative Hydrologie*. Gebrüder Borntraeger, Berlin, Stuttgart. 271–312.

Kuls, W. & A. Semmel (1965). Zur Frage pluvialzeitlicher Solifluktionvorgängeim Hochland von Godjam (Aethiopien). *Erdkunde*, 19, 292–297.

Kurowski, L. (1891). Die Höhe der Schneegrenze. *Geogr. Abh.*, 5, 1(124), 119–160.

Langdale Brown, I., H.A. Osmaston & J.G. Wilson (1964). *The Vegetation of Uganda and its Bearing on Landuse*. Govt. of Uganda, Entebbe, with maps.

Lauer, W. (1975). Vom Wesen der Tropen. Klimatologische Studien zum Inhalt und zur Abgrenzung eines irdischen Landschaftsgürtels. *Akademie der Wissenschaften und der Literatur. Abhandlungen der mathematisch–naturwissenschaftlichen Klasse*, 1975(3). Franz Steiner Verlag, Wiesbaden.

Lauer, W. & M.D. Rafiqpoor (1986). Geoökologische Studien in Ecuador. Bericht über eine Studienreise 1985. *Erdkunde* 40, 68–72.

Lauscher, F. (1966). Die Tagesschwankungen der Lufttemperatur auf Höhenstufen in allen Erdteilen. *60–62. Jahresbericht des Sonnblick-Vereines*, 1962–1964, 3–17.

Lautensach, H. (1952). Die Isanormalenkarte der Jahresschwankung der Lufttemperatur. *Petermann's Mitt.*, 96, 145–155.

Leibundgut, Ch., P. Berger, A. Brodbeck, R. Brunner, S. Decurtins, T. Kohler, T. Moeri, I. Mäller, U. Schotterer & M. Winiger (1986). Hydrogeographical Map of Mount Kenya Area. *Geographica Bernensia – African Studies Series*, A3, Institute of Geography, University of Berne.

Letreguilly, A. (1984). *Bilans de masses des glaciers Alpins: méthodes de mésure et répartition spacio-temporelle*. CNRS, Laboratoire de Glaciologie, Grenoble, Publ. No. 439.

Lettau, H. (1976). Dynamic and energetic factors which cause and limit aridity along South America's Pacific Coast. In: Schwerdtfeger, W. (ed.) *Climates of Central and South America*. vol. 13 of Landsberg, H.E. (ed.) *World Survey of Climatology*, Elsevier Scientific, Amsterdam. 188–192.

Lichtenegger, J. & B. Lichtenegger (1978). In der Wetterküche des Ruwenzori. *Die Alpen*, 54(3), 119–126.

Liljequist, G.H. & K. Cehak (1984). *Allgemeine Meteorologie*. Viehweg, Braunschweig.

Livingstone, D.A. (1962). Age of deglaciation in the Ruwenzori Range, Uganda. *Nature*, 194, 137–139.

Livingstone, D.A. (1967). Postglacial vegetation of the Ruwenzori Mountains in Equatorial Africa. *Ecological Monographs*, 37, 25–52.

Livingstone, D.A. (1975). Late Quaternary climatic change in Africa. *Annual Review of Ecology and Systematics*, 6, 249–280.

Livingstone, D.A. (1980). Environmental changes in the Nile Headwaters. In: M.A.Williams & H. Faure (eds.): *The Sahara and the Nile*, Balkema, Rotterdam, 339–359.

Lliboutry, L. (1954). The origin of penitentes. *J. Glaciol.*, 2(15), 331–338.

Lliboutry, L. (1975). La catastrophe du Yungay (Pérou). UGGI–IAHS–ICSI Symposium Moskow, 1971, *IAHS Publ.* No. 104, 353–363.

Lliboutry, L. (1977). Glaciological problems set by the control of dangerous lakes in Cordillera Blanca, Peru. II. Movement of a covered glacier embedded within a rock glacier. *J. Glaciol.*, 18(79), 255–273.

Lliboutry, L. (1998). Glaciers of the dry Andes: glaciers of Chile and Argentina. *USGS Prof. Paper*, 1386–I.

Lliboutry, L., B. Morales, A. Pautre & B. Schneider (1977¹). Glaciological problems set by the control of dangerous lakes in Cordillera Blanca, Peru. I. Historical failures of morainic dams, their causes and prevention. *J. Glaciol.*, 18(79), 239–254.

Lliboutry, L., B. Morales & B. Schneider (1977²). Glaciological problems set by the control of dangerous lakes in Cordillera Blanca, Peru. III. Study of moraines and mass balances at Safuna. *J. Glaciol.*, 18(79), 275–290.

Löffler, E. (1972). Pleistocene glaciation in Papua and New Guinea. *Zeitschrift für Geomorphologie*, suppl. vol. 13, 32–58.

Löffler, E. (1976). Potassium-Argon dates and pre-Wurm glaciations of Mount Giluwe volcano, Papua New Guinea. *Z. Gletscherk. Glazialgeo.*, 12, 55–62.

Löffler, E. (1977). *Geomorphology of Papua New Guinea*. CSIRO and Australian National University Press, Canberra.

Löffler, E. (1982). Pleistocene and present-day glaciations. In: Gressit, J.L. (ed.): *Biogeography and Ecology of New Guinea*. Monographiae Biologicae, 42, Junk, The Hague, 39–55.

MacLachlan, H. (1951). Air photographs – Rwenzori, 11.12.1951, inclined.

MacLachlan, H. (1952). Air photographs – Rwenzori, 17. and 27.9.1952, vertical.

Mahaney, W.C. (1977). Quaternary history of Mt. Kenya, East Africa. *Nat. Geogr. Soc. Research Reports, 1976 Projects*, 561–581.

Mahaney, W.C. (1982). Chronology of glacial deposits on Mt. Kenya. *Palaeoecology of Africa*, 14, 25–43.

Mahaney, W.C. (1989). Quaternary glacial geology of Mt. Kenya. In: Mahaney, W.C. (ed.): *Quaternary and Environmental Research on East African Mountains*. Balkema, Rotterdam, 121–140.

Mahaney, W.C. (1990). *Ice on the Equator: Quaternary Geology of Mt. Kenya*. Sister Bay, WI, USA. Caxton.

Mantovani, R. (1996). *The Ruwenzori Discovery – Luigi Amadeo di Savoia Duca degli Abruzzi*. Museo Nazionale della Montagna 'Duca degli Abruzzi', Club Alpino Italiano, Torino.

Meierding, T.C. (1982). Late Pleistocene glacial equilibrium-line altitudes in the Colorado Front Range: a comparison of methods. *Quaternary Research*, 18(3), 289–310.

Menzies, I.R. (1951¹). Some observations on the glaciology of the Ruwenzori Range. *J. Glaciol.*, 1(9), 511–512.

Menzies, I.R. (1951²). The Glaciers of Ruwenzori. *Uganda Journal*, 15(2), 177–181.

Mercer, J.H. (1979). Chronology of the Last Glaciation in Perú. *Boletín de le Sociedad Geológica del Perú*, 61, 113–120.

Mercer, J.H. & O. Palacios (1977). Radiocarbon dating of the last glaciation in Peru. *Geology*, 5, 600–604.

Messerli, B., H. Hurni, M. Kienholz & M. Winiger (1977). Bale Mountains. Largest Pleistocene mountain glacier system of Ethiopia. *X INQUA Congress Abstracts, Birmingham*, 300.

Messerli, B., M. Winiger & P. Rognon (1980). The Saharan and East African uplands during the Quaternary. In: M.A.J. Williams, H. Faure (eds.) *The Sahara and the Nile*. Balkema, Rotterdam, 87–132.

Michot, P. (1933). Les traits charactéristiques de la morphologie du Ruwenzori dans leur relations avec la tectonique du massif. *Soc. Roy. Belge Géogr.*, 57, 5–13.

Middendorf, E.W. (1895). *Perú – Beobachtungen und Studien über das Land und die Bewohner während eines 25 jährigen Aufenthalts*. Oppenheim–Schmidt, Berlin, Bd. III.

Mitchell, J.M. (1961). Recent secular changes of global temperature. *Ann. New York Acad. Sci.*, 95, 235–250.

Moore, J.E.S. (1901). *To the Mountains of the Moon*. Hurst and Blackett, London.

Moore, J.E.S. (1902). The first ascent of one of the Snow Ridges in the Mountains of the Moon. *Alpine Journal*, 21, 77–90.

Morales, B. (1966). The Huascarán avalanche in the Santa Valley, Peru. UGGI–IAHS–ICSI Symposium Davos 1965. *IAHS Publ.* No. 69, 304–315.

Morales, B. (1969). Las lagunas y glaciares de la Cordillera Blanca y su control. *Rev. Peruana de Andinismo y Glaciol.*, 1966–67–68, 8, 76–79.

Morales, B. (1979). Avalanchas y aluviones en el Departamento de Ancash. *Boletín Informativo del Instituto Geológico, Minero y Metalúrgico*, 2, 2–9.

Morales-Arnao, B. & S.L. Hastenrath (2000). Glaciers of Péru. In *Satellite Image Atlas of Glaciers of the World: Glaciers of S. America*. USGS P 1386-I.

Morales, B., M. Zamora & A. Ames (1979). Inventario de lagunas y glaciares del Perú. *Boletín de le Sociedad Geológica del Perú*, 62, 63–82.

Morton, W.H. (1968). *Expedition to the Ruwenzori Mountains, June–July 1968. Part 1: Glaciology and Geology*. Water Development Department, Geological Survey & Mines Department, Makarere University College.

Mosiño, P.A. & E. García (1974). The climate of Mexico. In: R.A. Bryson and F.C. Hare (eds.): Climates of North America. *World Survey of Climatology*, 11. Elsevier Scientific Publ. Comp., Amsterdam, London, New York, 345–404.

Müller, F. (1962). Zonation in the accumulation area of the glaciers of Axel Heiberg Island, N.W.T., Canada. *J. Glaciol.*, 4, 302–313.

Nakawo, M. & G.J. Young (1982). Estimate of glacier ablation under a debris layer from surface temperature and meteorological variables. *J. Glaciol.*, 28, 29–34.

Nakawo, M., H. Yabuki & A. Sakai (1999). Characteristics of Khumbu Glacier, Nepal Himalaya: recent change in the debris covered area. *Annals Glaciol.*, 28, 118–122.

Niedertscheider, J. (1990). *Untersuchungen zur Hydrographie der Cordillera Blanca (Perú)*. Diplomarbeit. Universität Innsbruck.

Nogami, M. (1976). Altitude of the modern snowline and Pleistocene snowline in the Andes. *Geographical Reports of Tokyo Metropolitan University*, 11, 71–86.

Noggler, B. (1992). *Neuzeitliche Gletscherschwankungen am Ruwenzori – Ostafrika*. Diplomarbeit, Inst. f. Geographie, Univ. Innsbruck.

Nye, J.F. (1960). The response of glaciers and ice-sheets to seasonal and climatic changes. *Proc. R. Soc. London, Ser. A*, 256(1287), 559–584.

Nye, J.F. (1963¹). On the theory of the advance and retreat of glaciers. *Geophys. J. R. Astron. Soc.*, 7(4), 431–456.

Nye, J.F. (1963²). The response of a glacier to changes in the rate of nourishment and wastage. *Proc. R. Soc. London, Ser. A*, 257(1360), 87–112.

Nye, J.F. (1965). A numerical method of inferring the budget history of a glacier from its advance and retreat. *J. Glaciol.*, 5(41), 589–607.

Nye, J.F. (1976). Water flow in glaciers: jökulhlaups, tunnels and veins. *J. Glaciol.*, 17(76), 181–207.

Ohmura, A., P. Kasser & M. Funk (1992). Climate at the equilibrium line of glaciers. *J. Glaciol.*, 38(130), 397–411.

Olago, D.O., F.A. Street-Perrott, R.A. Perrott, M. Ivanovich & D.D. Harkness (1999). Late Quaternary glacial-interglacial cycle of climatic and environmental change on Mount Kenya, Kenya. *J. African Earth Sciences*, 29, 593–618.

Oppenheim, V. (1945). Las glaciaciones en el Perú. *Boletín de le Sociedad Geológica del Perú*, 18, 37–43.

Oppenheim, V. & H. Spann (1946). *Investigaciones glaciológicas en el Perú 1944–1945*. Instituto Geológico del Perú, Boletín no. 5.

Osmaston, H.A. (1958). *Pollen Analysis in the Study of the Past Vegetation and Climate of Ruwenzori and its Neighbourhood*. B.Sc. thesis, Oxford.

Osmaston, H.A. (1961). Notes on the Ruwenzori Glaciers. *Uganda Journal*, 25, 99–104.

Osmaston, H.A. (1965). *The Past and Present Climate and Vegetation of Ruwenzori and its Neighbourhood*. D. Phil. thesis, University of Oxford.

Osmaston, H.A. (1967). The sequence of glaciations in the Ruwenzori and their correlation with glaciations of other mountains in East Africa and Ethiopia. *Palaeoecology of Africa*, 2, 26–28.

Osmaston, H.A. (1975). Models for the estimation of firnlines of present and Pleistocene glaciers. In: Peel. R.F., M.D.I. Chisholm & P. Haggett (eds.) *Processes in Physical and Human Geography, Bristol Essays*, 218–245.

Osmaston, H.A. (1986). The Siachen and Terong Glaciers, East Karakoram. *Himalayan J.*, 42, 87–96. Repr. as App. 5 in: S. Venables: *Painted Mountains*. London, Hodder & Stoughton.

Osmaston, H.A. (1989[1]). Glaciers, glaciations and equilibrium line altitudes on Kilimanjaro. In: Mahaney, W. C. (ed.) *Quaternary and Environmental Research on East African Mountains*. Balkema, Rotterdam / Brookfield, 7–30.

Osmaston, H.A. (1989[2]). Glaciers, glaciations and equilibrium line altitudes on the Ruwenzori. In: Mahaney, W.C., ed: *Quaternary and Environmental Research on East African Mountains*. Balkema, Rotterdam / Brookfield, 31–104.

Osmaston, H.A. (1998). The influence of the Quaternary history and glaciations of the Rwenzori on the present landscape and ecology. In: H. Osmaston, C. Basalirwa, J. Nyakaana (eds.): *The Rwenzori Mountains National Park: exploration, environment and biology; conservation, management and community relations*. Makerere Univ., Uganda, 49–65.

Osmaston, H.A. & G. Kaser (2001). *Glaciers and Glaciations, Rwenzori Mountains National Park*. 1:100,000 map., Henry Osmaston, Ulverston.

Osmaston, H.A. & D. Pasteur (1972). *Guide to the Ruwenzori*. Mountain Club of Uganda.

Osmaston, H.A., J. Tukahirwa, C. Basalirwa & J. Nyakaana (1998). *The Rwenzori Mountains National Park, Uganda*. Department of Geography, Makerere University, Kampala, Uganda.

Ota, Y. & J. Chappell (1999). Holocene sea-level rise and coral reef growth on a tectonically rising coast, Huon Peninsula, Papua New Guinea. *Quaternary International*, 55, 51–59.

Paffen, K. (1967). Das Verhältnis der tages- zur jahreszeitlichen Temperaturschwankung. Erläuterungen zu einer neuen Weltkarte als Beitrag zur allgemeinen Klimatologie. *Erdkunde*, 21, 94–111; map incl.

Paterson, W.S.B. (1994). *The Physics of Glaciers* (third ed.). Pergamon, Oxford.

Patzelt, G. (ed.) (1983). Die Berg- und Gletscherstürze vom Huascarán, Cordillera Blanca, Perú. *Hochgebirgsforschung – High Mountain Research*, 6, Arbeitsgemeinschaft für vergleichende Hochgebirgsforschung.

Patzelt, G. (1985). The period of glacier advances in the Alps, 1965 to 1980. *Z. Gletscherk. Glazialgeol.*, 21, 403–407.

Patzelt, G., E. Schneider & G. Moser (1984). Der Lewis Gletscher, Mount Kenya. Begleitwort zur Gletscherkarte 1983. *Z. Gletscherk. Glazialgeol.*, 20, 177–195.

Perrott, R.A. (1982[1]). A postglacial pollen record from Mt. Satima, Aberdare range, Kenya. *Amer. Quat. Assoc. 7th Conf.*, 153.

Perrott, R.A. (1982[2]). A high altitude pollen diagram from Mt. Kenya; its implications for the history of glaciation. *Palaeoecology of Africa*, 14, 77–83.

Peterson, J.A., G.S. Hope & R. Mitton (1973). Recession of snow and ice fields of Irian Jaya, Republic of Indonesia. *Z. Gletscherk. Glazialgeol.*, 9, 73–87.

Peterson, J.A. & L.F. Peterson (1994). Ice retreat from the Neoglacial maxima in the Puncak Jayakesuma area, Republic of Indonesia. *Z. Gletscherk. Glazialgeol.*, 30, 1–9.

Platt, C.M. (1966). Some observations on the climate of Lewis Glacier, Mount Kenya, during the rainy season. *J. Glaciol.*, 6(44), 267–287.

Porter, S.C. (1979[1]). Hawaian glacial ages. *Quat. Res.*, 12, 161–187.

Porter, S.C. (1979[2]). Quaternary stratigraphy and chronology of Mauna Kea, Hawaii: a 380,000 year record of mid-Pacific volcanism and ice-cap glaciation: summary. *Geol. Soc. Amer. Bull.*, I(90), 609–611.

Porter, S.C. (1986). Pattern and forcing of northern hemisphere glacier variations during the last millenium. *Quat. Res.*, 26(1), 27–48.

Potter, E.C. (1976). Pleistocene glaciation in Ethiopia. *J. Glaciol.*, 17, 148–150.

Pouyaud, B., B. Francou & P. Ribstein (1995). Un résau d'observation des glaciers dans les Andes tropicales. *Bull. Inst. fr. études andines*, 24(3), 707–714.

Prohaska, F. (1973). New evidence on the climatic controls along the Peruvian coast. In: D.H.K. Amiran & A.W. Wilson (eds.) *Coastal Deserts, Their Natural and Human Environments*. The Univ. of Arizona Press. Tucson Ariz., 91–107.

R.A.F. Mosquito (1952). Air photographs – Rwenzori, 4. and 20.4.1952. 82D/555 and 587.

Raggl, M. (1996). *Abschätzung von Wasserhaushaltsgrößen für das Ötztal mit Hilfe eines Geographischen Informationssystems*. Diplomarbeit, Institut für Geographie, Universität Innsbruck.

Raimondi, A. (1873). *El Departamento de Ancachs y sus Riquezas Minerales*. El Nacional, Lima, Peru.

Raimondi, A. (1876). *El Perú*. Tome II, Imprenta del Estado, Lima.

Rebmann, J. (1849[1]). Narrative of a journey to Jagga, the snow country of Eastern Africa. *Church Missionary Review*, 1, 12–23.

Rebmann, J. (1849[2]). Journal d'une excursion au Djagga. *Nouvelles Annales des Voyages et des Sciences Géographiques*, 2, 257–307.

Rebmann, J. (1855). Extracts of letter to Rev. Venn. *Report of the 24th meeting of the British Association of the Advancement of Science*, London, 123–124.

Rex, F.D. (1969). *Climate of the Free Atmosphere*. Elsevier, Amsterdam.

Reynaud, L. (1978). Glacier fluctuations in the Mont Blanc area (French Alps). *Z. Gletscherk. Glazialgeol.*, 13(1/2), 1977, 155–166.

Reynaud, L. (1983). Recent fluctuations of Alpine glaciers and their meteorological causes: 1880–1980. In: Street-Perrott, F.-A., M. Beran & R.A.S. Ratcliffe (eds): *Variations in the Global Water Budget*. D. Reidel Publishing Co., Dordrecht, 197–205.

Reynolds, J., C. Portocarrero, A. Ames, M. Zapata & G. Kaser (1988). Hazard assessment in the Cordillera Blanca, Peru. *Ice, News Bulletin of the Intern. Glaciol. Soc.*, 3, 26.

Riehl, H. (1954). *Tropical Meteorology*. McGraw–Hill, New York.

Rietti-Shati, M., A. Shenesh & W. Karlen (1998). A. 3000-year climatic record from biogenic silica oxygen isotopes in an equatorial high-altitude lake. *Science*, 281, 980–982.

Rind, D. & D. Peteet (1985). Terrestrial conditions at the last glacial maximum and CLIMAP sea surface temperature estimates: are they consistent? *Quat. Res.*, 24, 1–22.

Roccati, A. (1909). Übersicht der geologischen, petrographischen, mineralogischen, sowie der zoologischen und botanischen Beobachtungen. In: De Filippi, F. (ed.) (1909). *Der Ruwenzori. Erforschung und erste Ersteigung seiner höchsten Gipfel*. Anhang C, 433–452.

Rodbell, D.T. (1992[1]). Late Pleistocene equilibrium–line altitude reconstructions in the northern Peruvian Andes. *Boreas*, 21, 43–52.

Rodbell, D.T. (1992[2]). Lichonometric and radiocarbon dating of Holocene Glaciation, Cordillera Blanca, Perú. *The Holocene*, 2, 19–29.

Rosanski, K. (1995). Isotope Hydrology Section. International Atomic Energy Agency (IAEA), Wien. pers. comm..

Rosanski, K. & L. Araguás (1995). Spatial and temporal variability of stable isotope composition of precipitation over the South American continent. *Bul. Inst. fr. études andines*, 24(3), 379–390.

Rosqvist, G. (1990). Quaternary glaciations in Africa. *Quat. Sci. Revs.*, 9, 281–297.

Röthlisberger, F. (1987). *10,000 Jahre Gletschergeschichte der Erde*. Aarau Verlag, Sauerlande.

Rudloff, W. (1981). *World Climates*. With tables of climatic data and practical suggestions. Books of the Journal Naturwissenschaftliche Rundschau. Wissenschaftl. Verlagsges. mbH Stuttgart.

Sapper, K. (1923). *Die Tropen – Natur und Mensch zwischen den Wendekreisen*. Strecker und Schröder Verlag, Stuttgart.

Schubert, C. (1970). Glaciation of the Sierra Santo Domingo, Venezuela Andes, *Quaternaria*, 13, 225–246.

Schubert, C. (1972). Geomorphology and glacier retreat in the Pico Bolívar area, Sierra Nevada de Mérida, Venezuela. *Z. Gletscherk. Glazialgeol.*, VIII(1–2), 189–202.

Schubert, C. (1974). Late Pleistocene Mérida glaciation. *Boreas*, 3, 147–152.

Schubert, C. (1975). Glaciation and periglacial morphology in the north-western Venezuelan Andes. *Eiszeitalter und Gegenwart*, 6, 196–211.

Schubert, C. (1979). La zona del Páramo: morfologia glacial y periglacial de los Andes de Venezuela. In: M.L. Salgado-Labouriau (ed.) *El Medio Ambiente Páramo*. Caracas, I.V.I.C., 11–27.

Schubert C. & C.M. Clapperton (1990). Quaternary glaciations in the northern Andes, Venezuela, Colombia and Ecuador. *Quaternary Science Reviews*, 9, 123–136.

Schug, H.-J. (1987). *Der Schwarzmilzferner. Meteorologisch–glaziologische Untersuchung an einem Kleingletscher in den Allgäuer Alpen*. Diplomarbeit, Universität Innsbruck.

Schwartz, D.P. (1983). *Evaluation of Seismic Geology Along the Cordillera Blanca Fault Zone, Peru*. Hidrandina S.A., Huaraz. Woodward–Clyde Consultants.

Schweitzer, G. (1898). *Emin Pascha. Eine Darstellung seines Lebens und Wirkens mit Benutzung seiner Tagebücher, Briefe und Wissenschaftlichen Aufzeichnungen*. Verlag Hemann Walther, Berlin.

Schwerdtfeger, W. (1976). The atmospheric circulation over Central and South America. In: *World Survey of Climatology*, 12, The Climates of Central and South America, 1–12. Elsevier, Amsterdam.

Seltzer, G.O. (1990). Recent glacial history and palaeoclimate of the Peruvian–Bolivian Andes. *Quat. Sc. Rev.*, 9, 137–152.

Seltzer, G.O. (1993). Late-Quaternary glaciation as a proxy for climate-change in the Central Andes. *Mountain Research and Development*, 13, 129–138.

Seltzer, G.O. (1994). A lacustrine record of Late Pleistocene climatic-change in the Subtropical Andes. *Boreas*, 23, 105–111.

Shanahan T.M. & M. Zreda (2000). Chronology of quaternary glaciations in East Africa. *Earth and Planetary Science Letters*, 177, 23–42.

Shi Yafeng *et al.* (1979). The Batura Glacier in the Karakoram Mountains and its variations. *Scientia Sinica*, 22, 958–974.

Sievers, W. (1914). *Reise in Peru und Equador, ausgeführt 1909*. Wissenschaftliche Veröffentlichungen der Gesellschaft für Erdkunde zu Leipzig, Band 8. Duncker & Humbolt, München, Leipzig.

Sissons, J.B. (1974). A late glacial ice-cap in the Grampians, Scotland. *Trans. Inst. Brit. Geogr.*, 62, 95–114.

Sissons, J.B. (1980). The Loch Lomond advance in the Lake District, northern England. *Trans. Royal Soc. Edinburgh: Earth Sci.*, 71, 13–27.

Spann, H.J. (1947). Informe sobre el origen de la catástrofe de Chavín de Huantar. *Boletín de la Sociedad Geológica del Perú*, 20, 29–33.

Stadelmann, J. (1985). Zur Dokumentation der Bergsturzereignisse vom Huascarán. In: Patzelt, G. (ed.): Die Berg- und Gletscherstürze vom Huascarán, Cordillera Blanca, Perú. *Hochgebirgsforschung – High Mountain Research*, 6, Arbeitsgemeinschaft für vergleichende Hochgebirgsforschung.

Stanley, H.M. (1890[1]). *In Darkest Africa*. Sampson Low, London.

Stanley, H.M. (1890[2]). *Im dunkelsten Afrika. Aufsuchung, Rettung und Rückzug Emin Paschas*. Brockhaus, Leipzig.

Stearns, H.T. (1945). Glaciation of Mauna Kea, Hawaii. *Geol. Soc. Amer. Bull.*, 56, 267–274.

Steinacker, R. (1980). Über die Ursachen sommerlicher Schneefälle in Alpentälern. *Tagungsbericht XVIème congrès international de météorologie alpine – Aix Les Bains*. Societé Météorologique de France. 261–265.

Street, F.A. (1979[1]). *Late Quaternary lakes in the Ziway-Shala basin, Southern Ethiopia*. D.Phil. thesis, Univ. Oxford.

Street, F.A. (1979[2]). Late Quaternary precipitation estimates for the Ziway-Shala basin, southern Ethiopia. *Palaeoecology of Africa*, 11, 135–143.

Street-Perrott, F.A., Y.S. Huang, R.A. Perrott, G. Eglinton, P. Barker, L. Ben Khelifa, D.D. Harkness & D.O. Olago (1997). Impact of lower atmospheric carbon dioxide on tropical mountain ecosystems. *Science*, 278, 1422–1426.

Stuhlmann, F. (1894). *Mit Emin Pasha ins Herz von Afrika*. Reimer, Berlin.

Stumpp, A. (1952). Kartierungsarbeiten im Ruwenzorigebirge. *Allgemeine Vermessungsnachrichten*, 6, 142–147. *Zentralgruppe des Ruwenzori-Gebirges 1:25.000*.

Stute, M., J.F. Clark, P. Schlosser, W.S. Broecker & G. Bonani (1995[1]). A 30,000-yr continental paleotemperature record derived from noble-gases dissolved in groundwater from the San-Juan Basin, New-Mexico. *Quat. Res.*, 43, 209–220.

Stute, M., M. Forster, H. Frischkorn, A. Serejo, J.F. Clark, P. Schlosser, W.S. Broecker & G. Bonani (1995[2]). Cooling of tropical Brazil (5°C) during the Last Glacial Maximum. *Science*, 269, 379–383.

Sutherland, D.G. (1984). Modern glacier characteristics as a basis for inferring former climates with particular reference to the Loch Lomond stadial. *Quaternary Science Reviews*, 3, 291–309.

Synge, P.M. (1937). *Mountains of the Moon*. Drummond, London.

Talks, A. (1993). *East African Hot Ice 1993*. Sir Roger Manwood's School Expedition to Kenya and Uganda. Sir Roger Manwood's School, Kent, internal report. Copies held by Roy. Geogr. Soc.

Tanzer, G. (1986). *Berechnung des Wärmehaushaltes an der Gleichgewichtslinie des Hintereisferners*. Diplomarbeit, Universität Innsbruck.

Temple, P.H. (ed.) (1961). *A Final Report of the Makerere College I.G.Y. Expeditions to Ruwenzori*. Dept. Geography, Makerere.

Temple, P.H. (1968). Further observations on the glaciers of the Ruwenzori. *Geografiska Annaler*, Ser. A, 50(3), 136–150.

Thompson, B.W. (1966). The mean annual rainfall of Mt.Kenya. *Weather*, 21, 48–49.

Thompson, L.G. (1995). Late Holocene ice core records of climate and environment from the tropical Andes, Perú. *Bull. Inst. fr. études andines*, 24(3), 619–629.

Thompson, L.G., E. Mosley-Thompson, P.M. Grootes & M. Pourchet (1986). The Little Ice Age as recorded in the stratigraphy of the Quelccaya Ice Cap, Peru. *Science*, 234, 4774, 971–973.

Thompson, L.G., E. Mosley-Thompson, M.E. Davis, P.N. Lin, K.A. Henderson, J. Cole-Dai, J.F. Bolzan & K.B. Liu (1995). Late Glacial Stage and Holocene Tropical Ice Core Records from Huascarán, Peru. *Science*, 269, 46–50.

Thouret, J.C., T. van der Hammen & B. Salomons (1996). Paleoenvironmental changes and glacial stades of the last 50,000 years in the Cordillera Central, Colombia. *Quaternary Research*, 46, 1–18.

Times Atlas (1999). *Times Comprehensive Atlas of the World*, Millennium edition. London, Times Books.

Trapnell, C.G. & J.F. Griffiths (1960). The rainfall–altitude relation and its ecological significance in Kenya. *East African Agr. J.*, 25(4), 207–213.

Trask, P.D. (1953). El problema de los aluviones de la Cordillera Blanca. *Bolletín de la Sociedad Geográfica de Lima*, 70, 1–75.

Tricart, J. (1971). Pleistocene snowline and present periglacial processes in the Venezuelan Andes compared with Papua. *Australian Geogr. Studies*, 9, 85–86.

Troll, C. (1942). Büßerschenee (nieve de los penitentes) in den Hochgebirgen der Erde. Ein Beitrag zur Geographie der Schneedecke und ihrer Ablationsformen. *Ergänzungen Nr. 240 zu Petermann's Mitt.*

Troll, C. (1943). Thermische Klimatypen der Erde. *Petermann's Geogr. Mitt.*, 43(3/4), 81–89.

Troll, C. (1949). Schmelzung und Verdunstung von Eis und Schnee in ihrem Verhältnis zur geographischen Verbreitung der Ablationsformen. *Erdkunde*, III, 18–29.

Troll, C. (1959). Die tropischen Gebirge. Ihre dreidimensionale klimatische und pflanzengeographische Zonierung. *Bonner Geogr. Abh.*, 25.

Troll, C. & K. Wien (1949). Der Lewis Gletscher am Mount Kenya. *Geografiska Annaler*, 31, 257–274.

Umer, M. & R. Bonnefille (1998). A late glacial late Holocene pollen record from a highland peat at Tamaa, Bale Mountains, south Ethiopia. *Global & Planetary Change*, 17, 121–129.

USGS (1999). *Satellite Image Atlas of Glaciers of the World*, Prof. Paper 1386, (available on http://pubs.usgs.gov/prof/p 1386h/).

Verstappen, H.Th. (1964). Geomorphology of the Star Mts.: scientific results of the Netherlands New Guinea Expedition 1959. *Nova Guinea (Geology)*, 5, 101–158.

Wagner, H.P. (1979). Strahlungshaushaltsuntersuchungen an einem Ostalpengletscher während der Hauptablationsperiode. Teil 1: Kurzwellige Strahlung. *Arch. Met.*, Ser. B, 27, 297–324.

Walker, D. & J.R. Flenley (1979). Late Quaternary of Papua New Guinea. *Phil. Trans. Roy. Soc. B*, 286, 265–344.

Webster, P.J. & N.A. Streten (1978). Late Quaternary Ice Age climates of tropical Australasia: interpretations and reconstructions. *Quat. Res.*, 10, 289–309.

Weischet, W. (1988). Einführung in die Allgemeine Klimatologie. *Teubner Studienbücher Geographie*. 4. überarb. und erweit. Auflage, Teubner, Stuttgart.

Welsch, W. & H. Kinzl (1970). Der Gletschersturz vom Huascarán (Perú) am 31. Mai 1970, die größte Gletscherkatastrophe der Geschichte. *Z. Gletscherk. Glazialgeol.*, 6(1–2), 181–192.

WGMS (1989). *World Glacier Inventory, status 1988*. IAHS (ICSI)–UNEP–UNESCO.

White, S.E. (1962). Late Pleistocene glacial sequence for the west side of Iztaccihuatl, Mexico. *Geol. Soc. Amer. Bull.*, 73, 935–958.

White, S.E. (1981). Equilibrium line altitudes of Late Pleistocene and recent glaciers in central Mexico. *Geografiska Annaler*, 63A, 241–249.

White, S.E. & S. Velastro (1984). Pleistocene glaciation of volcano Ajusco, central Mexico, and comparison with the standard Mexican glacial sequence. *Quaternary Research*, 21, 21–35.

Whittow, J.B. (1959). The glaciers of Mount Baker, Ruwenzori. *Geogr. J.*, 125(3–4), 370–379.

Whittow, J.B. (1960). Some observations on the snowfall of Ruwenzori. *J. Glaciol.*, 3(28), 765–772.

Whittow, J.B. (1966). The landforms of the central Ruwenzori, East Africa. *Geogr. J.*, 132, 32–42.

Whittow, J.B. & A. Shepherd (1958). The Speke Glacier, Ruwenzori. *Uganda Journal*, 23(2), 153–161.

Whittow, J.B., A. Shepherd, J.E. Goldthorpe & P.H. Temple (1963). Observations on the glaciers of the Ruwenzori. *J. Glaciol.*, 4(35), 581–616.

Williams, M.A.J. (1975). Late Pleistocene tropical aridity synchronous in both hemispheres? *Nature* 253, 617–618.

Williams, R.S. & J.G. Ferrigno (eds) (1989). Glaciers of Irian Jaya, Indonesia, and New Zealand. *Satellite Image Atlas of Glaciers of the World*. U.S. Geological Survey professional paper 1386. G.U.S. Government Printing Office, Washington.

Williams, R.S. & J.G. Ferrigno (eds) (1991). Glaciers of the Middle East and Africa. *Satellite Image Atlas of Glaciers of the World*. U.S. Geological Survey professional paper 1386. G.U.S. Government Printing Office, Washington.

Wollaston, A.F.R. (1908). *From Ruwenzori to the Congo: a Naturalist's Journey Across Africa*. John Murray, London.

Wollaston, A.F.R. (1912). *Pygmies and Papuas. The Stone Age Today in Dutch New Guinea*. Smith and Elder, London.

Woosnam, R.B. (1907). Ruwenzori and its life zones. *Geogr. J.*, 30, 630–642.

Workman, W. H. (1914). Nieve penitente and allied formations in Himalaya, or surface-forms of névé and ice created or modelled by melting. *Z. Gletscherk.*, VIII, 289–330

Wright, H.E. (1983). Glaciation round the Junin Plain, central Peruvian Highlands, *Geografiska Annaler Ser.A*, 65, 35–43.

Xerez, F. de (1534). *Relación de la Conquista del Perú*. Biblioteca de la Cultura Peruana, T.2, Paris.

Index

Bold page numbers are important sections; italic page numbers are tables or figures; the summary has not been indexed.